Springer Proceedings in Mathematics & Statistics

Volume 362

Springer Proceedings in Mathematics & Statistics

This book series features volumes composed of selected contributions from workshops and conferences in all areas of current research in mathematics and statistics, including operation research and optimization. In addition to an overall evaluation of the interest, scientific quality, and timeliness of each proposal at the hands of the publisher, individual contributions are all refereed to the high quality standards of leading journals in the field. Thus, this series provides the research community with well-edited, authoritative reports on developments in the most exciting areas of mathematical and statistical research today.

More information about this series at http://www.springer.com/series/10533

Jan Awrejcewicz
Editor

Perspectives in Dynamical Systems I: Mechatronics and Life Sciences

DSTA, Łódź, Poland December 2–5, 2019

Springer

Editor
Jan Awrejcewicz
Department of Automation, Biomechanics
and Mechatronics
Lodz University of Technology
Lodz, Poland

ISSN 2194-1009 ISSN 2194-1017 (electronic)
Springer Proceedings in Mathematics & Statistics
ISBN 978-3-030-77305-2 ISBN 978-3-030-77306-9 (eBook)
https://doi.org/10.1007/978-3-030-77306-9

Mathematics Subject Classification: 28DXX, 34Cxx, 37-XX, 46LXX, 65-XX, 70-XX, 74-XX, 76-XX

This Springer imprint is published by the registered company Springer Nature Switzerland AG
The registered company address is: Gewerbestrasse 11, 6330 Cham, Switzerland

Preface

15th International Conference "Dynamical Systems – Theory and Applications" (DSTA 2019) took place in Lodz, Poland from the 2nd to 5th December of 2019. It was the 15th edition in the series of conferences organized every 2 years in Lodz by the Department of Automation, Biomechanics and Mechatronics of the Lodz University of Technology.

For this edition, the scientific committee composed of 64 scientists had to review over 360 submitted topics to choose 200 that were to be presented during the DSTA 2019 by participants representing 40 countries from all over the world.

It resulted in the program of conference that covered both theoretical and experimental approaches to widely understood dynamical systems, including topics devoted to bifurcations and chaos, control in dynamical systems, asymptotic methods in nonlinear dynamics, stability of dynamical systems, lumped mass and continuous systems vibrations, original numerical methods of vibration analysis, nonsmooth systems, dynamics in life sciences and bioengineering, as well as to the engineering systems and differential equations.

All papers included in the following book were submitted and presented during DSTA 2019. They contribute partially to the diverse approaches and topics covered by wide scope of dynamical systems.

In what follows a brief description of the book content is provided.

In Chap. 1, authors utilized Lagrange's principles and multiple scale technique to obtain governing equations of the vibrating motion of a cylinder over circular surface under the influence of an exciting force and its asymptotic solutions using Routh-Hurwitz criterion for systems stability determination.

Mykulyak and Skurativskyi (Chap. 2) considered problem of the system dynamics, when the friction is incorporated and the harmonic force is applied to the most upper level of the system. The bifurcations with respect to the structural parameter were investigated for the periodic, quasiperiodic and chaotic attractors regimes revealed by application of the numerical and qualitative analysis methods.

Awrejcewicz et al. (Chap. 3) constructed the mathematical model of the nonlinear dynamics of flexible mesh cylindrical panels in the field of additive white noise taking into account a Cosserat medium. Model obtained applying Pshenichniy continuum model and Kirchhoff-Love hypotheses was used to prove that the noise with intensity commensurate with the intensity of the normal load does not change oscillations character of the system.

A method for measuring the motion trajectory of a vehicle as well as results of computer simulation tests of the control system, which involves individual braking of one, two or three wheels of the vehicle in order to improve its directional stability are presented in Chap. 4.

Kim et al. (Chap. 5) studied nonlinear phenomenon through the experiments and the frequency response curves. Their research proved possibility of the occurrence of the super-harmonic resonance when the excitation frequency is three times of the natural one by performing experimental investigations.

Methodology for the development of a dynamic test bench is presented by Siqueira et al. (Chap. 6). For this purpose the commonly applied for high speed kinematic systems delta-robot configuration was adapted in a way to allow application of high transverse loads in three axis while keeping a considerably large range of movement.

Finite difference model of a robust universal heat exchanger applicable for prediction of the heat exchanger dynamics for condenser at normal operation level was developed by Zamojski et al. (Chap. 7). The carried out analysis included simulation and control design of multiphase fluid dynamics of an existing heat pump.

Mathematical approach to assess a human gait is proposed in Chap. 8. Modelling of a normal gait in sagittal and frontal anatomical planes of the body using Newton-Euler formulation yielded three multibody biomechanical models that can be used to model a single support phase and double support phase of the gait.

Carvalho and Pinto (Chap. 9) proposed a non-integer order model to describe the role of the immune system in cancer cells' growth in a HIV-infected individual. By considedratuon of the various orders for the fractional derivative the model for different values of biologically relevant parameters is simulated yielding biologically relevant results.

Using the vibrissae of rats as the model of mechanoreceptors Scharff (Chap. 10) investigated how an artificial vibrissa-like tactile sensor interacts with an object contour that is superimposed with macroscopic features. Simulation for a straight horizontal contour with superimposed sinus undulation was validated experimentally using a wavy contour with superimposed macroscopic features.

Grzelczyk et al. (Chap. 11) proposed and experimentally validated a design of lower limb exoskeleton driven by linear electric actuators. A new gait generator, which can be used to produce rhythmic movements in hip and knee joints of both limbs, was developed and tested using the time histories of human joint angles in normal gait as an articulation variables of individual joints of the investigated device.

Recorded dynamic variables of hexapod walker robots gait scenarios were used by Kecskés et al. (Chap. 12) for analyses of the model uncertainties. Five different methods were applied for both quantification and evaluation of the experimentally obtained results yielding important information for the robust control design research.

Augustynek and Urbaś (Chap. 13) proposed model of the revolute joint with the clearance for analysis of the linkage composed of the five rigid or flexible links which form a serial closed-loop kinematic chain. Applied numerical simulations allowed to investigate an interaction between the links' flexibility and clearance in the joint during the motion of the linkage.

In Chap. 14, Harlecki et al. present developed mathematical model for analysis of the truck with a trailer combination of vehicles. Model constituting it as a multibody system, for using formalism of Lagrange's equations, based on the joint coordinates and homogeneous transformations taken from robotics, can be treated as a virtual prototype of the system in question with application in truck trailers design.

Martowicz et al. (Chap. 15) applied artificial neural networks to simulate and experimentally identify complex behavior of the shape memory alloys type of smart materials. The constitutive models allowed to reliably model the hysteretic character of the stress-strain relationship observed for the experimentally tested material.

Chapter 16 is devoted to the synthesis of a mathematical model of the electrohydraulic servo-drive taking into account such nonlinearities as friction model, characteristics of the modulus of the elasticity, dependence of flow intensity on pressure drop at control edges of the valve slide, dependence of hydrodynamic force, and characteristics of the volumetric loss factor in the pump. Proposed model was experimentally through comparison with results obtained for real electrohydraulic servo-drive and can be used in fast prototyping of the nonlinear state-space control systems.

In Chap. 17, comparison of the performance of payload weighing systems involving neural networks is proposed. Kosiara et al. discussed also possibility of replacing the conventional models implemented in the most up to date payload weighing systems with the ones based on the neural networks and influence of training dataset size on the accuracy of the systems.

Stańczyk et al. (Chap. 18) studied degrees of freedom (DOFs), kinematics and drive systems of available constructions of lower limb exoskeletons. Results of those comparative analyses were used to design and construct a lower limb and spine exoskeleton that can be applied in gait rehabilitation of patients suffering from different mobility impairments.

Chapter 19 deals with the theory of solution of transverse shock wave propagation in thin plane elastic isotropic plate. Presented are both analytical solutions of transverse displacement, velocity and stress for various material and geometric models of the plate, as well as results of their experimental validation.

Cherkasov and Makieva in Chap. 20 proposed application of the Pontryagin maximum principle as a method of reduction of the optimal control problem to a boundary value problem for the initial variables in the classical differential game theory. Analyzed was the two-dimensional pursuit-evasion problem for the case of proportional navigation of the unmanned aerial vehicle.

Method of optimization of the geometry of aeroelastic energy harvester using a genetic algorithm that processes data from computational fluid dynamics calculations is proposed in Chap. 21. Results generated by applied algorithm to maximize the efficiency of the device were experimentally validated, and their efficiency was then compared with that of commonly used resonator implemented in the aeroelastic energy harvester.

Hedrih (Chap. 22) studied rolling homogeneous heavy ball over the surface with arbitrary shape, in the real Rn3 space. Proposed new methodology allowed to determine the current angular velocity of rolling the ball over the surface in the function of generalized coordinates and their direction using the velocity vector of the center of the ball.

In Chap. 23, results of applying the methodology of hyperbolic-elliptic models for surface wave field to the case of a half-space coated by a vertically inhomogeneous layer are presented. The study was focused on surface waves propagating in an isotropic elastic half-space coated with a thin, vertically inhomogeneous layer, subject to action of a prescribed normal surface stress.

Behn et al. (Chap. 24) analyzed eigenvalues in the first octant of the complex plane for two models of boundary damping for vibrissa-like sensors and focusing on bending beam vibrations. During their studies, authors found two systems with complementary spectra of eigenvalues that have alternative instead of common eigenvalues.

DSTA Conferences are aimed to provide a common platform for exchange of new ideas and results of recent research in the field of scientific and technological advances in modern dynamical systems. Over the last 25 years both approaches and understanding of sciences significantly evolved to include new ideas and trends, but the traditional views are still present and provide the basic understanding. Therefore, both as Head of Organizing and Scientific Committees of DSTA 2019 and as the Editor of volume of Springer Proceedings, I hope that this book will provide the readers with both answers to their problems and ideas for their novel approaches to study nonlinear dynamical systems.

I greatly appreciate the help of Springer Editor Dahlia Fisch, Springer Project Coordinators Murugesan Tamilsevan and Saveetha Balasundaram as well as T. Metilda Nancy Marie Rayan, the project manager at Straive – in publishing this volume in the Springer Proceedings in Mathematics and Statistics series. I would like also to express my gratitude to Scientific Committee of DSTA 2019 and all reviewers for their help and professional support during the book preparation.

Łódź, Poland Jan Awrejcewicz

Contents

On the Vibrational Analysis for the Motion of a Rotating Cylinder

M. A. Bek, Tarek Amer, and Mohamed Abohamer

Abstract The main purpose of this work is to study the motion of 2-DOF of an auto-parametric dynamical system attached with a damped system. The governing equations of motion are gained utilizing Lagrange's equations in terms of the generalized coordinates. The method of multiple scales (MS) is used to obtain the solutions of the governing equations up to the third order of approximation. The primary external resonance simultaneously with the internal one are investigated to establish the solvability conditions and the modulation equations. The graphical representations of the time histories together with the amplitude and phases of the dynamical system are represented in some plots to describe the motion of the system at any instance. The stability of the solution has been made with use of *Mathematica*.

Keywords Vibration · Nonlinear dynamics · Stability

1 Introduction

Rotating cylinder over circular surface is widely appeared in many engineering applications. For example, electric motors, vibrating buildings, aviation, missiles, and locomotive engines are increasingly used as an example of it. As a rule,

M. A. Bek (✉)
Faculty of Engineering, Physics and Engineering Mathematics Department, Tanta University, Tanta, Egypt

Faculty of Engineering/General Systems Engineering Department, Modern Sciences and Arts University, October, Egypt
e-mail: m.ali@f-eng.tanta.edu.eg

T. Amer
Faculty of Science, Mathematics Department, Tanta University, Tanta, Egypt

M. Abohamer
Faculty of Engineering, Physics and Engineering Mathematics Department, Tanta University, Tanta, Egypt

J. Awrejcewicz (ed.), *Perspectives in Dynamical Systems I: Mechatronics and Life Sciences*, Springer Proceedings in Mathematics & Statistics 362,
https://doi.org/10.1007/978-3-030-77306-9_1

a rotating cylinder might be one of the main vibration's sources. Hence, it is very important to fully understand the vibration motion in order to offer better engineering solutions to reduce the vibration using a good design. Thus, various type of frequencies and mode shapes of such vibrating structures are significant in the design stage. Hence, it is very important to carefully understand where the resonance occurs to avoid structural familiar. For instant the proposed model is a good example of such systems. Discussion of such models may be found in [1–3]. The auto parametric resonance phenomena is observed as a cause of the coupling occurring in the equations of motion. Dynamical analysis of nonlinear vibrations of a mass of a cylinder shape which is rotated over circular body of radius R was presented in the paper. With the use of the multiple scale method the solution up to the third order is achieved [4, 5]. The system stability is investigated using Routh-Hurwitz criterion. Mathematica was the selected software to solve the algebra system and to obtain the results and graphically present it [6].

2 Dynamical Modeling

Let us consider the planar motion of a cylinder where its mass is m_1 and its radius is r over a circular surface of radius R and mass M. The system of the two masses is attached with the ground with an elastic spring of k spring stiffness, a damper of c damping coefficient, g earth's acceleration, and θ generalized co-ordinates admitted according to Fig. 1. The motion is considered under the influence of an external force $F(t)$ in the vertical direction.

Lagrange's equations were used to obtain the governing equation of motion for the corresponding system, the following

$$\begin{aligned} \frac{d}{dt}\left(\frac{\partial L}{\partial \dot{u}}\right) - \left(\frac{\partial L}{\partial u}\right) = Q_u, \\ \frac{d}{dt}\left(\frac{\partial L}{\partial \dot{\theta}}\right) - \left(\frac{\partial L}{\partial \theta}\right) = Q_\theta \end{aligned} \tag{1}$$

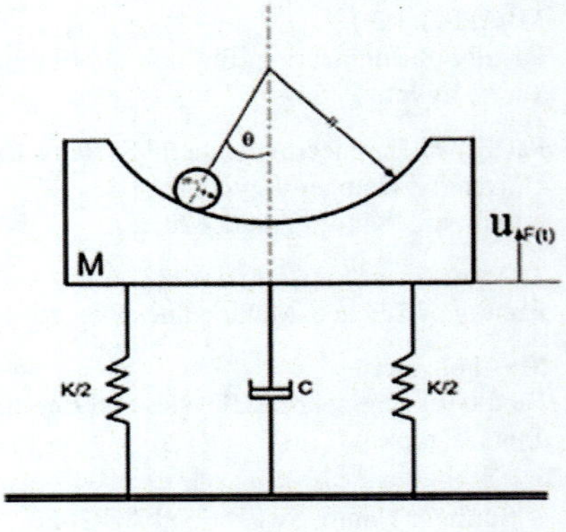

Fig. 1 The dynamical model

Where Q_θ and Q_u represent the general forces, which have the forms

$$Q_u = f(t),\ Q_\theta = 0$$

The equations of motion can be written as

$$\begin{aligned} &\ddot{u} + \omega_1^2 u + m\left(\ddot{\theta}\ \sin\theta + \dot{\theta}^2\ \cos\theta\ \right) + 2cu\ \dot{u} = f(t) \\ &\ddot{\theta} + \left(\omega_2^2 + 2\ddot{u}\right)\sin\ \theta = 0, \end{aligned} \tag{2}$$

where

$$u = \frac{y - y_c}{R - r},\ m = \frac{m_1}{(m_1 + M)},\ \omega_1^2 = \frac{k}{(m_1 + M)}, \omega_2^2 = \frac{2g}{3(R - r)}, f(t) = \frac{F(t)}{[(R - r)(m_1 + M)]} \tag{3}$$

The external force has the form $F(t) = F\cos\left(\Omega_1 t\right)$.

The higher order of the trigonometric of functions $\cos\theta = 1 - \frac{\theta^2}{2}$ and $\sin\theta = \theta - \frac{\theta^3}{6}$ are admitted.

3 The Proposed Method

The amplitudes of all oscillations are theoretic to be of the order of a small parameter ε. This can be expressed as

$$\theta(t) = \varepsilon\ \phi\ (t;\varepsilon)\,, \qquad u(t) = \varepsilon\ x\ (t;\varepsilon)\,, \tag{4}$$

where $0 \prec \varepsilon \prec\ \prec 1$. We seek the asymptotic solutions ϕ and x in the form of power series of ε as

$$\begin{aligned} \phi &= \textstyle\sum_{k=1}^{3} \varepsilon^k \phi_k\left(\tau_0, \tau_1, \tau_2\right) + O\left(\varepsilon^4\right), \\ x &= \textstyle\sum_{k=1}^{3} \varepsilon^k x_k\left(\tau_0, \tau_1, \tau_2\right) + O\left(\varepsilon^4\right), \end{aligned} \tag{5}$$

where $\tau_n = \varepsilon^n t$; $(n = 0, 1, 2)$ are different time scales.

The derivatives in terms of the new time scales will be written in the following form

$$\begin{aligned} \frac{d}{d\tau} &= \frac{\partial}{\partial\tau_0} + \varepsilon\frac{\partial}{\partial\tau_1} + \varepsilon^2\frac{\partial}{\partial\tau_2}, \\ \frac{d^2}{d\tau^2} &= \frac{\partial^2}{\partial\tau_0^2} + 2\varepsilon\frac{\partial^2}{\partial\tau_0\partial\tau_1} + \varepsilon^2\left(\frac{\partial^2}{\partial\tau_1^2} + 2\frac{\partial^2}{\partial\tau_0\partial\tau_2}\right) + O\left(\varepsilon^3\right). \end{aligned} \tag{6}$$

Terms of $O(\varepsilon^3)$ and higher order in Eq. (6) are neglected. Presumptuous that the amplitudes of generalized forces, the damping coefficients and the eccentricity to be small and have the form

$$f = \varepsilon^3\ \tilde{f}. \tag{7}$$

Substituting expressions (4), (5), (6), and (7) into Eq. (2) and equating coefficients of like powers of ε in both sides, one obtains three groups of partial differential equations according to the asymptotic solutions (5). Therefore, one obtains the following system that consists six partial linear differential equations as follows.

Order (ε)

$$\frac{\partial^2 x_1}{\partial \tau_0^2} + \omega_1^2 x_1 = 0. \tag{8}$$

$$\frac{\partial^2 \phi_1}{\partial \tau_0^2} + \omega_2^2 \phi_1 = 0, \tag{9}$$

Order of (ε^2)

$$\frac{\partial^2 x_2}{\partial \tau_0^2} + \omega_1^2 x_2 = -m\left(\frac{\partial \phi_1}{\partial \tau_0}\right)^2 - 2\frac{\partial^2 x_1}{\partial \tau_0 \partial \tau_1} - m\ \phi_1 \frac{\partial^2 \phi_1}{\partial \tau_0^2}, \tag{10}$$

$$\frac{\partial^2 \phi_2}{\partial \tau_0^2} + \omega_2^2 \phi_2 = -2\frac{\partial^2 \phi_1}{\partial \tau_0 \tau_1} - \frac{2}{3}\phi_1\ \frac{\partial^2 x_1}{\partial \tau_0^2}, \tag{11}$$

Order of (ε^3)

$$\begin{aligned}\frac{\partial^2 x_3}{\partial \tau_0^2} + \omega_1^2 x_3 = \tilde{F}_1 \cos\ \Omega_1\ \tau_0 - \frac{\partial^2 x_1}{\partial \tau_1^2} - 2\tilde{C}\frac{\partial x_1}{\partial \tau_0} - 2\left(\frac{\partial^2 x_1}{\partial \tau_0 \partial \tau_2} + \frac{\partial^2 x_2}{\partial \tau_0 \partial \tau_1}\right)\\ - 2\ m\left[\frac{\partial \phi_1}{\partial \tau_0}\frac{\partial \phi_1}{\partial \tau_1} + \frac{\partial \phi_1}{\partial \tau_0}\frac{\partial \phi_2}{\partial \tau_0} + \phi_1\frac{\partial^2 \phi_1}{\partial \tau_0 \partial \tau_1} + \phi_2\frac{\partial^2 \phi_1}{\partial {\tau_0}^2} + \phi_1\frac{\partial^2 \phi_2}{\partial {\tau_0}^2}\right],\end{aligned} \tag{12}$$

$$\begin{aligned}\frac{\partial^2 \phi_3}{\partial \tau_0^2} + \omega_2^2 \phi_3 = \tfrac{1}{6}\omega_2^2\ \phi_1^3 - \frac{\partial^2 \phi_1}{\partial \tau_1^2} - 2\left(\frac{\partial^2 \phi_1}{\partial \tau_0 \partial \tau_2} + \frac{\partial^2 \phi_2}{\partial \tau_0 \partial \tau_1}\right)\\ - \tfrac{4}{3}\ \phi_1\frac{\partial^2 x_1}{\partial \tau_0 \partial \tau_1} - \tfrac{2}{3}\left(\phi_1\frac{\partial^2 x_2}{\partial \tau_0^2} + \phi_2\frac{\partial^2 x_1}{\partial \tau_0^2}\right),\end{aligned} \tag{13}$$

the previous partial differential equations system can be solved successively. In order to achieve this purpose, we start with the general solutions of Eqs. (8) and (9) in the form

$$x_1 = \mathrm{A}_1 e^{i\omega_1 \tau_0} + \overline{\mathrm{A}}_1 e^{-i\omega_1 \tau_0}, \tag{14}$$

$$\phi_1 = A_2 e^{i\omega_2\tau_0} + \overline{A}_2 e^{-i\omega_2\tau_0}, \tag{15}$$

where (A_i; $i = 1, 2, 3$) represent unknown complex functions of τ_1 and τ_2 while $\overline{A}_i$ denotes to its complex conjugate.

Substitution of the solutions (14) and (15) into the equations of higher order (10) and (11) produce secular terms. In order to eliminate these terms, substituting (14) and (15) into (10) and (11) to obtain the required conditions for this purpose in the form

$$\frac{\partial A_1}{\partial \tau_1} = 0, \quad \frac{\partial A_2}{\partial \tau_2} = 0. \tag{16}$$

Consequently, the second order solutions become

$$x_2 = \frac{2\,m\omega_2^2 A_2\overline{A}_2}{\omega_1^2} + \frac{e^{2i\omega_2\tau_0}\,(\omega_2 - 2\,i)\,m\,\omega_2\,{A_2}^2}{\left(\omega_1^2 - 4\,\omega_2^2\right)} + CC, \tag{17}$$

$$\phi_2 = \frac{2\,e^{i(\omega_1+\omega_2)\tau_0}\omega_1 A_1 A_2}{3\,(\omega_1 + 2\omega_2)} + \frac{2\,e^{i(\omega_1-\omega_2)\tau_0}\omega_1 A_1\overline{A}_2}{3\,(\omega_1 - 2\omega_2)} + CC, \tag{18}$$

where CC refers to the complex conjugates of the preceding terms.

Referring to the above procedure, the elimination of the secular terms required that the functions (A_i; $i = 1, 2$) depend upon the time scale τ_2 only.

The solutions for Eqs. (12) and (13) of the third order approximations can be obtained as previously in a similar way. The elimination of the secular terms in (12) and (13) demands the following conditions

$$-2\,i\,\tilde{C}\,\omega_1\,A_1 - 2i\omega_1\frac{\partial A_1}{\partial \tau_2} - \frac{4\,m\omega_1^4 A_1 A_2\overline{A}_2}{3\left(\omega_1^2 - 4\,\omega_2^2\right)} = 0. \tag{19}$$

$$\begin{aligned} -2i\omega_2\frac{\partial A_2}{\partial \tau_2} &- \frac{4\,\omega_1^3\,A_1A_2\overline{A}_1}{9\,(\omega_1 + 2\omega_2)} \\ &- \frac{\omega_2^2\left(3\,\omega_1^2 + 4\,\omega_2\,(-8\,i\,m + (-3 + 4m)\,\omega_2)\right)A_2^2\overline{A}_2}{6\left(\omega_1^2 - 4\,\omega_2^2\right)} = 0. \end{aligned} \tag{20}$$

The solutions for Eqs. (12) and (13), after eliminating secular terms, take the form

$$\begin{aligned} x_3 = {} & \frac{\tilde{F}_1 e^{i\Omega_1\tau_0}}{2(\omega_1^2 - \Omega_1^2)} - \frac{2\,m\,\omega_1(\omega_1+2\omega_2)A_1A_2^2 e^{i\tau_0(\omega_1+2\omega_2)}}{3\left[\omega_1^2-(\omega_1+2\omega_2)^2\right]} \\ & + \frac{2\,m\omega_1(\omega_1-2\omega_2)A_1\overline{A}_2^2 e^{i\tau_0(\omega_1-2\omega_2)}}{3\left[\omega_1^2-(\omega_1-2\omega_2)^2\right]} + CC \end{aligned} \tag{21}$$

$$\phi_3 = -\frac{e^{3\,i\,\omega_2\tau_0}\left(\omega_1^2+4\,\omega_2(-8\,i\,m+(-1+4\,m)\omega_2)\right)A_2^3}{48\,\left(\omega_1^2-4\,\omega_2^2\right)} - \frac{4\,e^{i(2\omega_1+\omega_2)\tau_0}\omega_1^3A_1^2A_2}{9(\omega_1+2\,\omega_2)\left[\omega_2^2-(2\omega_1+\omega_2)^2\right]} - \frac{4\,e^{i(2\omega_1-\omega_2)\tau_0}\omega_1^3A_1^2\overline{A}_2}{9(\omega_1-2\,\omega_2)\left[\omega_2^2-(2\omega_1-\omega_2)^2\right]} + CC, \quad (22)$$

The unknown functions (A_i; $i = 1, 2$) can be estimated from the system with the aid of the following initial conditions $\phi(0) = z_{01}$, $\dot{\phi}(0) = z_{02}$, $u(0) = z_{03}$, $\dot{u}(0) = z_{04}$, .

4 Vibrations and Resonance Conditions

Referring to the above procedure of the analytical solutions, the values of resonance parameters can be determined. The resonance cases can be attained if any of the polynomials in the denominators tends to zero and can be classified as

Primary external resonance, at $\Omega_1 = \omega_1$
Internal resonance occurred, if $\omega_1 = \omega_2$ are satisfied.

If any resonance case is satisfied, in particular the case of incidence internal resonance, we can predict that the dynamical behaviour of the system will be very complicated, the previous approximated solutions considered in the above section are valid if oscillations run away from resonances. If any one of the above listed conditions are satisfying; which indicates the need to adjust the used method.

5 External Resonances

5.1 Solvability Conditions

Here, we discuss the simultaneously occurring three primary external resonances case. Therefore, we consider that the combinations $\Omega_1 \approx \omega_1$ is satisfied which qualitatively describes the nearness of Ω_1 to ω_1. If we introduce the detuning parameters ($\sigma_i; i = 1, 2$) as in the form

$$\begin{aligned} \Omega_1 &= \omega_1 + \sigma_1, \\ 2\omega_1 &= 2\omega_2 + \varepsilon\sigma_2\ . \end{aligned} \quad (23)$$

Then the effectiveness of resonance is reflected in the secular terms. The detuning parameters are considered a measure of distance of the vibrations from the strict resonance. Therefore, we express them in terms of the small parameter ε as

$$\sigma_i = \varepsilon\tilde{\sigma}_i; \qquad i = 1, 2. \tag{24}$$

Substituting about the detuning parameters from (23) and (24) into Eq. (2) and focusing attention on the secular terms, then the solvability conditions are obtained as a result of the elimination of secular terms, in which that can be written as for the 2nd order approximation

$$\frac{\partial A_1}{\partial \tau_1} = 0, \qquad \frac{\partial A_2}{\partial \tau_1} = 0, \tag{25}$$

-For the 3rd approximation

$$\frac{1}{2}\tilde{F}_1\, e^{i\tau_1\tilde{\sigma}_1} - 2\, i\, \tilde{C}\, \omega_1\, A_1 - 2i\omega_1 \frac{\partial A_1}{\partial \tau_2} - \frac{4\, m\omega_1^4 A_1 A_2 \bar{A}_2}{3(\omega_1^2 - 4\,\omega_2^2)} = 0,$$

$$- \frac{4\,\omega_1^3\, A_1^2 \bar{A}_2\, e^{i\tau_1\tilde{\sigma}_2}}{9(\omega_1 - 2\,\omega_2)} - 2i\omega_2 \frac{\partial A_2}{\partial \tau_2} - \frac{4\,\omega_1^3\, A_1 A_2 \bar{A}_1}{9(\omega_1 + 2\omega_2)} - \frac{\omega_2^2\big(3\,\omega_1^2 + 4\,\omega_2(-8\, i\, m + (-3+4m)\omega_2)\big)A_2^2 \bar{A}_2}{6(\omega_1^2 - 4\,\omega_2^2)} = 0. \tag{26}$$

6 Problem's Modulation Close to Resonances

Based on the above section, we can see that the solvability conditions of the considered model constitute a system consists of four nonlinear partial deferential equations in terms of unknown function A_i. It is worthwhile to notice from Eq. (25) that, the functions A_i depend upon the slow time scale τ_2 only. Therefore, we can express these functions in the polar notation as

$$A_i = \frac{\tilde{a}_i\,(\tau_2)}{2} e^{i\tilde{\psi}_i\tau_2}, \qquad a_i = \varepsilon\tilde{a}_i; \quad i = 1, 2. \tag{27}$$

Since A_i are independent functions of variable τ_0 andτ_1, then the first order derivative operator can be simplified to the form

$$\frac{\partial A_i}{\partial \tau} = \varepsilon^2 \frac{\partial A_i}{\partial \tau_2}; \quad (i = 1, 2)\,. \tag{28}$$

Bearing in mind the above formula (28), Eq. (26) turn into ordinary differential equations. In order to transform them into an autonomous ones, the modified phases can be introduced in the form

$$\begin{aligned} \theta_1\,(\tau_1, \tau_2) &= \tau_1\, \tilde{\sigma}_1 - \psi_1\,(\tau_2)\,, \\ \theta_2\,(\tau_1, \tau_2) &= \tau_1\, \tilde{\sigma}_2 + 2\, \left[\psi_1\,(\tau_2) - \psi_2\,(\tau_2)\right]. \end{aligned} \tag{29}$$

Substituting (27), (28), and (29) into (26), separating the real and imaginary parts, one obtains directly the following system that consists of four ordinary differential equations from first order in terms of θ_1, θ_2, a_1, and a_2

$$\begin{aligned}
a_1 \frac{d\theta_1}{d\tau} &= \frac{F_1}{2\omega_1}\cos\theta_1 + a_1\sigma_1 - \frac{m\, a_1 a_2^2\, \omega_1^3}{6(\omega_1^2 - 4\,\omega_2^2)},\\
\frac{da_1}{d\tau} &= \frac{F_1}{2}\sin\theta_1 - \omega_1\mu\, a_1,\\
a_2 \frac{d\theta_2}{d\tau} &= a_2\,(\sigma_2 + 2\sigma_1) - 2a_2\frac{d\theta_1}{d\tau} - \frac{\omega_1^3 a_1^3 a_2}{9\omega_2(\omega_1 + 2\omega_2)}\\
&\quad + \frac{a_2^3\,(3\,\omega_1^2\omega_2 + 4(-3+4m)\omega_2^3)}{24(\omega_1^2 - 4\,\omega_2^2)} - \frac{\omega_1^3 a_1^2 a_2\,\cos\theta_2}{9\omega_2(\omega_1 - 2\omega_2)},\\
\frac{da_2}{d\tau} &= -\frac{2\,m\,\omega_2^2\,a_2^3}{3(\omega_1^2 - 4\,\omega_2^2)} - \frac{\omega_1^3 a_1^2\,\sin\theta_2}{18\omega_2(\omega_1 - 2\omega_2)}.
\end{aligned} \tag{30}$$

With a view to solve this system, we can transform the previous initial conditions according to the new variables as

$$a_i(0) = 0.004, \quad \theta_i(0) = 0, \; where\; i = 1, 2.$$

Equations (30) have the solutions a_1 and θ_1 that govern both of amplitudes and phases modulation in terms of the slow time scale when two investigated resonances occur with each other. Therefore, the solutions of these equations can be plotted as in Figs. 2, 3, 4, and 5 after taking into consideration the following values of parameters where $M = 25\,kg$, $k = 50$, $c = 0.01$, $\sigma 1 = 0.001$, $\sigma 2 = 0.003$. We calculated the time histories curves in Figs. 6 and 7 of the desired solutions φ and u up to the third approximations.

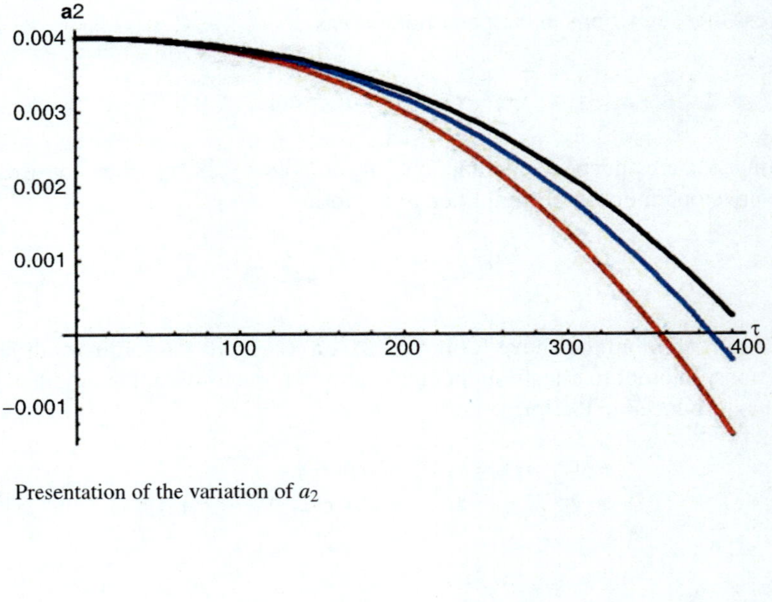

Fig. 2 Presentation of the variation of a_2

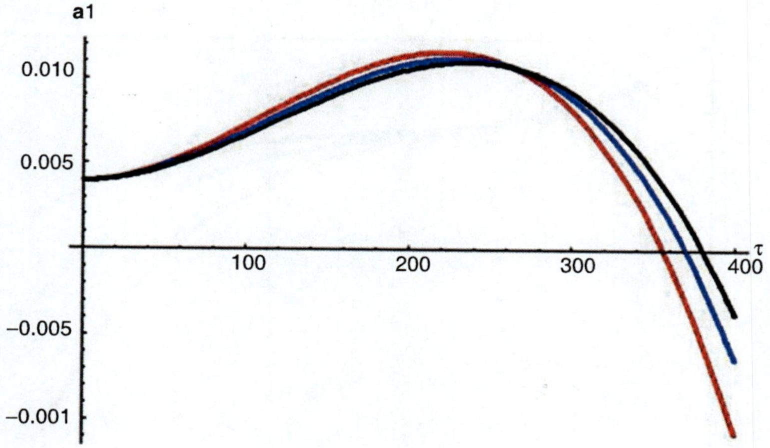

Fig. 3 Presentation of the variation of a_1

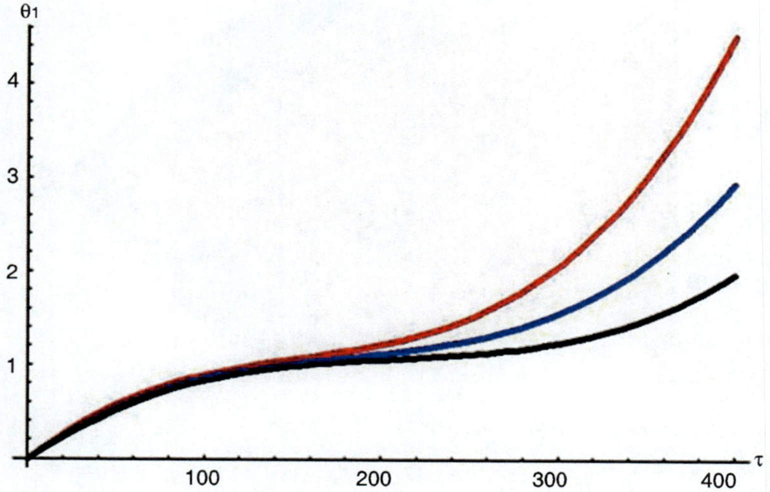

Fig. 4 Description of the variation of θ_1

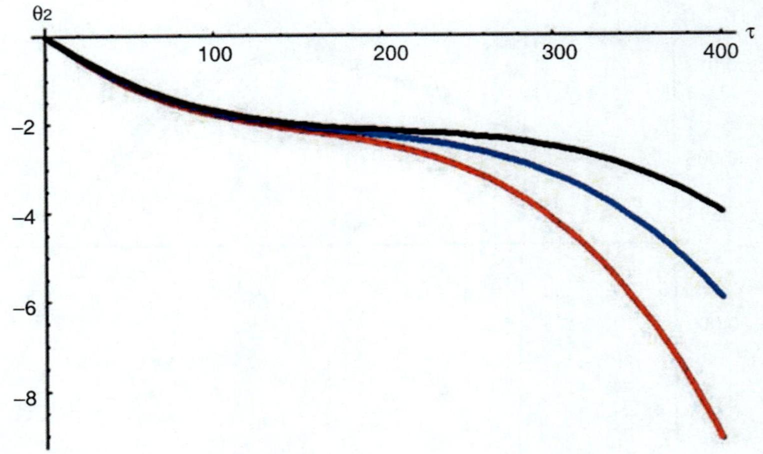

Fig. 5 Description of the variation of θ_2

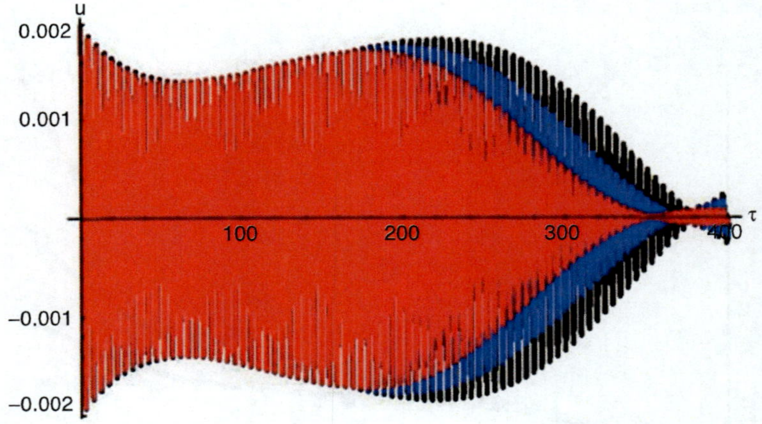

Fig. 6 Representation of the variation of u

7 Steady-State Solutions

The main objective of this section is to study the steady-state vibrations of the considered model. It is known that the steady state vibration arises if the behaviour of the transient processes disappears owing to the damping of the system. The amplitudes and modified phases of steady-state can be obtained from the Eq. (30), in order to explore such a case, let us assume that $\left(\frac{d\theta_i}{dt} = \frac{da_i}{dt} = 0; \;\; i = 1, 2, 3\right)$ equal

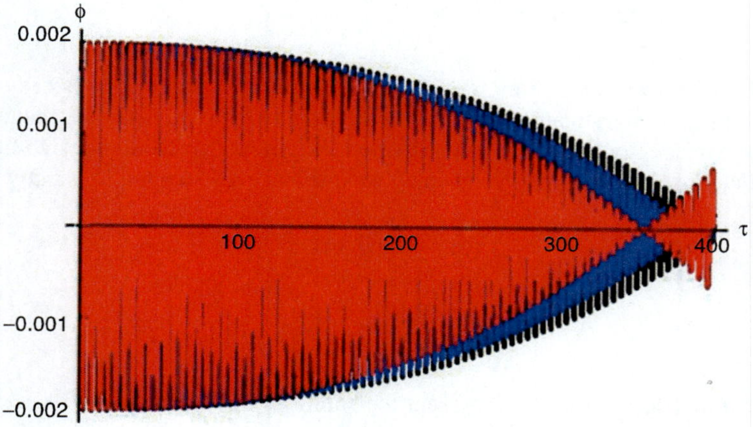

Fig. 7 Representation of the variation of φ

zero. Therefore, we have a system of algebraic equations in terms of the unknowns h_i and θ_i; $i = 1, 2$.

$$
\begin{aligned}
&\frac{F_1}{2\omega_1}\cos\theta_1 + a_1\sigma_1 - \frac{m\, a_1 a_2^2\, \omega_1^3}{6(\omega_1^2-4\,\omega_2^2)} = 0,\\
&\frac{F_1}{2}\sin\theta_1 - \omega_1\mu\, a_1 = 0,\\
&a_2\,(\sigma_2 + 2\sigma_1) - \frac{\omega_1^3 a_1^2 a_2}{9\omega_2(\omega_1+2\omega_2)}\\
&\quad + \frac{a_2^3\,(3\,\omega_1^2\omega_2+4(-3+4m)\omega_2^3)}{24(\omega_1^2-4\,\omega_2^2)} - \frac{\omega_1^3 a_1^2 a_2\,\cos\theta_2}{9\omega_2(\omega_1-2\omega_2)} = 0,\\
&-\frac{2\,m\,\omega_2^2\,a_2^3}{3(\omega_1^2-4\,\omega_2^2)} - \frac{\omega_1^3 a_1^2\,\sin\theta_2}{18\omega_2(\omega_1-2\omega_2)} = 0.
\end{aligned}
\tag{31}
$$

Elimination of the modified phases θ_1 and θ_2 from Eq. (31), produces the relationships between both of the amplitudes and the frequency clarified by the detuning parameters

$$
\begin{aligned}
F_1^2 &= \left[-\frac{m\,a_1 a_2^2\,\omega_1^4}{3(\omega_1^2-4\,\omega_2^2)} - 2a_1\omega_1\sigma_1\right]^2 + \omega_1^2\mu^2\,a_1^2,\\
\frac{\omega_1^6\,a_1^4}{81\,\omega_2^2(\omega_1-2\omega_2)^2} &= \left[(\sigma_2 + 2\sigma_1) - \frac{\omega_1^3 a_1^2}{9\omega_2(\omega_1+2\omega_2)} + \frac{a_2^3\,(3\,\omega_1^2\omega_2+4(-3+4m)\omega_2^3)}{24(\omega_1^2-4\,\omega_2^2)}\right]^2\\
&\quad + \left[\frac{4\,m\,\omega_2^2\,a_2^3}{3(\omega_1^2-4\,\omega_2^2)}\right]^2.
\end{aligned}
\tag{32}
$$

It is worthwhile to notice that Eq. (32) is considered as implicit nonlinear algebraic equation with respect to the variables a_1 and a_2.

8 Steady State Analysis

One of the important factors for the mentioned problem of the steady-state oscillations is to investigate their stability. For this task, we analyse the manner of the system in a region that is very close to the fixed points. To discuss the stability for the particular solution of the steady state, we introduce the substitutions into (30).

$$
\begin{aligned}
a_1 &= a_{10} + a_{11}, \\
a_2 &= a_{20} + a_{21}, \\
\theta_1 &= \theta_{10} + \theta_{11}, \\
\theta_2 &= \theta_{20} + \theta_{21}.
\end{aligned} \tag{33}
$$

Here a_{10}, θ_{10}, a_{20} and θ_{20} represent the solutions of (34) and a_{11}, θ_{11}, a_{21} and θ_{21} denote perturbations which are assumed to be very small, compared to the predecessors. Then the linearized equations take the form

$$
\begin{aligned}
a_{10}\frac{d\theta_{11}}{d\tau} &= -\frac{F_1}{2\omega_1}\theta_{11}\sin\theta_{10} + a_{11}\left[\sigma_1 - \frac{m\ \omega_1^3 a_{10}^2}{6(\omega_1^2-4\ \omega_2^2)}\right] - \frac{m\ \omega_1^3 a_{10}a_{20}a_{21}}{3(\omega_1^2-4\ \omega_2^2)}, \\
\frac{da_{11}}{d\tau} &= \frac{F}{2}\theta_{11}\cos\theta_{10} - \omega_1\mu\ a_{11}, \\
a_{20}\frac{d\theta_{21}}{d\tau} &= a_{21}\left[\sigma_2 + 2\sigma_1 - \frac{\omega_1^3 a_{10}^2}{9\omega_2(\omega_1+2\omega_2)} + \frac{a_2^3\ (3\ \omega_1^2\omega_2+4(-3+4m)\omega_2^3)}{24(\omega_1^2-4\ \omega_2^2)}\right] \\
&\quad - 2a_{20}\frac{d\theta_{11}}{d\tau} - \frac{2\omega_1^3 a_{20}a_{10}a_{11}}{9\omega_2(\omega_1+2\omega_2)}, \\
\frac{da_{21}}{d\tau} &= -\frac{2m\ \omega_2^2\ a_{20}^2\ a_{21}}{(\omega_1^2-4\ \omega_2^2)} - \frac{\omega_1^3\ a_{20}\ \theta_{21}\cos\theta_{20}}{18\omega_2(\omega_1-2\ \omega_2)}.
\end{aligned} \tag{34}
$$

Take into consideration that the small perturbations a_{11}, θ_{11}, a_{21} and θ_{21} are unknown functions. Every solution is a linear combination of $k_i\, e^{\lambda\tau}$, where k_i $i = 1$, 2, 3, 4 are constants and λ is the eigenvalue corresponding to the unknown perturbation, counted from the real parts of the roots. In this analysis, if the steady-state solutions (fixed points) a_{10}, θ_{10}, a_{20} and θ_{20} are asymptotically stable, the real parts of the roots of the following characteristic equation

$$
\lambda^4 + \Gamma_1\lambda^3 + \Gamma_2\lambda^2 + \Gamma_3\lambda + \Gamma_4 = 0. \tag{35}
$$

of the set of (34), must be negative. Here Γ_1, Γ_2, Γ_3 and Γ_4 take the form

$$
\begin{aligned}
\Gamma_1 =\ & \tfrac{F\ \sin\theta_{10}}{2a_{10}\omega_1} + \tfrac{\mu\ \omega_1^3 + 2(m\ a_{20}^2 - 2\mu\ \omega_1)\ \omega_2^2}{\omega_1^2 - 4\ \omega_2^2},\\
\Gamma_2 =\ & \tfrac{1}{1296\ a_{10}a_{20}\ \omega_1\omega_2^2(\omega_1 - 2\omega_2)(\omega_1 + 2\omega_2)} \big[108\ a_{20}\ (\omega_1 - 2\omega_2)\ \omega_2^2\big(24m\mu\ a_{10}a_{20}^2\omega_1^2\omega_2^2\\
& + 6\ F \sin\theta_{10}\left(\mu\omega_1^3 + 2\left(ma_{20}^2 - 2\mu\omega_1\right)\omega_2^2\right)\\
& - F\ \omega_1 \cos\theta_{10}\left(ma_{20}^2\omega_1^3 - 6\ \sigma_1\left(\omega_1^2 - 4\ \omega_2^2\right)\right)\big)\\
& + a_{10}^2\ \omega_1^4 \cos\theta_{10}\left(-8a_{10}^2\ \omega_1^3\ (\omega_1 - 2\omega_2) + 3\omega_2\right)\\
& \times \big(48\sigma_1\left(\omega_1^2 - 4\ \omega_2^2\right) + 24\sigma_1\left(\omega_1^2 - 4\ \omega_2^2\right)\\
& + a_{20}^2\left(-16m\omega_1^3 + 9\omega_1^2\omega_2 + 12\,(-3 + 4m)\ \omega_2^3\right)))\big],\\
\Gamma_3 =\ & \tfrac{1}{2592\ a_{10}a_{20}\omega_2^2\left(\omega_1^2 - 4\ \omega_2^2\right)^2}\left[a_{10}\omega_1^2\ \cos\theta_{20}\ (\omega_1 + 2\omega_2)\big(-\ 8a_{10}^2\omega_1^4\big(F\ \sin\theta_{10}\right.\\
& + 2\mu a_{10}\omega_1^2\big) + 8\omega_1^2\big(F\ \sin\theta_{10}\left(18\sigma_1 + 9\sigma_2 + 2a_{10}^2\ \omega_1\right) + 2\mu a_{10}\omega_1^2\\
& \left(18\sigma_1 + 9\sigma_2 + 2\left(a_{10}^2 - 3ma_{20}^2\right)\omega_1\right)\big)\omega_2 + 27a_{20}^2\omega_1^2\left(F\ \sin\theta_{10} + 2\mu a_{10}\omega_1^2\right)\omega_2^2\\
& - 288\,(2\sigma_1 + \sigma_2)\ \left(F\ \sin\theta_{10} + 2\mu a_{10}\omega_1^2\right)\omega_2^3\\
& + 36\,(-3 + 4m)\,a_{20}^2\left(F\ \sin\theta_{10} + 2\mu a_{10}\omega_1^2\right)\omega_2^4\\
& - 432\ F\ ma_{20}^2\ \omega_2^4\left(-6\mu \sin\theta_{10}\left(\omega_1^2 - 4\ \omega_2^2\right)\right.\\
& \left.\left. + \cos\theta_{10}\left(ma_{20}^2\ \omega_1^3 - 6\sigma_1\left(\omega_1^2 - 4\ \omega_2^2\right)\right)\right)\right],\\
\Gamma_4 =\ & \tfrac{F \cos\theta_{20}\ \omega_1^3}{5184\ a_{20}\omega_2^2(\omega_1 - 2\omega_2)^3(\omega_1 + 2\omega_2)^3}\ [-3\omega_2(8\,(2\sigma_1 + \sigma_2)\,\omega_1^2 + 3a_{20}^2\omega_1^2\omega_2\\
& - 32\,(2\sigma_1 + \sigma_2)\,\omega_2^2 + 4\,(-3 + 4m)\,a_{20}^2\omega_2^3\,)(-\ 6\mu \sin\theta_{10}\left(\omega_1^2 - 4\ \omega_2^2\right)\\
& + \cos\theta_{10}\left(ma_{10}^2\omega_1^3 - 6\sigma_1\left(\omega_1^2 - 4\ \omega_2^2\right)\right))\\
& - 8a_{10}^2\omega_1^3\ (\omega_1 - 2\omega_2)\left(2\mu \sin\theta_{10}\left(\omega_1^2 - 4\ \omega_2^2\right)\right.\\
& \left. + \cos\theta_{10}\left(ma_{10}^2\omega_1^3 - 6\sigma_1\left(\omega_1^2 - 4\ \omega_2^2\right)\right)\right)].
\end{aligned}
\tag{36}
$$

However, according to the Routh-Hurwitz criterion, the fundamental conditions of the stability for the particular steady-state solutions will be

$$
\begin{aligned}
& \Gamma_1 > 0,\\
& \Gamma_3\left(\Gamma_1\Gamma_2 - \Gamma_3\right) - \Gamma_4\Gamma_1^2 > 0,\\
& \Gamma_1\Gamma_2 - \Gamma_3 > 0,\\
& \Gamma_4 > 0.
\end{aligned}
\tag{37}
$$

The stability of the system amplitudes are varying with different spring stiffness value. As presented in Figs. 8 and 9 the dashed line (on the lift side) represent the unstable region. Where, the solid line represents the stable region. The solid red line represents the stable region of a_1 and the blue color represents the stable region of a_2.

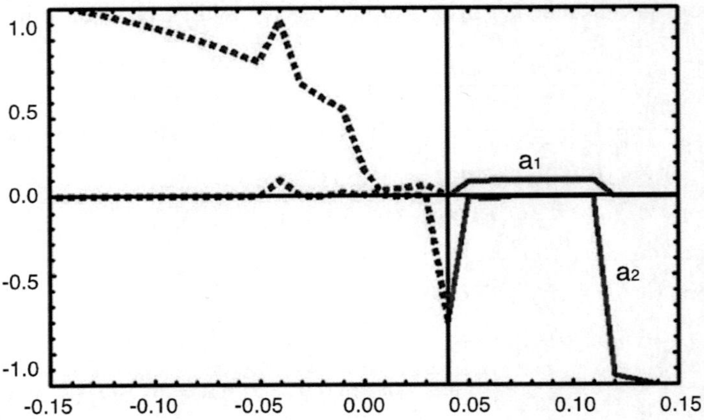

Fig. 8 stability for a_1 *and* a_2 *with* σ_1 *at* $k = 2000$

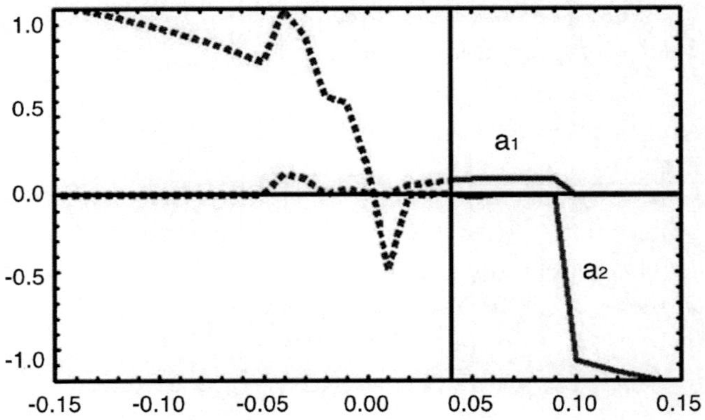

Fig. 9 stability for a_1 *and* a_2 *with* σ_1 *at* $k = 1500$

9 Conclusions

This work outlines the vibrating motion of a cylinder over circular under the influence of an exciting force. Lagrange's principles were utilized obtain the governing equation of the system's motion taking into account the presence of external forces acting on the vertical direction. The multiple Scale technique is used to obtain the asymptotic solutions up to third order and to gain the modulation equations in frame work of the solvability conditions. The various resonance cases, primary external resonance and internal one, are studied. The system stability is checked according to Routh-Hurwitz criterion condition. The graphical representations of time history of motion, resonance cases are presented through some plots to highlight the effectiveness of different physical parameters on the motion. The importance of this work is due to its direct applications in the fields of engineering machines which needs insight investigation in order to reduce the system vibrations.

References

1. Amer, T.S., Bek, M.A., Abouhmr, M.K.: On the vibrational analysis for the motion of a harmonically damped rigid body pendulum. Nonlinear Dynamics (2018)
2. Amer, W.S., Bek, M.A., Abohamer, M.K.: On the motion of a pendulum attached with tuned absorber near resonances. Results in Physics. **11**, 291–301 (2018)
3. Legeza, V.: Dynamics of vibration isolation system with a quasi-isochronous roller shock absorber. Int. Appl. Mech. **47**(3), 329 (2011)
4. Kevorkian, J., Cole, J., Nayfeh, A.H.: Perturbation methods in applied mathematics. Bull. Am. Math. Soc. **7**, 414–420 (1982)
5. Nayfeh, A.H.: Introduction to Perturbation Techniques. Wiley, London (2011)
6. Wolfram, S.: The mathematica book. Assembly Automation (1999)

Nonlinear Dynamics of the Hierarchic System of Oscillators

Sergiy Mykulyak and Sergii Skurativskyi

Abstract A significant part of materials under appropriate conditions manifests their internal structure. In particular, this concerns the geomedia which are endowed with discrete and hierarchic structure. To examine the dynamics of such systems, we develop the mathematical model on the basis of Hamiltonian formalism. This model describes the motion of the hierarchically connected oscillators interacting with each other via the power law. For certain simplifying constraints, we reduce the model to the three level strongly nonlinear system of ODE. The problem considered is the analysis of the system dynamics, when the friction is incorporated and the harmonic force is applied to the most upper level of the system. Using the numerical and qualitative analysis methods, the existence of periodic, quasiperiodic and chaotic attractors are revealed. The bifurcations of these regimes with respect to the structural parameter are studied in more detail. The statistical properties of chaotic attractors are considered as well.

Keywords Hierarchic structures · Coupled oscillators · Quasiperiodicity · Dynamic chaos · Tsallis index

1 Introduction

The rocks forming the lithosphere are significantly heterogeneous. The heterogeneity is related to both the heterogeneity of rock compositions and the existence of crack and fault networks. Such defect networks cause the selection of rock fragments that can be considered as separate discrete elements. Discreteness is observed in a wide range of levels: from rock pieces that can be observed in

S. Mykulyak · S. Skurativskyi (✉)
Subbotin Institute of Geophysics of the NAS of Ukraine, Kyiv, Ukraine
e-mail: mykulyak@ukr.net

J. Awrejcewicz (ed.), *Perspectives in Dynamical Systems I: Mechatronics and Life Sciences*, Springer Proceedings in Mathematics & Statistics 362,
https://doi.org/10.1007/978-3-030-77306-9_2

quarries or mountains with centimeters or meters in size to tectonic blocks and plates having the dimensions from kilometers to thousands kilometers. Sadovskiy et al. [1] performed a statistical analysis of discrete rock fragments and concluded that their distribution over a wide range of scales has a modal nature: there are allocated sizes of fragments and the relationships between adjacent sizes (hierarchical levels) are in a narrow interval $L_{i+1}/L_i = K$, where $K \in [2; 5]$ [1]. The process of energy redistribution between hierarchical levels plays an important role in the dynamics of discrete hierarchical media, in particular in the processes of earthquake preparation, seismic energy release during earthquakes themselves, and seismic wave propagation. To study the energy redistribution processes in hierarchical discrete media, the model of embedded oscillators has been proposed [2, 3]. The oscillators forming the hierarchic layers interact via the power law with the oscillators placing on the higher layers. The equations of motion for this model derived within the framework of Hamiltonian formalism possess complicated solutions including periodic, quasiperiodic, and chaotic ones.

There is also an important problem concerning the behavior of hierarchic media under the external excitation applied to the upper level of hierarchic system. In the case of weak dissipative processes and small harmonic loading, the resonant phenomena in the three-layer hierarchic system have been considered in detail [3, 4].

Now the question arises how the hierarchical system behaves when the small quantities are not expected? To elucidate this problem, the three-layer model incorporating the linear viscous friction and harmonic loading is studied by the qualitative analysis methods accompanied by the Fourier spectra analysis, Lyapunov and Tsallis index derivations.

2 Construction of the Mathematical Model for a Hierarchic System

The general approach to hierarchic system description has been developed in [2–4]. Now we are going to deal with the simplified model when the number of structural parameters is reduced as much as possible.

We thus consider the model schematically depicted in Fig. 1 and consisted of $n = 3$ layers. Each oscillator is characterized by the coordinate x_{jk}, where j is the layer number, k is the oscillator's position in this layer. All oscillators of the jth layer have identical masses m_j, $j = 1, 2, 3$. Pair of oscillators belonging to the adjacent layers interacts with the force having the power potential. The bond stiffness does not depend on the oscillator position in the layer, i.e. $C_{jk} = C_j$.

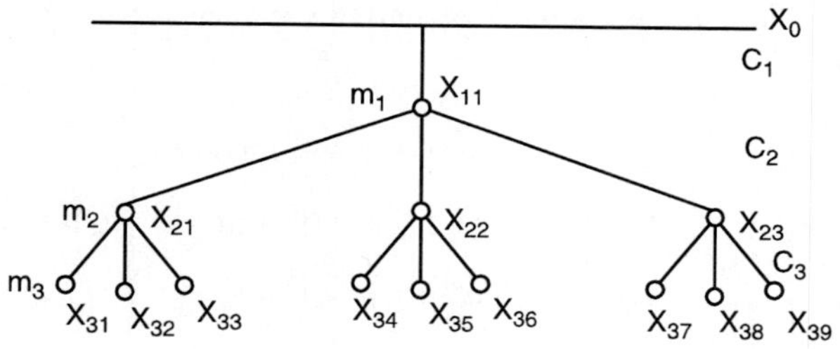

Fig. 1 The schematic representation of the model for hierarchical medium

Thus, the corresponding Hamiltonian is as follows

$$H = \frac{p_{11}^2}{2m_1} + \frac{C_1}{\beta+1}|x_{11} - x_0|^{\beta+1} + \sum_{k=1}^{3} \frac{p_{2k}^2}{2m_2} + \frac{C_2}{\beta+1}|x_{2k} - x_{11}|^{\beta+1} +$$
$$\sum_{k=1}^{3} \frac{p_{3k}^2}{2m_3} + \frac{C_3}{\beta+1}|x_{3k} - x_{21}|^{\beta+1} + \sum_{k=4}^{6} \frac{p_{3k}^2}{2m_3} + \frac{C_3}{\beta+1}|x_{3k} - x_{22}|^{\beta+1} +$$
$$\sum_{k=7}^{9} \frac{p_{3k}^2}{2m_3} + \frac{C_3}{\beta+1}|x_{3k} - x_{23}|^{\beta+1},$$

where p_{jk} is the momentum of kth oscillator placed on the jth level, $x_0 = \text{const}$.

Corresponding equations of motion

$$\dot{x}_{nk} = \frac{\partial H}{\partial p_{nk}}, \qquad \dot{p}_{nk} = -\frac{\partial H}{\partial x_{nk}}$$

lead us to the system of 13 ordinary differential equations. For instance, the equations with respect to x_{11}, x_{21}, and x_{31} can be written in the following form

$$\begin{gathered}
\ddot{x}_{11} = -\omega_1^2 |x_{11} - x_0|^\beta \chi_{11,0} + \\
\omega_2^2 \tfrac{m_2}{m_1} \left(|x_{21} - x_{11}|^\beta \chi_{21,11} + |x_{22} - x_{11}|^\beta \chi_{22,11} + |x_{23} - x_{11}|^\beta \chi_{23,11}\right), \\
\ddot{x}_{21} = -\omega_2^2 |x_{21} - x_{11}|^\beta \chi_{21,11} + \\
\omega_3^2 \tfrac{m_3}{m_2} \left(|x_{31} - x_{21}|^\beta \chi_{31,21} + |x_{32} - x_{21}|^\beta \chi_{32,21} + |x_{33} - x_{21}|^\beta \chi_{33,21}\right), \\
\ddot{x}_{31} = -\omega_3^2 |x_{31} - x_{21}|^\beta \chi_{31,21},
\end{gathered} \tag{1}$$

where $\omega_i^2 = C_i/m_i$, $\chi_{a,b} = \text{sgn}(x_a - x_b)$.

Let us assume that system (1) admits the solution when oscillators in the specified layer move synchronously. Then the components of solution are identical, i.e. $x_{3j} = x_3$, $x_{2j} = x_2$, $x_{11} = x_1$. The proof of such a regime existence in the general model is not a trivial problem, but in the simplified model such regime can occur due to the coincidence of the corresponding equations of the system and specifying the proper initial conditions for these equations.

Thus, the movement of partially synchronized oscillators is described by the system

$$\begin{aligned} \ddot{x}_1 &= -\omega_1^2 |x_1 - x_0|^\beta \chi_{1,0} + \omega_2^2 \varphi_1 |x_2 - x_1|^\beta \chi_{2,1}, \\ \ddot{x}_2 &= -\omega_2^2 |x_2 - x_1|^\beta \chi_{2,1} + \omega_3^2 \varphi_2 |x_3 - x_2|^\beta \chi_{3,2}, \\ \ddot{x}_3 &= -\omega_3^2 |x_3 - x_2|^\beta \chi_{3,2}, \end{aligned} \tag{2}$$

where $\varphi_1 = 3\frac{m_2}{m_1}$, $\varphi_2 = 3\frac{m_3}{m_2}$.

Suppose that the set m_i is the geometric sequence, i.e. $m_i = m_0 h^{i-1}$, $h < 1$. Then $\varphi_i = sh = \varphi = \text{const}$. Using the new variables $q_n = x_n - x_{n-1}$—displacements from the steady state, let us write model (2) in the form

$$\ddot{q}_1 = -F_1 + \varphi F_2, \qquad \ddot{q}_2 = F_1 - F_2(1+\varphi) + \varphi F_3, \qquad \ddot{q}_3 = F_2 - F_3(1+\varphi), \tag{3}$$

where $F_i = \omega_i^2 |q_i|^\beta \operatorname{sgn}(q_i)$.

Note that doing in a similar manner we can construct the hierarchic model with n layers obeying the system [2]

$$\begin{aligned} \ddot{q}_1 &= -F_1 + \varphi_1 F_2, \quad \ddot{q}_n = F_{n-1} - F_n(1+\varphi_{n-1}) + \varphi_n F_{n+1}, \\ \ddot{q}_N &= F_{N-1} - F_N(1+\varphi_{N-1}), \end{aligned} \tag{4}$$

where $\varphi_i = s m_{i+1}/m_i$. From the analysis of the last system it follows that at $\varphi_n = 1$ it reduces to the equations of motions for the chain of masses with pair interactions. Such a case can be realized when $s = 1$ and all masses are identical in particular, i.e. the chain without hierarchic structure. If $s \neq 1$, the system is hierarchic.

Even in the simple case of system (3) the strong nonlinearity of the system does not allow one to carry out the complete its investigation. But using the Poincaré section technique some typical model's solutions were revealed [2]. In particular, the quasiperiodic (Fig. 2a), periodic (Fig. 2b, the inset contains the phase portrait of complex periodic trajectory), and chaotic (Fig. 2c) regimes were distinguished. Note that to prove the chaoticity of trajectory presented in Fig. 2c the Lyapunov spectrum is suitable. It should be mentioned that for Hamiltonian systems the sum of Lyapunov exponents is equal to zero. Therefore, the periodic orbits are characterized by zero exponents, whereas among the Lyapunov spectrum of chaotic trajectory there is a pair of exponents with opposite signs. Using the classical numerical method [5], the Lyapunov spectrum $\lambda = \{0.0013, -0.0015, 0.00068, -0.00061, 0.00079, -0.00063\}$ was evaluated. Due to numerical errors, we should introduce some threshold [6] to distinguish zero exponents. We thus assume that $\lambda_{1,2} \neq 0$ while others are zero. Since the spectrum contains the positive index, the observed regime should be classified as a chaotic.

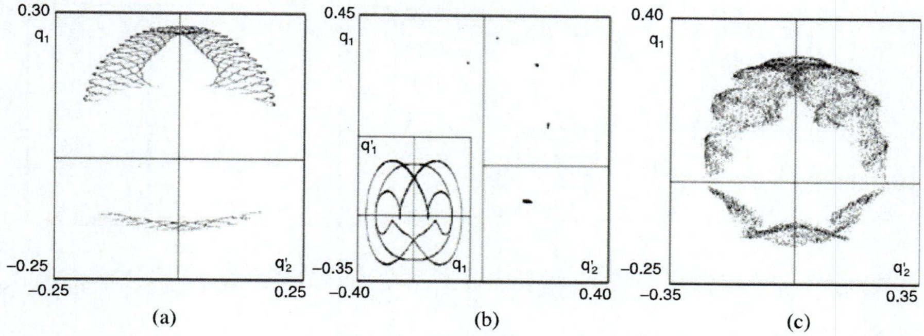

Fig. 2 The poincaré sections of system (3) for the quasiperiodic regime at $\beta = 1.05$ (**a**), periodic orbit at $\beta = 1.5$ (**b**), and chaotic trajectory at $\beta = 1.15$ (**c**)

3 Dynamics of Three-layer Dissipative Hierarchic System Under the Harmonic Loading

Incorporating the dissipative processes description and harmonic force $\gamma \sin \alpha t$ applied to the uppermost layer of three-layer dissipative hierarchic system (3), we lead to the problem of forced oscillations of hierarchic medium. In this case, after the substitution $t \to \alpha t$, system (3) can be written in the following form

$$\begin{aligned}
\alpha^2 \ddot{q}_1 &= -F_1 + \varphi F_2 - \alpha \mu_1 \dot{q}_1 + \gamma \sin t, \\
\alpha^2 \ddot{q}_2 &= F_1 - F_2(1+\varphi) + \varphi F_3 - \alpha \mu_2 \dot{q}_2 + \alpha \mu_1 \dot{q}_1, \\
\alpha^2 \ddot{q}_3 &= F_2 - F_3(1+\varphi) - \alpha \mu_3 \dot{q}_3 + \alpha \mu_2 \dot{q}_2.
\end{aligned} \tag{5}$$

Now the period of external loading is 2π. When the model is close to the linear one and force amplitude is small enough, to consider the resonant oscillating regimes, the small parameter method can be applied to the model [3]. Now parameters' smallness is not assumed and dynamical system (5) is studied by the qualitative analysis methods and numerically.

To do this, we fix the parameters $\omega = \{1, 1.1, 1.21\}$, the friction coefficients $\mu = \{0.01, 0.03, 0.05\}$, the Hertz law index $\beta = 3/2$, the amplitude of loading $\gamma = 0.18$ and its frequency $\alpha = 1.4$. It turned out that at these parameters the system possesses an especially rich set of solutions which we are going to discuss further. Namely, we will describe the regimes and their bifurcations when the parameter φ is varied.

The periodic trajectory can be derived numerically starting from the initial data $\{q_{1,2,3}, q'_{1,2,3}\} = \{-0.0009, 0.0015, -0.0006, -0.1087, 0.0183, -0.0014\}$ at $\varphi = 0.5$. The profile of q_2 component solution in Fig. 3a allows one to suppose that its period is 2π. To proof this, the Fourier spectrum (Fig.3b) of this trajectory is derived. Since it contains only one essential maximum at $\omega = 1$, the observed

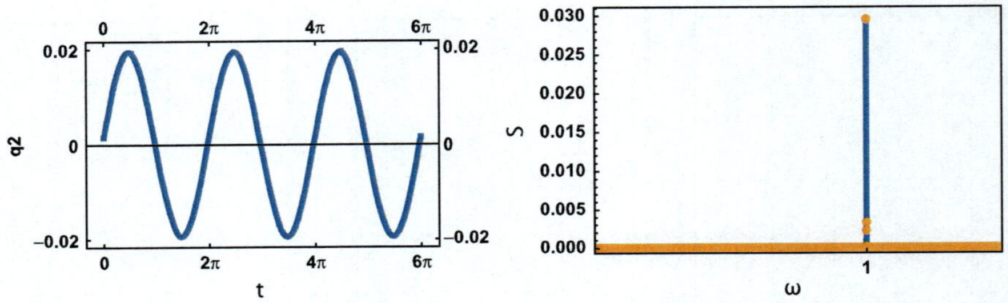

(a) (b)

Fig. 3 The profile of $q_2(t)$ and its Fourier spectrum at $\varphi = 0.5$

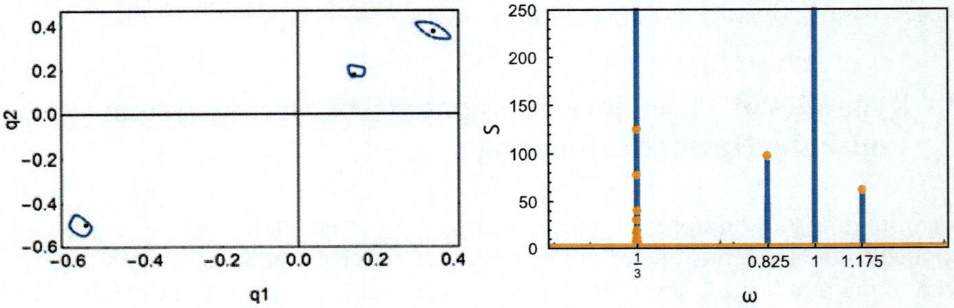

(a) (b)

Fig. 4 The Poincaré sections (**a**) at $\varphi = 0.6$ (separate black points) and $\varphi = 0.61$ (blue points forming closed curves). Panel (**b**) shows the Fourier spectrum of the regime depicted by closed curves in the left panel

regime corresponds to the oscillations with frequency 1 and presents the movement with the external frequency.

In addition to this mode, we reveal more complex periodic orbit at $\varphi = 0.6$ and initial conditions $\{q_{1,2,3}, q'_{1,2,3}\} = \{-0.5536, -0.4969, -0.3425, -0.0612, -0.0821, 0.2492\}$. To study this solution and others, the Poincaré section technique is used. Since we deal with the nonautonomous system, the Poincaré section consists of the trajectory points extracted at $t_n = 2\pi n$, $n = 1, 2, \ldots$. The resulting Poincaré section depicted in Fig. 4a contains three points that testifies about the existence of solution with period 6π. The Fourier spectrum of this solution contains two maxima at $\omega = 1/3$ and $\omega = 1$ (Fig.4b). Since these frequencies are commensurate, we can state that the observed mode presents the resonant torus in the system's phase space. The stability of trajectory and its type can be examined with the help of Lyapunov spectrum. Integrating system (5) together with its linearization during 3000 time units (and for all calculations below), the spectrum for the periodic trajectory $\lambda = \{-0.0035, -0.0036, -0.0051, -0.0046, -0.0236, -0.0239\}$ is evaluated. Accounting for the Fourier spectrum analysis and assuming that $\lambda_{1,2}$ correspond to zero Lyapunov exponents, we can conclude that the observed trajectory is the two-dimensional torus [6]. Note that such a regime is also recognized at lower values of φ till 0.465.

When the parameter $\varphi = 0.61$, the periodic trajectory loses its stability. Then the toroidal surface causing the creation of closed curves in the Poincaré section (Fig. 4a) appears. This torus corresponds to a quasiperiodic solution with incommensurate frequencies as it can be seen in the Fourier spectrum (Fig. 4b). Note that two auxiliary maxima 1 ± 0.175 appear in a vicinity of unit frequency. Deriving the Lyapunov spectrum, we obtain $\lambda = \{-0.0002, -0.0032, -0.0034, -0.0117, -0.0211, -0.0248\}$. Let us take into account that $\lambda_{1,2,3}$ are regarded as zero. Then from this it follows that the trajectory relates to the 3D torus.

Another quasiperiodic solution is observed at $\varphi = 0.48$ and initial data $\{0.0069, 0.0252, 0.0202, -0.1036, 0.0114, 0.0083\}$ the phase portrait of which is depicted in Fig. 5. It is worth noting that the amplitude of this regime is essentially smaller than in the solutions mentioned above. The Fourier spectrum possesses two main frequencies $\Omega_1 = 0.215$, $\Omega_2 = 0.570$, and the combinational frequency $3\Omega_1$. The Lyapunov spectrum is $\lambda = \{-0.0077, -0.0077, -0.0116, -0.0116, -0.0133, -0.0123\}$ and contains two zero exponents providing the 2D torus existence proof.

When we fix $\varphi = 0.50$ and initial data $\{0.0311, 0.0502, 0.01707, -0.0889, 0.0286, -0.0320\}$, the solution's Poincaré section (Fig. 5) is qualitatively similar to existing at $\varphi = 0.48$ but it is symmetrically mapped with respect to the origin.

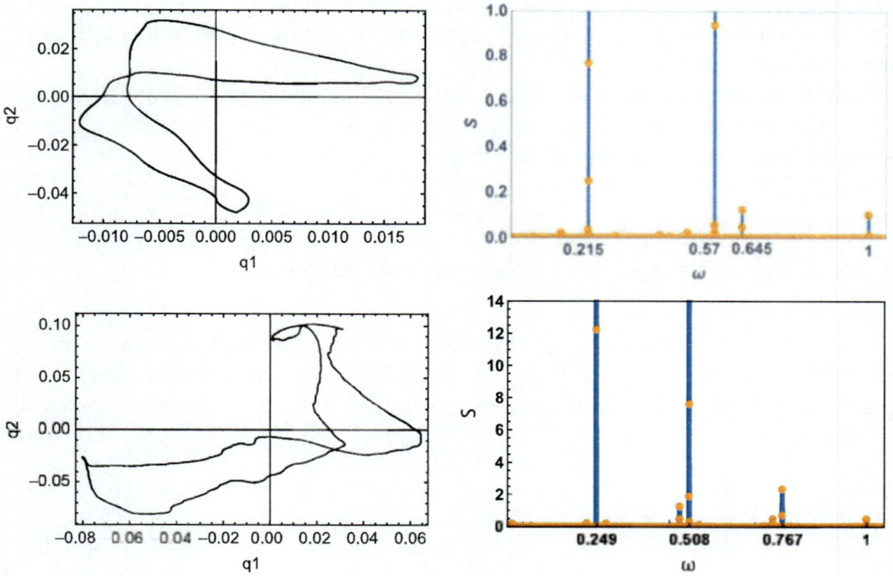

Fig. 5 The Poincaré sections and Fourier spectra at $\varphi = 0.48$ (top) and $\varphi = 0.50$ (bottom)

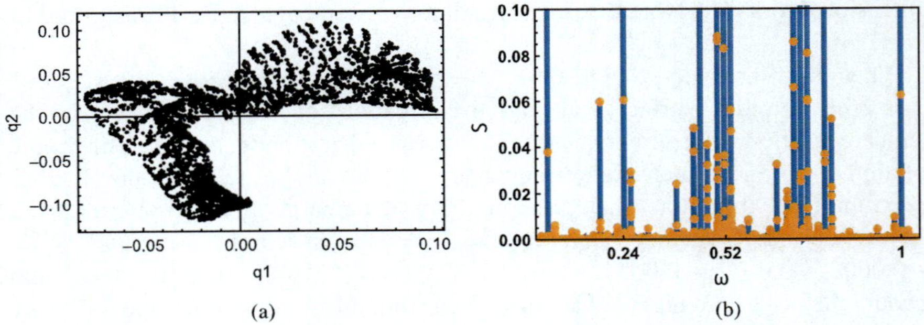

Fig. 6 The Poincaré sections at $\varphi = 0.561$ (**a**) and its Fourier spectrum (**b**)

Moreover, the spectrum of the signal contains two main frequencies $\Omega_1 = 0.249$ and $\Omega_2 = 0.508$, in addition, the combinational frequency $2\Omega_2 - \Omega_1 = 0.767$ is distinguished as well. The evaluation of the Lyapunov spectrum gives us the following exponents $\lambda = \{-0.0005, -0.0004, -0.0132, -0.0148, -0.0171, -0.0182\}$ among which two first quantities are assumed to be zero. This convinces in the 2D torus existence.

Choosing $\varphi = 0.561$ and initial data $\{-0.0965, -0.0275, 0.0027, -0.1227, 0.0536, -0.0178\}$, the Poincaré section of resulting attractor is of irregular structure (Fig. 6a). The corresponding Fourier spectrum contains a large number of exited frequencies that tells us about the possibility to observe a chaotic regime. According to the Lyapunov spectrum $\lambda = \{0.0025, -0.0009, -0.0051, -0.0174, -0.0202, -0.0231\}$ the top Lyapunov exponent is positive (λ_2 vanishes). From this it follows the attractor chaoticity.

The aforementioned results, in particular, show that in the phase space of system (2) can coexist at least three attractors with separated basins of attraction. Let us examine some statistical properties of these attractors.

4 Statistical Properties of System's Attractors

The statistical properties of the observed attractors, especially chaotic, can be examined by using the generalized entropy conception, introduced by Tsallis [7] and dealing with the nonergodic systems. Instead of classical entropy, the quantity S_T is defined

$$S_T = \frac{1 - \sum_{i=1}^{Q} p_i^q}{q - 1}, \qquad q \in R,$$

where q is the Tsallis (or entropy) index, $\sum_{i=1}^{Q} p_i = 1$. Note that the limit $q \to 1$ corresponds to the Boltzmann-Gibbs entropy $S_T \to -\sum_{i=1}^{Q} p_i \ln p_i$ and the deviation of q from 1 points to the appearance of long-range correlations. It has been shown that at $q < 1$ the physical system behavior depends on rare events, whereas at $q > 1$ the frequent events have more weight [8].

To estimate the index q, the approach proposed in [9] is used. According to this method, the maximum entropy principal on the base of S_T is applied. The auxiliary function known as redundancy R_T is defined

$$R_T(q) = 1 - \frac{S_T}{S_{T\,\max}},$$

where $S_{T\,\max} = \frac{1-Q^{1-q}}{q-1}$ is the maximum of the function S_T which is reached on the equiprobable microstates $p_i = 1/Q$. It is evident that for the nonrandom quantities, when $p_1 = 1$ and $p_2 = 0$, $R_T = 1$. For the equiprobable distribution when $p_1 = p_2 = 1/2$, we have $R_T = 0$.

Thus, to construct the functions S_T and R_T we need some discrete distribution. To construct it, let us consider the solution of system (5) at $\varphi = 0.561$ during total time $4 \cdot 10^3 \cdot 2\pi$. Next, we extract from the q_1-component of solution the sequence of intervals $R_i = (\tau_i; \tau_{i+1})$ (Fig. 7a) between its zeros τ_i such that $q_1'(\tau_i) > 0$ [10]. This sequence of intervals R_i allows one to compose the sequence of their length $\ell_i = \tau_{i+1} - \tau_i$. Then the required sampling distribution of this sequence ℓ_i (Fig. 7b) can be constructed. The numerical estimation of extremum coordinates gives $q = 0.643 < 1$. This means that the process described by this distribution possesses long-range correlations and the system dynamics is defined by the mutual influence of a large number of rare events.

Using the procedure outlined above we evaluate the Tsallis index q for other attractors. In particular, for the quasiperiodic regime depicted in Fig. 5 at $\varphi = 0.48$ the parameter $q = 0.756$ and for the similar regime shown in Fig. 4 at $\varphi = 0.61$ the parameter $q = 0.825$. The Tsallis index can provide the identification of attractors and their bifurcations, thus, the variation of q with respect to φ requires additional studies.

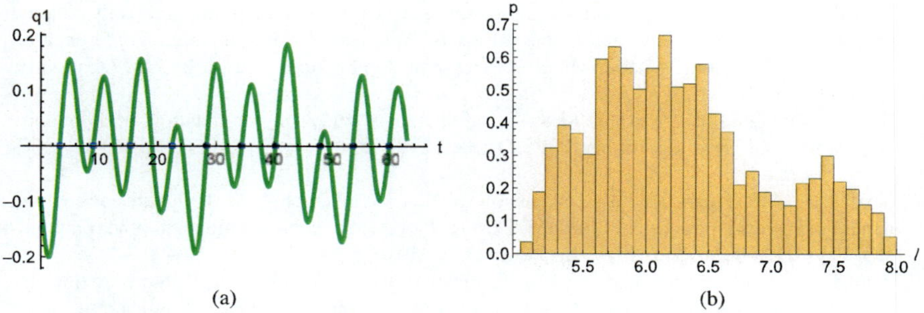

Fig. 7 Construction of the sequence of intervals (**a**) and the corresponding distribution of relative frequencies (**b**) at $\varphi = 0.561$

5 Concluding Remark

Thus, we have developed the three-layered hierarchic system in the form of embedded oscillators with strongly nonlinear bonds. Among possible dynamic regimes of the system, we focused on the mode when all oscillators in the same layer move synchronously. Although this allows one to reduce the number of equations of motion, the resulting equations are still strongly nonlinear. Therefore, the qualitative and numerical analysis methods were applied. According to the studies, the variation of structural parameter φ stimulates developing a wide variety of nonlinear phenomena in the considered system. Notably, the periodic, quasiperiodic, and chaotic attractors were discovered. Moreover, the coexistence of periodic and quasiperiodic modes with different partial frequencies was revealed.

The types of attractors and their bifurcations were studied via the Lyapunov exponent spectra, Poincaré sections, and Fourier spectra. In particular, we have shown that the two-frequency, three-frequency tori, and their bifurcations occur in the phase space of the system.

The findings mentioned above can be useful for understanding the processes of energy redistribution in the natural structured geomedia and other hierarchical discrete systems.

References

1. Sadovskiy, M.A., Pisarenko, V.F., Rodionov, V.N.: From seismology to geomechanics. On the model of the geophysical medium. Bull. USSR Acad. Sci. **1**, 82–88 (1983)
2. Danylenko, V.A., Mykulyak, S.V., Skurativskyi, S.I.: Energy redistribution in hierarchical systems of oscillators. Eur. Phys. J. B **88**, 143–151 (2015). https://doi.org/10.1140/epjb/e2015-60225-0
3. Mykulyak, S.V., Skurativska, I.A., Skurativskyi, S.I.: Forced nonlinear vibrations in hierarchically constructed media. Int. J. Non-Linear Mech. **98**, 51–57 (2018). https://doi.org/10.1016/j.ijnonlinmec.2017.10.005
4. Mykulyak, S.V., Skurativskyi, S.I.: Nonlinear dynamics of the system of hierarchically coupled oscillators with power law interactions. Book of Abstracts. In: International Conference on Differential Equations, Mathematical Physics and Applications (DEMPhA-2017,Cherkasy, Ukraine), pp.42–43 (2017)
5. Ramasubramanian, K., Sriram, M.S.: A comparative study of computation of Lyapunov spectra with different algorithms. Phys. D **139**, 72–86 (2000). https://doi.org/10.1016/S0167-2789(99)00234-1
6. Kuznetsov, A.P., Kuznetsov, A.P., Shchegoleva, N.A., Stankevich, N.V.: Dynamics of coupled generators of quasiperiodic oscillations: Different types of synchronization and other phenomena. Phys. D **398**, 1–12 (2019). https://doi.org/10.1016/j.physd.2019.05.014
7. Tsallis, C. : Nonextensive statistical mechanics: construction and physical interpretation. In: Gell-Mann, M., Tsallis, C. (eds.) Nonextensive Entropy: Interdisciplinary Applications, Santa Fe Institute Studies on the Sciences of Complexity, pp. 1–53. Oxford University, Oxford (2004)
8. Boghosian, B.M.: Thermodynamic description of the relaxation of two-dimensional turbulence using Tsallis statistics. Phys. Rev. E Oxford **53**(5), 4754–4763 (1996). https://doi.org/10.1103/physreve.53.4754

9. Ramirez-Reyes, A., Raul Hernandez-Montoya, A., Herrera-Corral, G., Dominguez-Jimenez, I.: Determining the entropic index q of Tsallis entropy in images through redundancy. Entropy **18**(14), 299 (2016). https://doi.org/10.3390/e18080299
10. Skurativskyi, S., Kudra, G., Wasilewski, G., Awrejcewicz, J.: Properties of impact events in the model of forced impacting oscillator: experimental and numerical investigations. Int. J. Non-Linear Mech. **113**, 55–61 (2019). https://doi.org/10.1016/j.ijnonlinmec.2019.03.013

Nonlinear Dynamics of Flexible Meshed Cylindrical Panels in the White Noise's Field

Jan Awrejcewicz, Ekaterina Krylova, Irina Papkova, and Vadim Krysko

Abstract The mathematical model of the nonlinear dynamics of flexible mesh cylindrical panels in the field of additive white noise is constructed in this paper. To account for size-dependent behavior, a nonclassical continual model based on a Cosserat medium is considered. Thus, along with the classical stress field, the moment voltages are also taken into account. It is also assumed that the fields of displacements and rotations are not independent. The equilibrium equations for the plate element and the boundary conditions are obtained from the Ostrogradskiy-Gamilton variation principle on the basis of Kirchhoff-Love kinematic hypotheses and Karman's geometric nonlinearity. In accordance with a continual model, a mesh panel consisting of a regular system of often located same material's ribs is replaced by an equivalent continuous layer having some averaged stiffness depending on the layout of the ribs and their stiffness. The system of differential equations in partial derivatives is reduced to a system of ODE using the finite difference method of the second order of accuracy. The resulting system is solved by the fourth-order Runge-Kutta methods.

Keywords Meshed panel · Micropolar theory · Buckling · Geometric nonlinearity

J. Awrejcewicz (✉)
Department of Automation, Biomechanics and Mechatronics Lodz University of Technology, Lodz, Poland
e-mail: jan.awrejcewicz@p.lodz.pl

E. Krylova
Saratov State University, Saratov, Russia

I. Papkova · V. Krysko
Yuri Gagarin State Technical University of Saratov, Saratov, Russia
e-mail: tak@san.ru

J. Awrejcewicz (ed.), *Perspectives in Dynamical Systems I: Mechatronics and Life Sciences*, Springer Proceedings in Mathematics & Statistics 362,
https://doi.org/10.1007/978-3-030-77306-9_3

1 Formulation of the Problem

Micro- and nano-sized plates and shells are widely used in electromechanical systems. High growth rates of microsystem technologies cause the interest of scientists not only to the dynamics of full-size mechanical systems in the form of plates and shells [1–3], but also the need to create mathematical models that take into account the scale effects at the micro and nano level [4–6]. In most works on this subject linear models are used for numerical analysis [7–11]. However, there are experimental data confirming the need to take into account the nonlinearity in modeling the behavior of the objects under consideration [12].

Subsequent paragraphs, however, are indented. Mesh panels and shells due to their lightness and increased strength are used in many important industries, such as the military and space industries, robotics and medicine. Despite the large number of works devoted to the size-dependent behavior of mechanical objects in the form of plates, panels and shells [13–16], studies of the behavior of mesh plates and shells based on theories that take into account the effects of scale is very small [17,18].

In this work the mathematical model of the geometrically nonlinear micropolar mesh cylindrical panel behavior is constructed. The panel is exposed to the distributed normal load in the normal white noise field. The panel material is the Cosserat pseudo-continuum with constrained particle rotation. The mesh structure is taken into account according to the G.I. Pshenichnov theory [19]. In this work, the vibrations of mesh panels as a system with many freedom's degrees are studied with the addition of an external chaotic component in the form of white noise. White noise is the generalized stationary random process X(t) with constant spectral density. The term "white" was assigned by analogy with white light, which in the visible part of the spectrum has the entire set of frequencies. The correlation (generalized) function of the white noise process has the form: $B(t) = \sigma^2\delta(t)$, where σ^2 is some positive constant, and $\delta(t)$ is the delta function. Gaussian white noise as a model is well suited for the mathematical description of many natural processes, in particular sound pressure.

The investigation object is a rectangular plate occupying in space $\mathbb{R}^3$ a region $\Omega = \left\{0 \le x \le c; 0 \le y \le b; -\frac{h}{2} \le z \le \frac{h}{2}\right\}$ The non-zero components of the strain tensor in the case of the Kirchhoff–Love hypotheses and the Karman theory can be written as:

$$\begin{aligned} e_{xx} &= \frac{\partial u}{\partial x} + \frac{1}{2}\left(\frac{\partial w}{\partial x}\right)^2 - z\frac{\partial^2 w}{\partial x^2}; \\ e_{yy} &= \frac{\partial v}{\partial y} + \frac{1}{2}\left(\frac{\partial w}{\partial y}\right)^2 - k_y w - z\frac{\partial^2 w}{\partial y^2}; \\ e_{xy} &= \frac{1}{2}\left(\frac{\partial u}{\partial y} + \frac{\partial v}{\partial x}\right) - z\frac{\partial^2 w}{\partial x \partial y}, \end{aligned} \tag{1}$$

Here w, u, v – are the axial displacements of the plate middle surface in the directions z, x, y, respectively, k_y – is the geometric parameter of panel's curvature.

The panel material is considered as a Cosserat pseudo-continuum with constrained particle rotation where, along with the usual stress field, moment stresses are also taken into account. It is assumed here that the displacement and rotation fields are not independent.

The components of the symmetric bending-torsion tensor taking into account the accepted hypotheses and assumptions will take the form:

$$\begin{gathered}\chi_{xx}=\frac{\partial^2 w}{\partial x\partial y};\quad \chi_{yy}=-\frac{\partial^2 w}{\partial y\partial x};\quad \chi_{xy}=\frac{1}{2}\left(\frac{\partial^2 w}{\partial y^2}-\frac{\partial^2 w}{\partial x^2}\right);\\ \chi_{xz}=\frac{1}{4}\left(\frac{\partial^2 v}{\partial x^2}-\frac{\partial^2 u}{\partial x\partial y}\right);\quad \chi_{yz}=\frac{1}{4}\left(\frac{\partial^2 v}{\partial y\partial x}-\frac{\partial^2 u}{\partial y^2}\right).\end{gathered}\tag{2}$$

For the plate material the defining relations are taken in the form:

$$\begin{gathered}\sigma_{xx}=\frac{E}{1-\nu^2}\left[e_{xx}+\nu e_{yy}\right],x\rightleftarrows y,\quad \sigma_{xy}=\frac{E}{(1+\nu)}e_{xy},\\ \left(m_{xx},m_{xy},m_{zx}\right)=\frac{El^2}{1+\nu}\left(\chi_{xx},\chi_{xy},\chi_{zx}\right),\end{gathered}\tag{3}$$

Here σ_{ij} – are the components of the stress tensor, m_{ij} – are the components of the moment tensor of higher order, E – is the Young's modulus, ν – is the Poisson's ratio, l – is the additional independent material length parameter.

The equations of motion of an element of a smooth plate equivalent to a mesh, the boundary and initial conditions were obtained from the Hamilton-Ostrogradskiy energy principle.

The subject of study is a mesh plate under the action of a normal distributed load. The panel consists n families of densely spaced edges of the same material. The regular system of edges can be replaced by a continuous layer according to the continuum model of G. I. Pshenichnov [19]. In this case, the stresses arising in the equivalent smooth plate, associated with the stresses in the ribs making up the angles φ_j with the x-axis, will have the form:

$$\begin{gathered}\sigma_{xx}=\sum_{j=1}^{n}\frac{\sigma_x^j\delta_j\mathrm{Cos}^2\varphi_j}{a_j},\quad \sigma_{yy}=\sum_{j=1}^{n}\frac{\sigma_x^j\delta_j\mathrm{Sin}^2\varphi_j}{a_j},\quad \sigma_{xy}=\sum_{j=1}^{n}\frac{\sigma_x^j\delta_j\mathrm{Cos}\varphi_j\mathrm{Sin}\varphi_j}{a_j},\\ m_{xx}=\sum_{j=1}^{n}\frac{m_x^j\delta_j\mathrm{Cos}^2\varphi_j}{a_j}\quad m_{yy}=\sum_{j=1}^{n}\frac{m_x^j\delta_j\mathrm{Sin}^2\varphi_j}{a_j},\quad m_{xy}=\sum_{j=1}^{n}\frac{m_x^j\delta_j\mathrm{Cos}\varphi_j\mathrm{Sin}\varphi_j}{a_j},\\ m_{xz}=\sum_{j=1}^{n}\frac{m_{zx}^j\delta_j\mathrm{Cos}\varphi_j}{a_j},\quad m_{yz}=\sum_{j=1}^{n}\frac{m_{zx}^j\delta_j\mathrm{Sin}\varphi_j}{a_j},\end{gathered}\tag{4}$$

where a_j – is the distance between edges of j-th set, δ_j – is the edge thickness of the j-th set, φ_j – is the angle between the x-axis and the edge axis of the j-th set, stresses with index j refer to rods. The physical relationships for the mesh plate are determined based on the Lagrange multipliers method:

$$\begin{gathered}\sigma_x^j=\sigma_{xx}\mathrm{Cos}^2\varphi_j+\sigma_{yy}\mathrm{Sin}^2\varphi_j+\sigma_{xy}\mathrm{Cos}\varphi_j\mathrm{Sin}\varphi_j;\quad \tau^j=\sigma_{xz}\mathrm{Cos}\varphi_j+\sigma_{yz}\mathrm{Sin}\varphi_j;\\ m_x^j=m_{xx}\mathrm{Cos}^2\varphi_j+m_{yy}\mathrm{Sin}^2\varphi_j+m_{xy}\mathrm{Cos}\varphi_j\mathrm{Sin}\varphi_j;\quad m_z^j=m_{xz}\mathrm{Cos}\varphi_j+m_{yz}\mathrm{Sin}\varphi_j.\end{gathered}\tag{5}$$

The obtained physical relations (5) and expressions connecting the stresses arising in the equivalent smooth plate with the stresses in the edges (4) will make it possible to write the relations for the forces and moments of the smooth plate to the equivalent mesh plate. Substituting the latter into the equations of motion of a smooth plate, we obtain the equations of motion of a micropolar plate of a mesh structure in mixed form.

In what follows, we will consider a panel with two families of edges (Fig. 1) $\varphi_1 = 45^o$, $\varphi_2 = 135^o$, , $a_1 = a_2 = a$.

Additive noise added to the system in the form of a random term with constant intensity: $q_{nois} = q_{n0}(2.0 *$ rand ()/(RAND _ MAX + 1.0) − 1.0), here q_{n0} – is the noise intensity. The (2.0 * rand ()/(RAND _ MAX + 1.0) − 1.0) expression takes arbitrary fractional values in the range (−1; 1).

We introduce the following dimensionless parameters: $x = c\overline{x}$, $y = b\overline{y}$, $w = h\overline{w}$, $u = \frac{h^2}{c}\overline{u}$, $v = \frac{h^2}{b}\overline{v}$, $\delta = h\overline{\delta}$, $a = h\overline{a}$, $l = h\overline{l}$, $k_y = \frac{1}{b}\overline{k_y}$, $t = \frac{cb}{h}\sqrt{\frac{\rho}{E}}\overline{t}$, $\varepsilon = \frac{h}{cb}\sqrt{\frac{E}{\rho}}\overline{\varepsilon}$, $q = \frac{Eh^4}{c^2b^2}\overline{q}$, $q_{nois} = \frac{Eh^4}{c^2b^2}\overline{q_{nois}}$, here $q = q_0\mathrm{Sin}(\omega_p t)$ – is the external normal load, q_0 – its intensity, ω_p – its frequency, c, b – are the plate's linear dimensions in x and y direction, ρ – is the panels material density, ε – is the dissipation coefficient.

The equilibrium equations of the element of the considered micropolar mesh cylindrical panel in displacements take the form. The line over dimensionless variables is omitted.

$$
\begin{aligned}
&2(\nu-1)\frac{\partial^2 u}{\partial y^2}-2(3+\nu)\frac{\partial^2 v}{\partial x\partial y}-4(\nu+1)\frac{b^2}{c^2}\frac{\partial^2 u}{\partial x^2}-4k_y(\nu+1)\frac{b}{h}\frac{\partial w}{\partial x}+4(\nu-1)\frac{\partial^2 w}{\partial y^2}\frac{\partial w}{\partial x}-8\frac{\partial w}{\partial y}\frac{\partial^2 w}{\partial x\partial y}-\\
&-4(1+\nu)\frac{b^2}{c^2}\frac{\partial w}{\partial x}\frac{\partial^2 w}{\partial x^2}+l^2(\nu-1)\left(-\frac{h^2}{b^2}\frac{\partial^4 u}{\partial y^4}+\frac{h^2}{b^2}\frac{\partial^4 v}{\partial x\partial y^3}-\frac{h^2}{c^2}\frac{\partial^4 u}{\partial x^2\partial y^2}+\frac{h^2}{c^2}\frac{\partial^4 v}{\partial x^3\partial y}\right)=\frac{8h^2a\left(\nu^2-1\right)}{c^2\delta}\frac{\partial^2 u}{\partial t^2},\\
&-4k_y(1+\nu)\frac{c^2}{bh}\frac{\partial w}{\partial y}-4(1+\nu)\frac{c^2}{b^2}\frac{\partial^2 v}{\partial y^2}-2(3+\nu)\frac{\partial^2 u}{\partial x\partial y}-2(1-\nu)\frac{\partial^2 v}{\partial x^2}-4(1+\nu)\frac{c^2}{b^2}\frac{\partial w}{\partial y}\frac{\partial^2 w}{\partial y^2}-8\frac{\partial w}{\partial x}\frac{\partial^2 w}{\partial y\partial x}+\\
&+4(\nu-1)\frac{\partial w}{\partial y}\frac{\partial^2 w}{\partial x^2}+l^2(\nu-1)\left(\frac{h^2}{b^2}\frac{\partial^4 u}{\partial x\partial y^3}-\frac{h^2}{b^2}\frac{\partial^4 v}{\partial x^2\partial y^2}+\frac{h^2}{c^2}\frac{\partial^4 u}{\partial x^3\partial y}-\frac{h^2}{c^2}\frac{\partial^4 v}{\partial x^4}\right)=\frac{h^2 8a\left(\nu^2-1\right)}{b^2\delta}\frac{\partial^2 v}{\partial t^2},\\
&\left([1+\nu]+6l^2[1-\nu]\right)\frac{c^2}{bh}\frac{\partial^4 w}{\partial y^4}+4\left(1+l^2[\nu-1]\right)\frac{c^2}{b^2}\frac{\partial^4 w}{\partial x^2\partial y^2}+\left([1+\nu]+6l^2[1-\nu]\right)\frac{b^2}{c^2}\frac{\partial^4 w}{\partial x^4}+\\
&+12k_y^2(1+\nu)\frac{c^2}{b^2}w+12k_y(1+\nu)\frac{c^2}{bh}\frac{\partial v}{\partial y}+12k_y(1+\nu)\frac{b^2}{h^2}\frac{\partial u}{\partial x}-6k_y(1+\nu)\frac{c^2}{bh}\left(\frac{\partial w}{\partial y}\right)^2-\\
&-12(1+\nu)\frac{c^2}{b^2}\frac{\partial w}{\partial y}\frac{\partial^2 v}{\partial y^2}-12k_y(1+\nu)\frac{c^2}{bh}w\frac{\partial^2 w}{\partial y^2}-12(1+\nu)\frac{c^2}{b^2}\frac{\partial v}{\partial y}\frac{\partial^2 w}{\partial y^2}-18(1+\nu)\left(\frac{\partial w}{\partial y}\right)^2\frac{\partial^2 w}{\partial y^2}-\\
&-12(\nu+1)\frac{\partial^2 w}{\partial y^2}\frac{\partial u}{\partial x}+12(\nu-1)\frac{\partial^2 u}{\partial y^2}\frac{\partial w}{\partial x}-6k_y(\nu+1)\frac{b}{h}\left(\frac{\partial w}{\partial x}\right)^2+6(-5+3\nu)\frac{\partial^2 w}{\partial y^2}\left(\frac{\partial w}{\partial x}\right)^2-\\
&-24\frac{\partial w}{\partial y}\frac{\partial^2 u}{\partial x\partial y}-24\frac{\partial w}{\partial x}\frac{\partial^2 v}{\partial x\partial y}+24(\nu-1)\frac{\partial u}{\partial y}\frac{\partial^2 w}{\partial x\partial y}+24(\nu-1)\frac{\partial v}{\partial x}\frac{\partial^2 w}{\partial x\partial y}+24(3\nu-5)\frac{\partial w}{\partial y}\frac{\partial w}{\partial x}\frac{\partial^2 w}{\partial x\partial y}-\\
&-12(1+\nu)\frac{b^2}{c^2}\frac{\partial w}{\partial x}\frac{\partial^2 u}{\partial x^2}+12(\nu-1)\frac{b}{c}\frac{\partial w}{\partial y}\frac{\partial^2 v}{\partial x^2}-12k_y(\nu+1)\frac{b}{h}w\frac{\partial^2 w}{\partial x^2}-12(\nu+1)\frac{c^2}{b^2}\frac{\partial v}{\partial y}\frac{\partial^2 w}{\partial x^2}+\\
&+6(3\nu-5)\left(\frac{\partial w}{\partial y}\right)^2\frac{\partial^2 w}{\partial x^2}-12(1+\nu)\frac{\partial u}{\partial x}\frac{\partial^2 w}{\partial x^2}-18(1+\nu)\frac{b^2}{c^2}\left(\frac{\partial w}{\partial x}\right)^2\frac{\partial^2 w}{\partial x^2}=\\
&=\frac{24a\left(\nu^2-1\right)}{\delta}\left[\frac{\partial^2 w}{\partial t^2}-\varepsilon\frac{\partial w}{\partial t}-q-q_{nois}\right].
\end{aligned}
$$

To the equations we add the initial zero conditions and the boundary conditions of the rigidly clamped edge:

$x = \pm 1, y = \pm 1 : u = v = w = 0, \frac{\partial u}{\partial x} = 0, \frac{\partial u}{\partial y} = 0, \frac{\partial v}{\partial x} = 0, \frac{\partial v}{\partial y} = 0, \frac{\partial w}{\partial x} = 0, \frac{\partial w}{\partial y} = 0.$

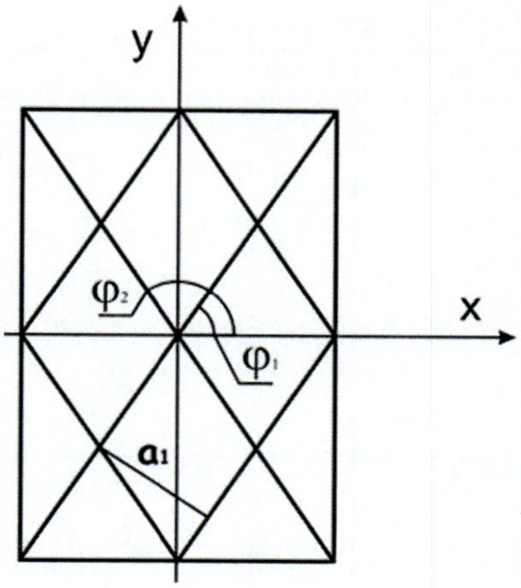

Fig. 1 Plate grid geometry

2 Numerical Results

The system of differential equations in partial derivatives is reduced to the ODE system using the finite difference method with second-order approximation. The Cauchy problem is solved by the Runge–Kutta method of the fourth order of accuracy.

Experiment Parameters: $\nu = 0.3$, $\delta = h = a = 0.002$, $\omega_p = 5$, $q_0 \in [0; 20]$.

The Ruelle-Tackens–Newhouse scenario was obtained in the experiment. The oscillation transition was carried out through two linearly independent frequencies and their linear combinations. Table 1 shows the Fourier spectra and phase portraits for some values of the external normal load amplitude. The phase portrait is a torus, which corresponds to the classic Ruelle-Tackens-Newhouse scenario.

The following frequency dependencies are available: ω_p and ω_1 – are the independent frequencies, $\omega_2 = 3\omega_1$, $\omega_3 = 6\omega_1$, $\omega_4 = \omega_p - \omega_1$, $\omega_5 = \omega_p - 4\omega_1$, $\omega_6 = \omega_p - 8\omega_1$, $\omega_7 = 2\omega_p - 18\omega_1$.

The effects of external additive white noise $q_{no} \in [0.1; 9]$ on the behavior of a shallow cylindrical mesh micropolar panel are investigated. A numerical experiment showed that the noise of intensity commensurate with the intensity of the external normal load does not significantly affect the nature of its oscillations and does not change the scenario of the transition of system vibrations into chaos.

Table 2 shows the Fourier spectra for the intensity of the external load $q_0 = 2$ and the noise intensity $q_{nois} = 0.1$ and $q_{nois} = 9$. An increase in the noise intensity to $q_{nois} = 9$ accelerated the transition of system oscillations to chaotic ones.

Table 1 Transition scenario

q_0	Fourier spectrum	Phase portrait
0.1	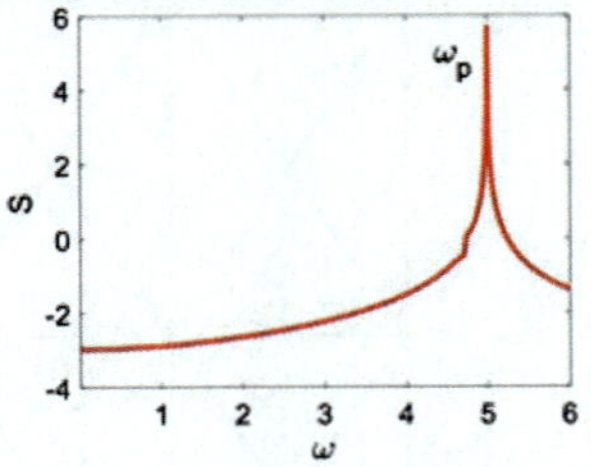	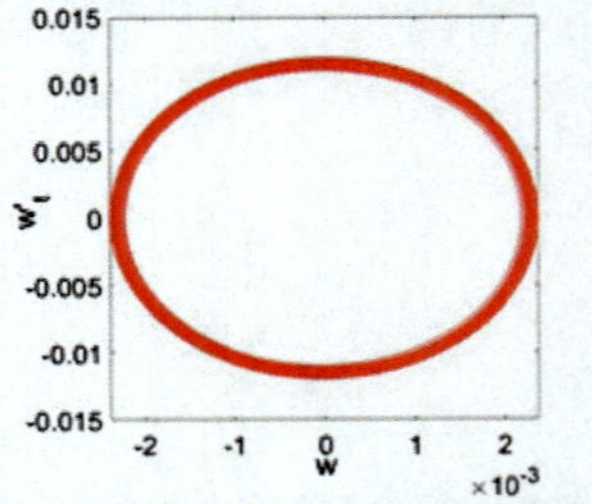
2	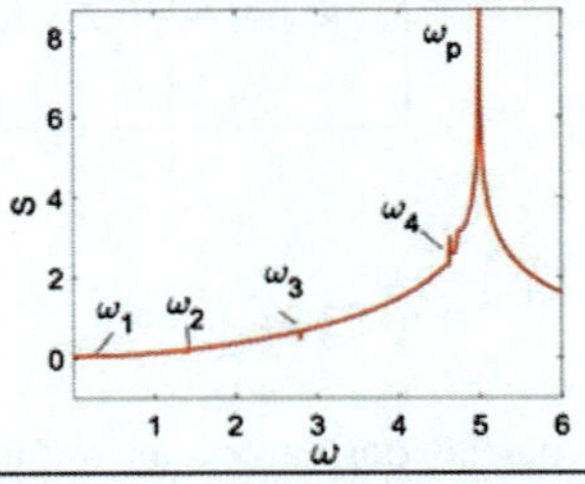	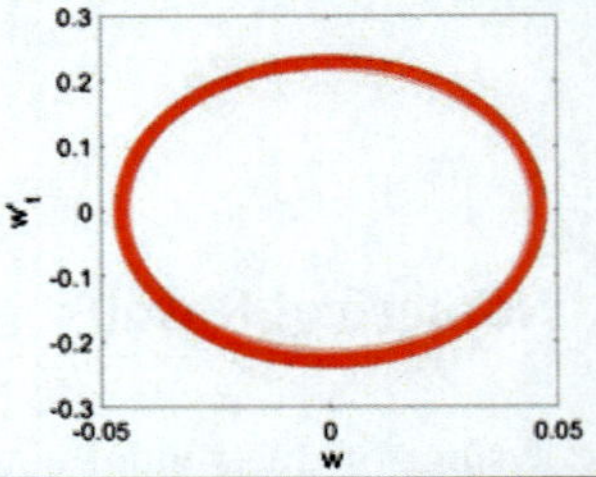
9	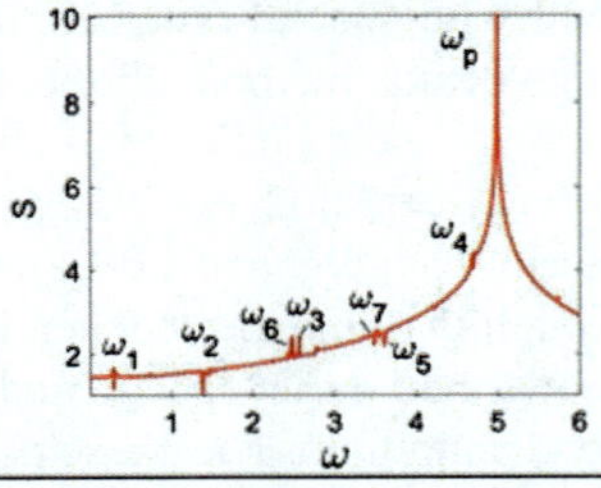	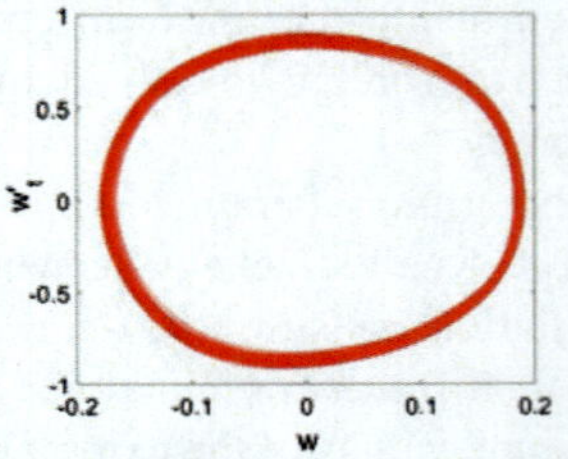

Table 2 The effect of white noise intensity on the panel vibrations character ($q_0 = 2$)

$q_{nois} = 0.1$	$q_{nois} = 9$
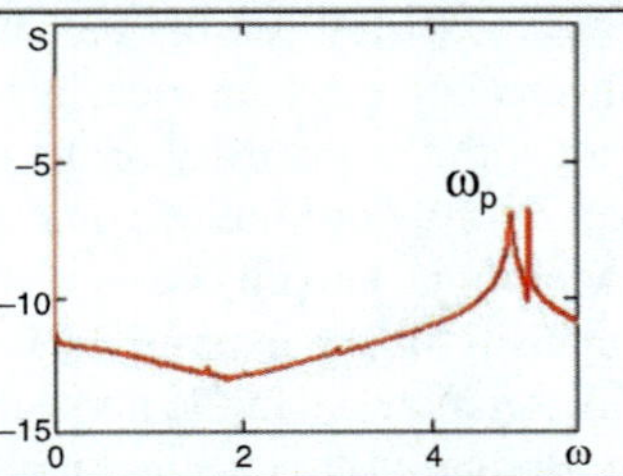	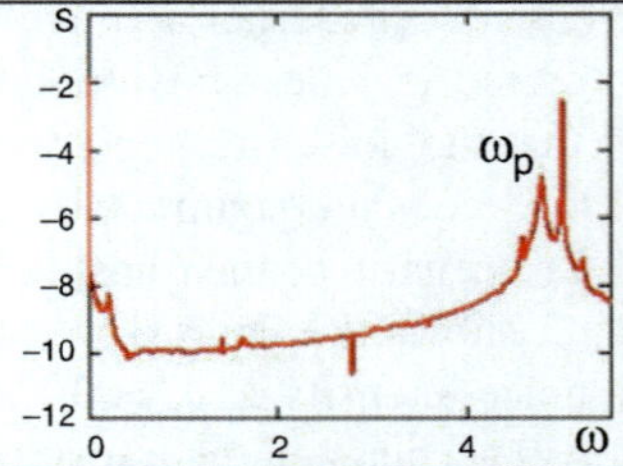

3 Conclusions

On the basis of Pshenichniy continuum model and Kirchhoff-Love hypotheses, the mathematical model of vibrations of flexible micropolar cylindrical mesh panels in the additive white noise field is constructed. The panel consists of two families of mutually orthogonal edges. Ruele–Takens–Newhouse scenario of transitions vibration from harmonic to chaotic for the meshed panels was obtained and studied. It was shown that the noise with intensity commensurate with the intensity of the normal load does not change oscillations character of the system.

Acknowledgments The work was supported by the RFBR, № 18-01-00351a, №18-41-700001 r_a.

References

1. Krylova, E.Y., Papkova, I.V., Erofeev, N.P., Zakharov, V.M., Krysko, V.A.: Complex fluctuations of flexible plates under longitudinal loads with account for white noise. J. Appl. Mech. Tech. Phys. **57**(4), 714–719 (2016)
2. Awrejcewicz, J., Mrozowski, J., Krysko, A.V., Papkova, I.V., Zakharov, V.M., Erofeev, N.P., Krylova, E.Y., Krysko, V.A.: Chaotic dynamics of flexible beams driven by external white noise. Mech. Syst. Signal Process. **79**, 225–253 (2016)
3. Krysko, V.A., Papkova, I.V., Awrejcewicz, J., Krylova, E.Y., Krysko, A.V.: Non-symmetric forms of non-linear vibrations of flexible cylindrical panels and plates under longitudinal load and additive white noise. J. Sound Vib. **423**, 212–229 (2018)
4. Sargsyan, S.H., Zhamakochyan, K.A.: Applied theory of micropolar elastic thin plates with constrained rotation and the finite element method. Mater. Phys. Mech. **35**(1), 145–154 (2018)
5. Nikabadze, M.U.: Some versions of equations of micropolar shell theories. Prikladnaya matematika i matematicheskaya fizika. **1**(1), 101–118 (2015)
6. Awrejcewicz, J., Krysko, V.A., Sopenko, A.A., Zhigalov, M.V., Kirichenko, A.V., Krysko, A.V.: Mathematical modelling of physically/geometrically non-linear micro-shells with account of coupling of temperature and deformation fields. Chaos, Solitons Fractals. **104**, 635–654 (2017)
7. Safarpour, H., Kianoosh Mohammadi, K., Ghadiri, M.: Temperature-dependent vibration analysis of a FG viscoelastic cylindrical microshell under various thermal distribution via modified length scale parameter: a numerical solution. J. Mech. Behav. Mater. **26**, 9–24 (2017)
8. Sahmani, S., Ansari, R., Gholami, R., Darvizeh, A.: Dynamic stability analysis of functionally graded higher-order shear deformable microshells based on the modified couple stress elasticity theory. Compos. Part B Eng. **51**, 44–53 (2013)
9. Zhou, X., Wang, L.: Vibration and stability of micro-scale cylindrical shells conveying fluid based on modified couple stress theory. Micro Nano Lett. **7**(7), 679–684 (2012)
10. Varygina, M.: Numerical modeling of micropolar cylindrical shells on supercomputers with GPUs. AIP Conf. Proc. **1895**, 080005 (2017)
11. Krylova, E.Y., Papkova, I.V., Sinichkina, A.O., Yakovleva, T.B., Krysko-yang, V.A.: Mathematical model of flexible dimension-dependent mesh plates. J. Phys. Conf. Ser. **1210**, –012073 (2019)
12. Scheible, D.V., Erbe, A., Blick, R.H.: Evidence of a nanomechanical resonator being driven into chaotic response via the Ruelle–Takens route. Appl. Phys. Lett. **81**, 1884–1886 (2002)

13. Belikov, G.I.: The general case of bending of a rectangular mesh plate taking into account tensile forces acting in the middle surface Vestnik Volgogradskogo gosudarstvennogo arhitekturno-stroitel'nogo universiteta. Seriya: Stroitel'stvoiarhitektura. **37**, 121–128 (2014)
14. Trushin, S.I., Zhuravleva, T.A., Sysoeva, E.V.: Dynamic stability loss of nonlinearly deformable mesh plates made of composite material with various lattice configurations. Nauchnoe obozrenie. **4**, 44–51 (2016)
15. Azikov, N.S., Pavlov, E.A.: The study of the stability of the mesh composite plate. Aviacionnaya promyshlennost. **3**, 46–50 (2016)
16. Wu, Q.L., Zhang, W., Dowell, E.H.: Detecting multi-pulse chaotic dynamics of high-dimensional non-autonomous nonlinear system for circular mesh antenna. Int. J. Non Linear Mech. **102**, 25–40 (2018)
17. dell'isola, F., Steigman, D.A.: Two-dimensional gradient-elasticity theory for woven fabrics. J. Elast. **118**(1), 113–125 (2015)
18. Krylova, E.Y., Papkova, I.V., Yakovleva, T.V., Krysko, V.A.: Theory of vibration of carbon nanotubes like flexible micropolar mesh cylindrical. Izv. Saratov Univ. (N. S.), Ser. Math. Mech. Inf. **19**(3), 305–316 (2019)
19. Pshenichnov, G.I.: The Theory of Thin Elastic Net Shells and Plates. Nauka, Moscow (1982)

A System for Improving Directional Stability Involving Individual Braking of 1, 2, or 3 Wheels of Articulated Rigid Body Vehicles

Aleksander Skurjat and Andrzej Kosiara

Abstract Road-safety of wheeled vehicles depends on the systems used to assist the driver while operating the vehicle. For commercial vehicles, i.e. cars and trucks, numerous systems supporting the driver and influencing the trajectory of vehicle motion are developed. Stiffness of the articulated vehicles' steering systems is relatively low. Consequently, in order to meet normative requirements for the steering system the maximum velocity of vehicles of this type is very limited. The article presents the results of computer simulation tests of the control system, which involves individual braking of one, two or three wheels of the vehicle in order to improve its directional stability. The principles of operation of various motion stabilization systems were also compared. Furthermore, the article presents a method for measuring the motion trajectory of a vehicle.

Keywords Snaking · Articulated vehicle · Braking system · Control system

1 Introduction

The rapid development of the automotive industry has contributed to an increase in the number of vehicles travelling on public roads at ever higher speeds. For this reason, safety during driving has become an extremely important factor in the development of new machines on the road. The largest number of safety systems can be seen in the design of passenger cars. The development of electronics and microprocessor systems has led to the emergence of quick driving assistance systems. These include active braking systems, slip control during acceleration, power steering, traction control during steering and many others. It should be noted that such systems are not used in a earth-working machines and the location of the

A. Skurjat (✉) · A. Kosiara
Faculty of Mechanical Engineering, Department of Off-Road Machine and Vehicle Engineering, Wrocław, Poland
e-mail: aleksander.skurjat@pwr.edu.pl; andrzej.kosiara@pwr.edu.pl

J. Awrejcewicz (ed.), *Perspectives in Dynamical Systems I: Mechatronics and Life Sciences*, Springer Proceedings in Mathematics & Statistics 362,
https://doi.org/10.1007/978-3-030-77306-9_4

tool, e.g. excavator bucket or loader, can be extremely dangerous in the event of a collision with a pedestrian.

One of the phenomena that prevent these machines from moving at speeds exceeding 50 km/h is snaking phenomenon considered as spontaneous change of a vehicle path. The systems supporting the braking process in heavy commercial vehicles are used more and more often. In this group of vehicles, the tractor/trailer combination is often a rotary pair with no additional components. Vehicles with articulated steering system (e.g. earth working machines), apart from connecting both frames by a rotary pair, contain a power element in the form of e.g. a hydraulic cylinder. Due to the different design of these vehicles, it is not possible to reproduce the already proven and improved systems for the enhancement of trajectory and shortening braking distances, already known in automotive vehicles or heavy commercial vehicles.

In the paper [1] authors tests the possibility of usage vehicle brakes as a method for diminishing snaking behavior. In proposed method control algorithm allows only to brake by a repeatable short pulse of one front wheel. A method for braking torque calculation was proposed and method for comparing results was discussed. The algorithm uses articulated angle as a signal for controlling brakes. In the paper [2] authors propose a control algorithm which allows for braking only one front wheel, but in this method as a control signal vehicle trajectory were used. The author finds it imprecise. A braking torque was chosen to get the best results without proposing a method for calculation and brakes was used until the threshold was achieved. In the paper [3] authors discussed factors affecting steering stiffness. Their work is focused on the geometry of a steering system and its structural stiffness. In the paper [4] a method for driver-assist is proposed to achieve inline motion. In the paper [5] a mathematical model of a hydraulic system is proposed. The dynamic characteristic of the steering system is analyzed. Obtained results are verified on a real machine with good compliance. In this study [6] a model coupled with the kinematic and dynamic properties of the steering struts is formulated to identify objective measures of the AFSV under steering inputs. The results suggest that the vehicle yaw oscillation/stability, steering power efficiency and maneuverability can be objectively measured in terms of the strut length, yaw oscillation frequency and damping ratio, steering gain, and steering response rate and overshoot. In the paper [7, 8] author proposes a control system for optimum distribution of longitudinal and lateral forces of the four tires of a towing vehicle. The system is designed to stabilize the motion of vehicle utilizing the tires entire ability in both longitudinal and lateral directions as well as to make the handling characteristics of an articulated vehicle similar to those of a single one. In this study [9] a control system was applied for an articulated vehicle to turn in a smaller radius.

An active brake control system was developed to generate additional yaw-moment and was controlled by the articulation angle determined by the driver. The paper [10] relates to the diminishing snaking behavior by braking caravan wheels. Obtained results are very promising in many road situations. Snaking vibrations are most effectively suppressed by trailer braking forces in phase with the trailer roll displacement, with the consequence that the braking is required when the tire load is greatest. In this study [11] a controller was proposed to modify the slip control braking strategy. The controller compares observed vehicle states to a linear yaw-plane reference model, then attenuates the demanded wheel slips as necessary at each axle to restore directional control to the driver during emergency braking.

2 Basic Vehicle Parameters and Braking Control System Principles

The model of a wheel articulated loader was used for simulated testing. The testing is conducted simultaneously in the integrated MSC Adams and Matlab/Simulink environments. MBS tests provide information on the loads that occur in kinematic pairs of cooperation, wheel – surface contact, steering angle in the articulation, while Matlab/Simulink complements the model with the driving torque necessary to accelerate the vehicle and maintain the set speed, and also introduces the braking torque values into the MBS model, which are calculated from the Matlab package control system. The vehicle model basic parameters gathered from CAD software are shown in Table 1. Depending on the design the control algorithm enables the braking of one, two or three vehicle wheels, and allows the change of the motion trajectory and affects steering joint oscillation time. The proposed system delivers the same braking torque to wheels. Braking torque individual wheel command depends on an angle measured between front and rear frame. A different three braking criterion is tested and principles of working are presented on Fig. 1. In order to prevent continuous and alternate braking of the wheels, a sensitivity threshold γ_{th}

Table 1 Basic parameters of masses, moments of inertia and gravity centers for the vehicle frames

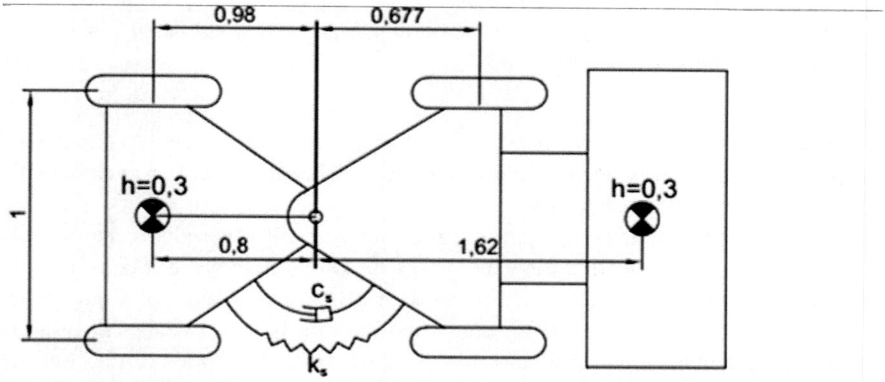

Tires radial stiffness = 50 [N/mm]	Steering system stiffness = 1600 [Nm/deg]
Tires radial damping = 2 [Ns/m]	Steering system damping = 1 [Nms/deg]
Tires diameter = 0.55 [m]	Front frame mass = 445 [kg]
Tires aspect ratio = 0.7	Rear frame mass = 1170 [kg]

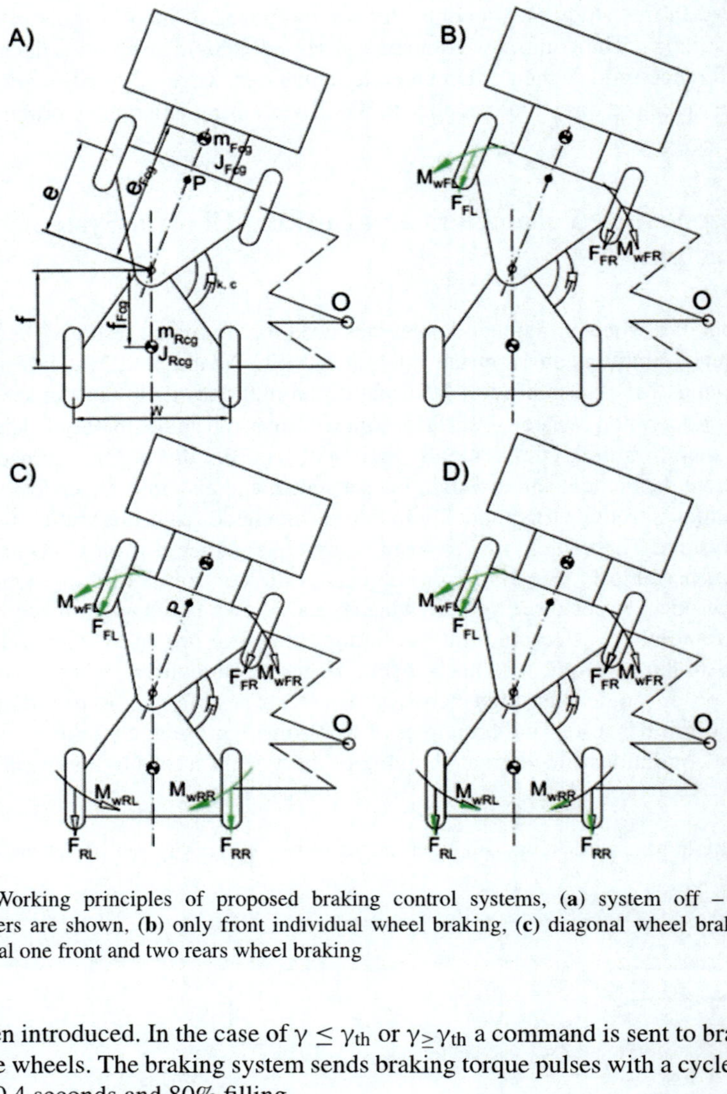

Fig. 1 Working principles of proposed braking control systems, (**a**) system off – a basic parameters are shown, (**b**) only front individual wheel braking, (**c**) diagonal wheel braking, (**d**) individual one front and two rears wheel braking

has been introduced. In the case of $\gamma \leq \gamma_{th}$ or $\gamma \geq \gamma_{th}$ a command is sent to brake one or more wheels. The braking system sends braking torque pulses with a cycle length of $t = 0.4$ seconds and 80% filling.

The braking process is completed when the steering angle is reduced below the value specified by the sensitivity threshold. The braking torque value is constant for each front and rear wheel. Rear wheels brake with 50% efficiency of front ones. Time parameters, values of braking torque and the threshold γ_{th} when the system is switched on and off are identical for all tested solutions. Operation of the control system shall be enforced by introducing a constant torque pulse between the front and rear of the vehicle for all tested solutions.

The system shown in Fig. 1b allows only the front wheel brakes to operate. Its operation is based on the braking of this wheel, which has a higher rotational speed than the opposite wheel. It is obvious that if a vehicle turns to the right, for example, its left wheel has a higher speed, for this reason it has been decided to use the angle between the frames of the vehicle as a control signal. The function of the system is to introduce M_{wFL} or M_{wFR} moment which will allow the front part of the vehicle to be rotated to turn vehicle back to straight path running.

Figure 1c shows the system in which the diagonal, front and rear wheels are braked. The function of the system is to introduce the moment of M_{wFL} lub M_{wFR} and M_{wRR} or M_{wRL} which will allow the front and rear part of the vehicle to be rotated. The purpose of the test is to determine whether the extra torque at the rear part of the vehicle will permit greater damping in the steering system and its effect on the trajectory of motion of the whole vehicle.

Figure 1d shows a system in which three wheels are used to minimize the snaking phenomenon. In this solution, one of the front wheels is braked – as in the case of the system shown in Fig. 1b – and the rear wheels are braked simultaneously. This always results in a situation in which one of the front wheels and both rear wheels brake. In this situation, it is not possible to achieve the torque to rotate rear part of the vehicle because equal braking forces of the wheels and the same length of the arms of their action in relation to the symmetry of the rear part of the vehicle reduce the resulting moment. What remains, however, is a rear-wheel braking force which is directed in the opposite direction of motion and which tightens the vehicle in the steering joint. This results in an additional damping effect. In this analysis, the rear wheel braking torque is half of that of the front wheels. This is due to the fact that the rear axle is relieved as a result of the whole vehicle's deceleration. These tests are intended to determine which of the above methods will help to eliminate oscillations in the articulated joint more quickly and will have a positive effect on straightness of movement.

In the proposed system the brakes are enabled when threshold γ_{th} has been exceeded and the system sends braking pulses. Soon an author tests a control system with variable braking torque, continuous tires braking and different braking command. Simulation testing using the MBS environment do not show computational equations and the user only receives the mathematical solution of the problem. Authors could propose equations for describing vehicle motion but further solution needs for oversimplifying. For this reason, to check the effectiveness of the stabilizing system, W_{SD} performance index has been developed. The dimensionless W_{SD} index combines both the behavior of the steering system, i.e. the amplitudes and duration of the oscillation and the vehicle resulting motion trajectory with side slip on wheels. The value W_{SD} is determined by the formula:

$$W_{SD} = \int_{t0}^{t1} \tan\left|\frac{x}{y}\right| dt \int_{t0}^{t1} \tan|\gamma|\, dt \tag{1}$$

where:

x – lateral displacement [m]
y – displacement of the vehicle in the direction of travel [m]

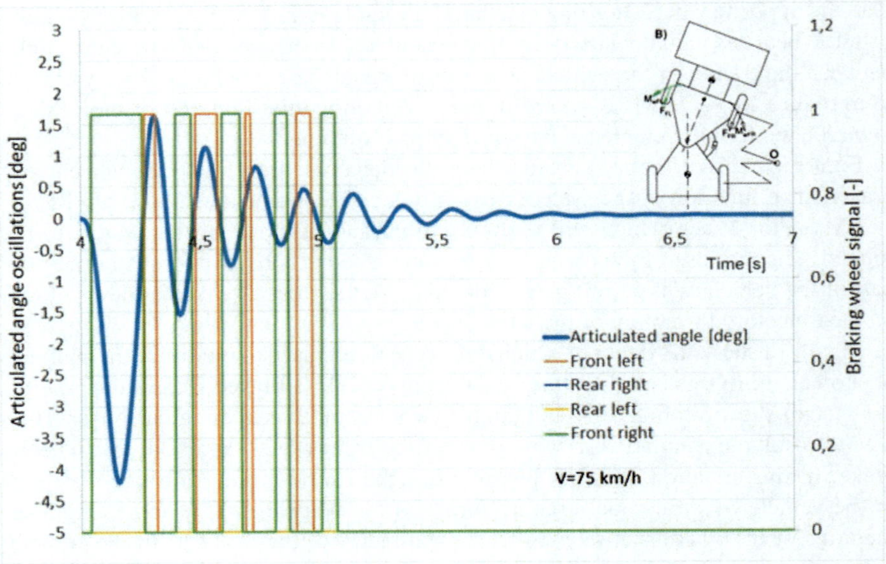

Fig. 2 Principle of working of a control system using front individual wheel braking (Fig. 1b)

γ – angle of turn of elements [rad].

The first part of Eq. 1 increases value in the case of snaking. The second part describes articulated joint oscillations. Articulated angle oscillations always influence vehicle trajectory, so both of equation parts increase. To compare the results it is important to limit covered by vehicle distance by adding a time limit.

3 Simulation Testing of the Anti-Oscillation System in Steering System

Comparative tests are aimed at determining which of the tested solutions: (a) will allow to obtain the settling time (Ts) of oscillations in a steering system (the quickest damping of oscillation) (b) to obtain a in-line direction of motion (c) will have the lowest W_{SD} index. The studies were conducted for three velocities: V = 75, V = 50 and V = 25 km/h. The action of the system braking the front wheels (Fig. 1b) is shown in Fig. 2. Figure 3 shows the action of the system using diagonally one front and one rear wheel with the same value of braking torques. In Fig. 4, the control algorithm allows one front and two rear wheels to be braked.

During simulation tests, the following results were obtained for the brake control system at the velocity of v = 75 km/h – Figs. 5. and 6.

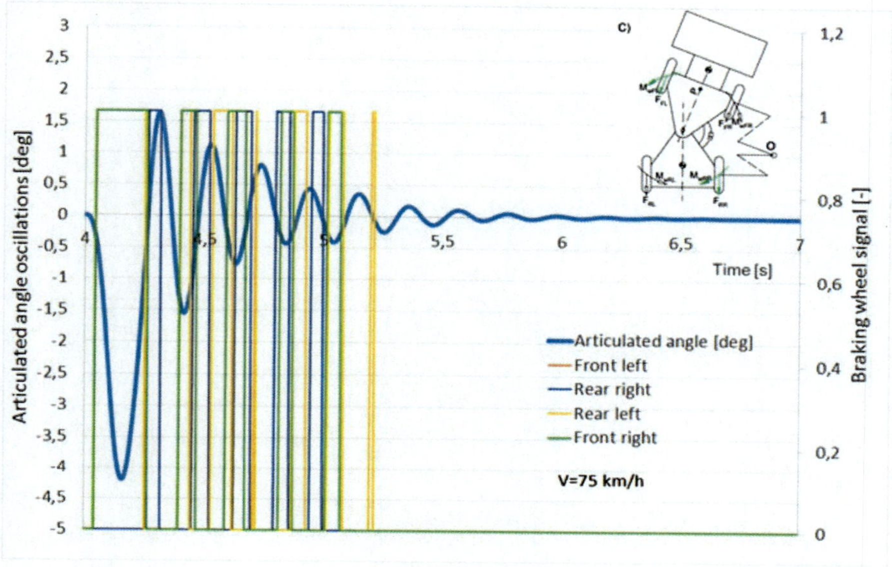

Fig. 3 Principle of working of a control system using diagonal wheels braking (Fig. 1c)

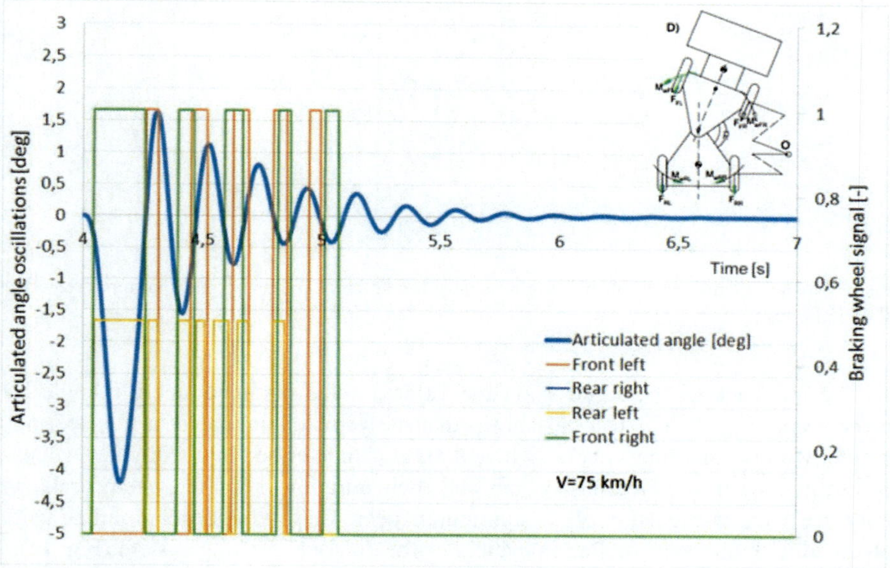

Fig. 4 Principle of working of a control system using one front and both rears wheels braking (Fig. 1d)

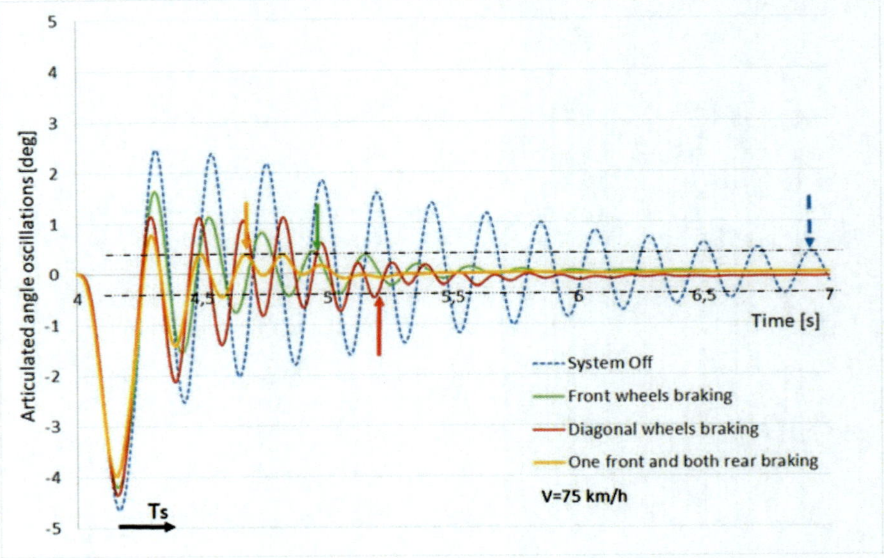

Fig. 5 Steering angle for different braking control system. Velocity V = 75 km/h

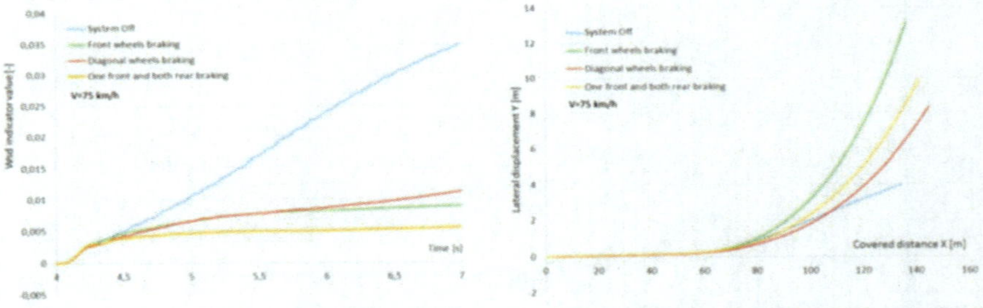

Fig. 6 W_{SD} indicator value and vehicle trajectory for different braking control system. Velocity V = 75 km/h

The influence of the solution on the values of the obtained steering angle is presented in Fig. 5. The test has shown that the vehicle stimulated in the steering system with the steering system switched off is characterized by the highest values of oscillations in the steering system and their duration. For the system braking only the front wheels (Fig. 1b), a continuous decrease in vibration amplitude was obtained with the next oscillation cycle. For the diagonal braking system (Fig. 1c), initially (for about four cycles), a constant and lower amplitude value was obtained than for the system from Fig. 1b, and then the value began to decrease until the complete extinction of the oscillation. The system having the ability to brake the front wheel and the rear wheel (Fig. 1d) is definitely characterized by the highest damping coefficient.

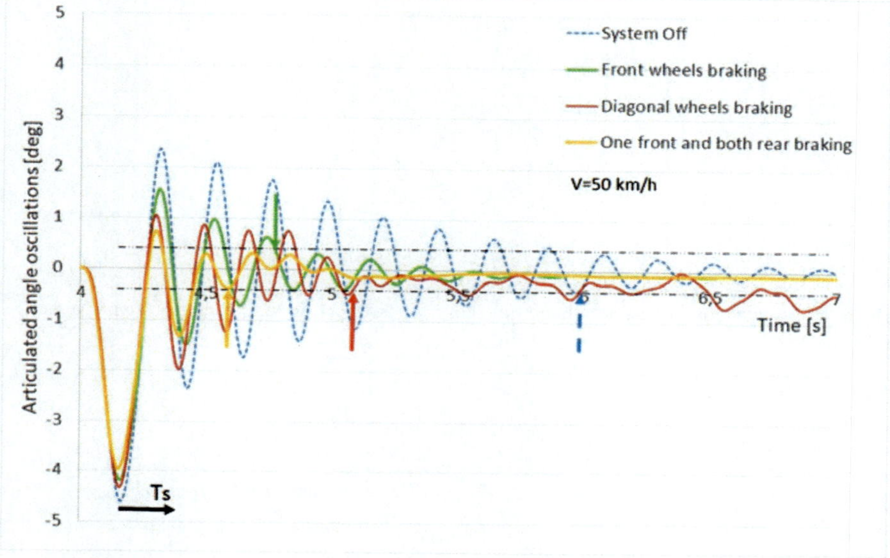

Fig. 7 Steering angle for different braking control system. Velocity V = 50 km/h

Tests of the W_{SD} indicator showed that the vehicle with the brake assist system switched off achieves the highest values. The algorithms of the systems presented in Fig. 1b, c can be compared with each other due to small differences between them. The best solution was the three-wheel braking system (Fig. 1d).

When examining the trajectory of motion – Fig. 6b, it can be observed that a vehicle with the braking system switched off travels a distance which is closest to a straight line due to the lowest lateral displacement Y values achieved. Next is the diagonal braking system, followed by the one with the braking force of both rear wheels. Only the front wheel braking system achieves the worst result. The result is explained by the tires slip angle. When the system is switched off, all wheels rotate freely, so that the values are low. In the diagonal brake system, the slip angle is partially compensated for by braking the left and right wheels (on different axles). For the braking system of both rear wheels, one of the rear wheels partially compensates for the slip angle of one of the front wheels. It is not possible to generate compensation for the braking control of the front wheels alone on other wheels because they rotate freely and result in a significant displacement in the transverse direction to the direction of travel.

The following results were obtained during simulation tests for V = 50 km/h in Figs. 7 and 8.

The influence of the solution on the values of the obtained steering angle is shown in Fig. 7. In this case, similar results were obtained with respect to the velocity V = 75 km/h both for the testing of steering angle oscillations, the W_{SD} indicator –

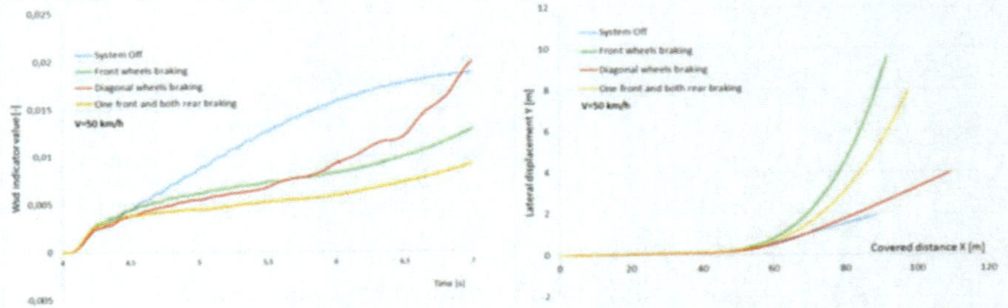

Fig. 8 W_{SD} indicator value and vehicle trajectory for different braking control system. Velocity V = 50 km/h

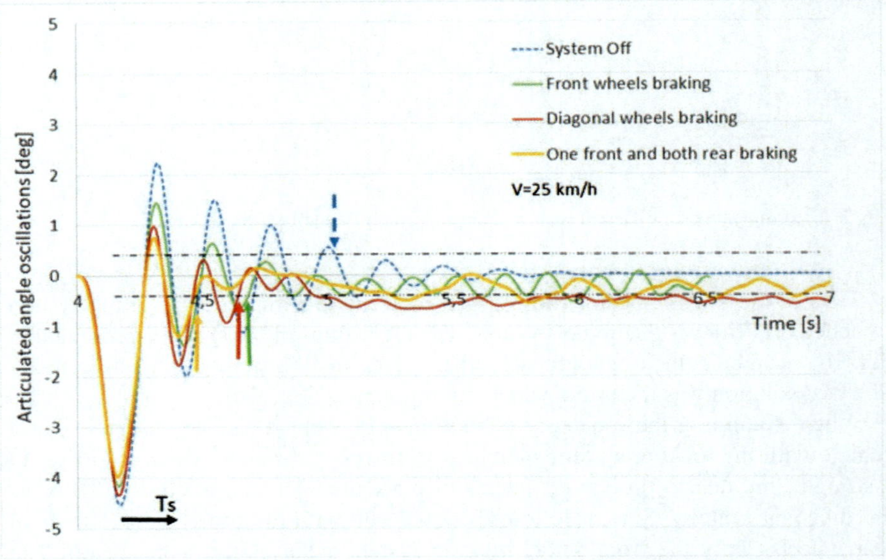

Fig. 9 Steering angle for different braking control system. Velocity V = 25 km/h

Fig. 8a and for the obtained motion trajectory – Fig. 8b. For lower speed of motion, the braking system diagonally decreased its efficiency.

During simulation tests for velocity V = 25 km/h the results presented in Figs. 9 and 10 were obtained. It can be observed that the effectiveness of reducing the oscillation amplitude for each of the proposed control algorithms is similar to that presented in Fig. 9. Similar values are also achieved for W_{SD} of each solution – Fig. 10a and for trajectory of motion – Fig. 10b.

In control theory the settling time of a dynamical system such as an amplifier or other output device is the time elapsed from the application of an ideal instantaneous step input to the time at which the amplifier output has entered and remained within a specified error band. The settling time is shown on Figs. 5, 7 and 9. To

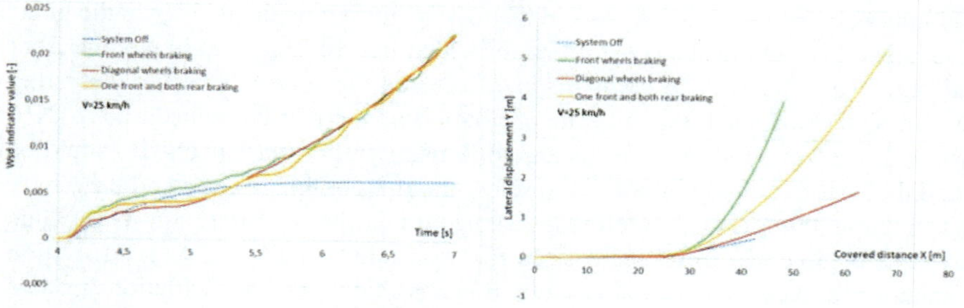

Fig. 10 W_{SD} indicator value and vehicle trajectory for different braking control system. Velocity V = 25 km/h

Table 2 Comparing results by calculating reducing time oscillation ratio

Braking mode:	V = 75 km/h	V = 50 km/h	V = 25 km/h
System off	0.0%	0.0%	0.0%
Front wheels	71.2%	66.3%	40.2%
Diagonal wheels	62.0%	50.0%	46.0%
Front and rear wheels	82.0%	76.1%	60.9%

compare obtained results an reducing time oscillation ratio (RTOR) is introduced and calculated for error band of ±0.4 [deg]. RTOR compares an efficiency of control system different modes with system switched off. T_s^{mode} is a time needed for reaching lower error band than 0.4 [deg] from disturbing input function when the system is switched off or system uses brakes. Obtained results are shown in Table 2.

Again, a control system efficiency reaches very high values. The best results could be observed for the highest vehicle velocity and for braking one front and both rear wheels. Braking only front wheels offer good efficiency. Diagonal wheels braking results the worst shortening of oscillation time but it reaches, still, very promising value. Method for diminishing snaking by using vehicle brakes brings great benefits with low cost of a new system.

$$RTOR_{0,4} = \frac{\left(T_s^{off} - T_s^{mode}\right)}{T_s^{off}} \cdot 100\% \tag{2}$$

4 Conclusions

Research shows that the use of the proposed algorithms for braking articulated rigid frame vehicle brings different results. The speed of the vehicle is extremely important. For velocities below V = 25 km/h the application of any algorithm has a similar effect. With the increase in speed, the benefits of the use of systems supporting the operator while driving are increasing. The best results in eliminating oscillations in the steering system and the lowest (most beneficial) W_{SD} indications have been obtained for the control system which has the ability to brake one front and both rear wheels. The comparison of the settling time allows for indicating efficiency of the control system. The oscillation time could be reduced up to 82% by using one front and rear wheels and 40% using only front wheels. It should be noted that it is extremely important to select the right braking forces and the length of a single braking cycle, which is the subject of further research. Too high value of torque or time causes an increase in fuel consumption and the wear of friction elements of brakes. In a critical situation it is possible to induce oscillations instead of suppressing them.

References

1. Dudziński, P., Skurjat, A.: System for improving directional stability for articulated vehicles. Computational Technologies in Engineering (TKI'2018) Conference Proceedings, ISSN 0094-243X (2019)
2. Skurjat, A., Serwatka, K.: Badanie możliwości eliminacji zjawiska wężykowania pojazdów przegubowych poprzez przyhamowanie kół. Logistyka. **5**, 1363–1370 (2015)
3. Dudziński, P., Skurjat, A.: Research on the influence of geometric parameters on the phenomenon of snaking of articulated vehicles. General Tadeusz Kosciuszko Military University of Land Forces, s. 294–301 (2017)
4. Dudziński, P., Skurjat, A., Dacko, P.: Sposób aktywnej poprawy stateczności kierunkowej pojazdu z wieloczłonkowym układem podwoziowym. Patent: 421682 (2019)
5. Yin, Y.M., Rakheja, S., Yang, J., Boileau, P.E.: Effect of articulated frame steering on the transient yaw responses of the vehicle. Proc. Inst. Mech. Eng. Part D J. Automob. Eng. (2018). https://doi.org/10.1177/0954407017702987
6. Yin, Y.M., Rakheja, S., Yang, J., Boileau, P.E.: Design optimization of an articulated frame steering system. Proce. Inst. Mech. Eng. Part D J. Automob. Eng. (2018). https://doi.org/10.1177/0954407017729052
7. Mokhiamar, O.: Stabilization of car-caravan combination using independent steer and drive/or brake forces distribution. Alex. Eng. J. **54**(3), 315–324 (2015)
8. Mokhiamar, O., Abe, M.: Examination of different models following types of yaw moment control strategy for improving handling safety of a car–caravan combination. Proc. Inst. Mech. Eng. Part D J. Automob. Eng. (2003). https://doi.org/10.1243/095440703322114942
9. Iida, M., Tomiyama, H., Oh, T., Nakashima, H.: Steering/braking control of articulated vehicle for small turning. IFAC Proc. Vol. **43**(26), 197–202 (2010)
10. Sharp, R.S., Alonso Fernandez, M.A.: Car–caravan snaking part 2: active caravan braking. Proc. Inst. Mech. Eng. Mech. Eng. Sci. (2002). https://doi.org/10.1243/09544060260128779
11. Graeme, M., David, C.: Combined emergency braking and turning of articulated heavy vehicles. Veh. Syst. Dyn. **55**(5), 725–749 (2017)

An Experimental Observation of the Spatial Motions of Strings in Resonance Points Under the Planar Excitation

Sungyeup Kim, Hiroshi Yabuno, and Kohei Mitaka

Abstract In general, strings are resonated when the excitation frequency is in the neighborhood of natural frequency by planar excitation. It is a primary resonant phenomenon. And they are also resonated by three times frequency of the natural one under the external excitation. This phenomenon is called super-harmonic resonance. In this study, we consider the case when the lower end of a string is excited periodically by shaker in a direction which is perpendicular to the longitudinal one and the other upper end is fixed. Then, we show experimentally nonlinear phenomena in strings by frequency response curves. As a result, we found out that spatial motion can occur by super-harmonic resonance. Finally, we observe the occurrence of the out of plane or spatial motions through the experiments. These phenomena are caused by the coupling effect of the stiffness due to the characteristic of the geometrical cubic nonlinear restoring force in strings.

Keywords String · Out-of-plane motion · Nonlinear stiffness

1 Introduction

There are many particular oscillations such as nonlinear vibrations which include nonplanar motion and super-harmonic resonance, in recent. Most of the mechanical systems we use for construction field are able to regard as continuous systems. Among them, strings are very interested in the engineering fields and the most fundamental isotropic elements of the continuous systems, as well. It is well-known

S. Kim (✉) · K. Mitaka
University of Tsukuba, Graduate School of Systems and Information Engineering, Department of Intelligent Interaction Technologies, Tsukuba city, Ibaraki, Japan
http://yabuno.iit.tsukuba.ac.jp/

H. Yabuno
University of Tsukuba, Graduate School of Systems and Information Engineering, Tsukuba city, Ibaraki, Japan

J. Awrejcewicz (ed.), *Perspectives in Dynamical Systems I: Mechatronics and Life Sciences*, Springer Proceedings in Mathematics & Statistics 362,
https://doi.org/10.1007/978-3-030-77306-9_5

that the nonplanar motion in strings is produced in previous studies [1–3]. This nonlinear phenomenon is caused by the coupling effect of the stiffness due to the characteristic of the geometrical cubic nonlinear restoring force in strings. The stability of the motion with respect to the nonlinear phenomenon in the strings was investigated [4]. Analizing the nonlinear dynamical characteristics of the continuous systems including strings, beams and plates is very important problem to design the mechanical systems and control the motion with respect to the infinite degrees of freedom. In this study, we consider the nonplanar motion caused by nonlinearity of the strings and also super-harmonic resonance [5] by showing the frequency response curves through the experiments.

2 Analytical Model and Equations of Motion

2.1 *Analytical Model for the Nonplanar Motion*

We introduce the analytical model for the nonplanar motion in the string as shown in Fig. 1.

where ρ is the density of the string, A is a area of cross-section of the string, l is a natural total length of the string, and N_0 is an initial tension of the string. We introduce the $x - y - z$ coordinate system with respect to the time for this study.

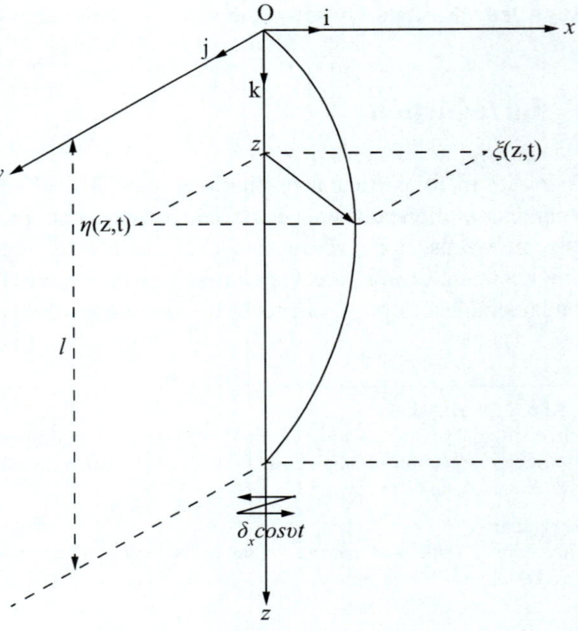

Fig. 1 Analytical model of the string related to the equations of motion for the nonplanar motion

Fixed upper end O is the origin of this coordinate system. And the other lower fixed end is excited periodically and harmonically. The stiffness of the string is followed the Hook's law. We consider the characteristic of the geometrical cubic nonlinear restoring force under the harmonic external excitation of the string.

2.2 Nonlinear Equations of Motion for the Nonplanar Motion

The nonlinear equations of motion for the nonplanar motion was analized in a preceding study [6] as follows:

$$\frac{\partial^2 \xi}{\partial t^2} + 2\mu \frac{\partial \xi}{\partial t} - c^2 \frac{\partial^2 \xi}{\partial z^2} = 0, \tag{1}$$

$$\frac{\partial^2 \eta}{\partial t^2} + 2\mu \frac{\partial \eta}{\partial t} - c^2 \frac{\partial^2 \eta}{\partial z^2} = 0, \tag{2}$$

where μ is a damping ratio of the system, ξ is a displacement of the excitation direction in the string, η is a displacement of orthogonal direction in the string. c^2 is a nonlinear coefficient and expressed as follows:

$$c^2 = 1 + \frac{\beta}{2} \int_0^1 \left\{ \left(\frac{\partial \xi}{\partial z} \right)^2 + \left(\frac{\partial \eta}{\partial z} \right)^2 \right\} dz, \tag{3}$$

where β is a constant and expressed as follows:

$$\beta = EA/N_0, \tag{4}$$

where E is a Young's modulus of the material property. A and N_0 are area of cross-section and density of the string. The boundary conditions of the system are as follows:

$$\begin{cases} \xi(0,t) = 0, \ \xi(1,t) = \delta_x \cos \nu t, \\ \eta(0,t) = 0, \ \eta(1,t) = 0, \end{cases} \tag{5}$$

where δ_x and ν are the excitation amplitude and frequency, respectively. In this study, we consider the case when the lower end of the string is periodically excited by external shaker in a direction which is perpendicular to the longitudinal one and the other upper end is fixed. It is well-known that the nonplanar motion occurs around the natural frequency of the string due to the nonlinearity of the equations including the coupling effect terms.

3 Experiments

3.1 Experimental Apparatus

We carried out the experiments with a simple apparatus as shown in Fig. 2. The diameter of the string is 0.54×10^{-3} m and the total length of the string is 1.45 m. The string is made by stainless steel wire. We use two laser displacement sensors for excitation and orthogonal directions to measure the displacement each direction of the string as shown in Fig. 3, and the load cell for the tension variation of the string. The initial tension of the string is 12.5 N. Parameters of the experimental apparatus are shown in Table 1. The natural frequency of the string we use in this study was obtained by free vibration.

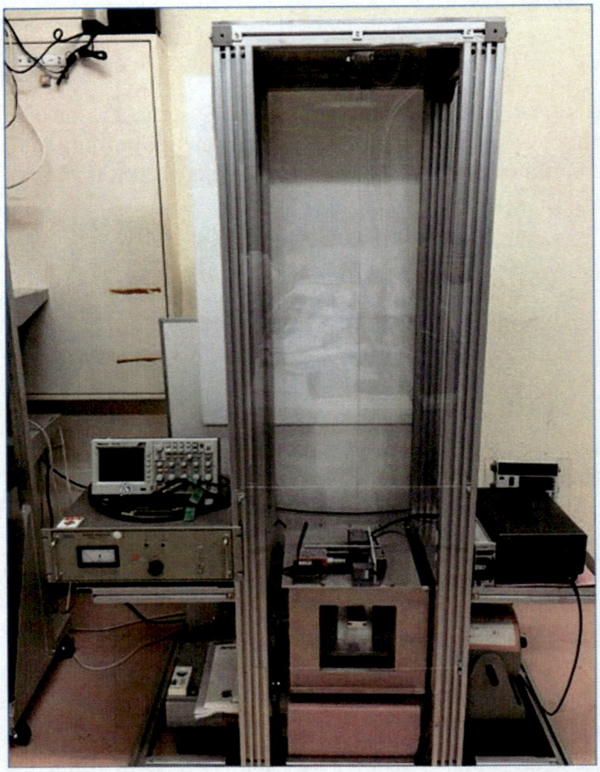

Fig. 2 String externally excited at the lower fixed end by shaker

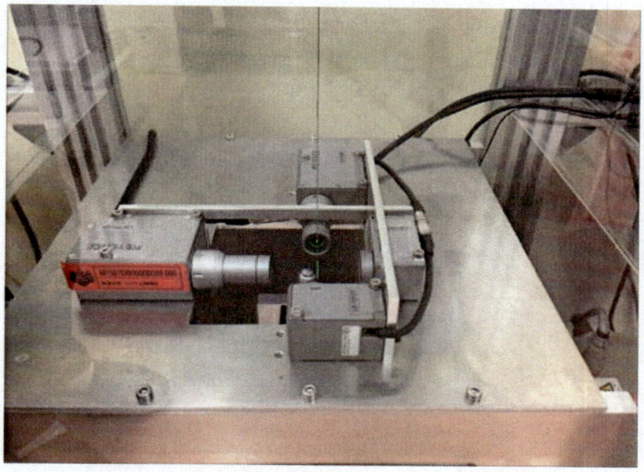

Fig. 3 Enlarged view of the measuring point in the string by laser displacement sensors

Table 1 Parameters of the experimental apparatus

Parameter	Value
First natural frequency of string, $\omega/2\pi$	39.4 Hz
Amplitude of excitation, δ_x	0.05×10^{-3} m
Length of string, l	1.45 m
Initial tension of string, N_0	12.5 N
Nonlinear coefficient, β	3.61×10^3
Area of cross-section of string, A	2.29×10^{-7} m^2
Young's modulus of string, E	1.97×10^{11} N/m^2
Density of string, ρ	5.24×10^3 kg/m^3

3.2 Experimental Results

We conducted the three types of experiments with respect to the nonlinear phenomena in the string. We invesitigated the characteristic of the frequency response curves by using a F.F.T. analizer. At first, we carried out the experiments when the excitation frequency is twice of the natural frequency. The result of the experiment is shown as in Fig. 4. In this figure, the horizontal axis and vertical axis denote the detuned excitation frequency and response amplitude, respectively. And the circle plot and rectangular one mean the vibration of the excitation direction and orthogonal one, respectively. Detuned excitation frequency 0 express the natural frequency of the first mode in the string. We observed the vibration with respect to the out-of-plane motion in the neighborhood of the natural frequency. And we also observed the jumping phenomenon. Figure 5 is the result of the experiment when the excitation frequency is three times of the natural frequency. These phenomena are caused by nonlinear coupling effect of the equations of motion in the string. We can also observed the out-of-plane motion which occur near the twice of the natural frequency through the experiment. Finally, we carried out the experiments when the

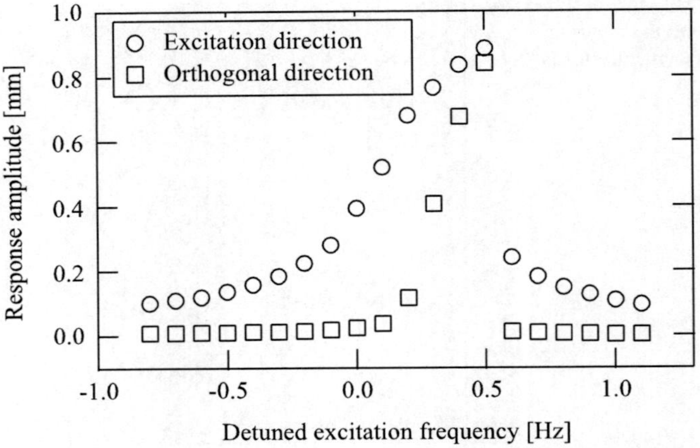

Fig. 4 Experimental frequency response curves when the excitation frequency is twice of the natural frequency: circle plot and rectangular one mean in-plane motion and out-of-plane one, respectively. And horizontal axis and vertical axis denote the detuned excitation frequency and response amplitude, respectively

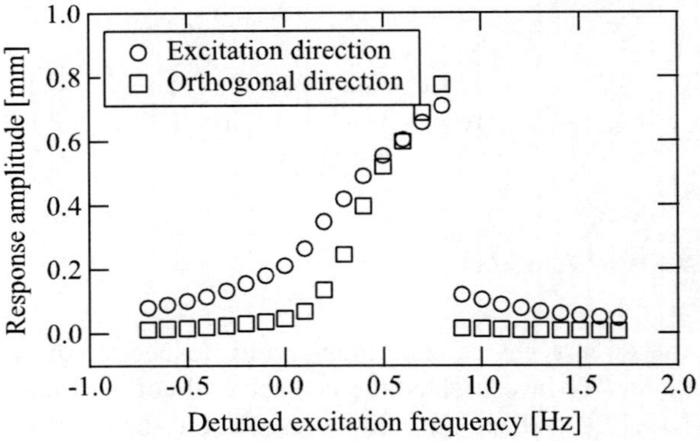

Fig. 5 Experimental frequency response curves when the excitation frequency is three times of the natural frequency: circle plot and rectangular one mean in-plane motion and out-of-plane one, respectively. And horizontal axis and vertical axis denote the detuned excitation frequency and response amplitude, respectively

excitation frequency is one third of the natural frequency. In this case, the frequency response curve is as shown in Fig. 6. The response amplitude which is the vertical axis express the amplitude subjected to natural frequency of the first mode. As a results, we can observed the nonplanar motion in three cases.

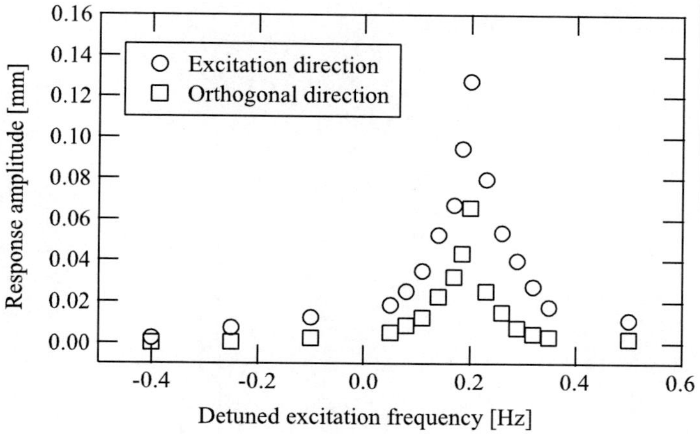

Fig. 6 Experimental frequency response curves when the excitation frequency is one third of the natural frequency: circle plot and rectangular one mean in-plane motion and out-of-plane one, respectively. And horizontal axis and vertical axis denote the detuned excitation frequency and response amplitude, respectively

4 Conclusions

In this study, we showed the nonlinear phenomenon through the experiments and the frequency response curves by using a F.F.T. analyzer. We observed the nonplanar motion around the primary resonant region experimentally. And we also observed the occurrence of the super-harmonic resonance when the excitation frequency is three times of the natural one by doing experiments.

References

1. Anand, G.: Nonlinear resonance in stretched strings with viscous damping. J. Acoust. Soc. Am. **40**(6), 1517–1528 (1966)
2. Harrison, H.: Plane and circular motion of a string. J. Acoust. Soc. Am. **20**(6), 874–875 (1948)
3. O'Reilly, O., Holmes, P.J.: Non-linear, non-planar and non-periodic vibrations of a string. J. Sound Vib. **153**(3), 413–435 (1992)
4. Miles, J.W.: Stability of forced oscillations of a vibrating string. J. Acoust. Soc. Am. **38**(5), 855–861 (1965)
5. Raghunandan, C., Anand, G.: Superharmonic vibrations of order 3 in stretched strings. J. Acoust. Soc. Am. **64**(4), 1093–1100 (1978)
6. Anand, G.: Large-amplitude damped free vibration of a stretched string. J. Acoust. Soc. Am. **45**(5), 1089–1096 (1969)

A Hydraulic Delta-Robot-Based Test Bench for Validation of Smart Products

Renan Siqueira, Osman Altun, Paul Gembarski, and Roland Lachmayer

Abstract With the development of new technologies, such as smart components, additive manufacturing or multi-materials, product performance tests play a decisive role on supporting effective design and product reliability. However, test machines are mostly designed to attend norms and perform standard tests, which requires a need for development of new machines when dealing with new cutting-edge technologies or reliability of a specific product. Therefore, these test benches must be designed to be flexible and robust, in order to attend the highest number of possibilities for a certain kind of test and a range of different components. With this intent, an innovative test bench for high loads was designed and constructed based on a delta-robot configuration. This configuration, which is commonly applied for high speed kinematic systems, was adapted to apply high transverse loads in three axis while keeping a considerably large range of movement. Thereunto, dynamic simulations were conducted considering a hydraulic actuation and the robust control approach of Sliding Mode Control (SMC), which delivered satisfactory results. Finally, after mechanical design, construction and calibration, first tests were performed for a self-sensing suspension arm, where the load prediction ability of the component can be analyzed and the ability of the developed system to test complicated components under multi-axial load was evaluated.

Keywords Delta-robot · Test-bench · Sliding mode control

1 Introduction

New technologies, such as smart components, additive manufacturing or multi-materials, require product performance tests in its development phase, being a crucial step to guarantee product reliability. When dealing with new cutting-edge

---

R. Siqueira (✉) · O. Altun · P. Gembarski · R. Lachmayer
Leibniz University Hannover, Hanover, Germany
e-mail: siqueira@ipeg.uni-hannover.de

J. Awrejcewicz (ed.), *Perspectives in Dynamical Systems I: Mechatronics and Life Sciences*, Springer Proceedings in Mathematics & Statistics 362,
https://doi.org/10.1007/978-3-030-77306-9_6

technologies, specific performance tests must be executed in structural components, such as dynamic loads, random loads, or even different load cases. Standard test machines are mostly designed to attend norms and perform standard tests, which are not suitable for those cases. Therefore, a test bench must be always newly designed to attend these needs. Ideally, the test bench should be flexible and robust, being able to execute the highest number of possibilities and configurations for a certain kind of test and for a range of different components.

This document presents the design process for the construction of a dynamic hydraulic test bench. The motivation for constructing this equipment is originally in the context of a collaborative research center (SFB 1153) [10], where Tailored Forming multi-material demonstrator components must be tested, as foreseen in the method presented in [9]. However, to allow possibilities of future research, the idea of this test bench was amplified, so that it can test a bigger variety of mechanical components, such as additive manufacturing parts and smart components to be implemented in vehicles [5].

For this reason, a more robust and generic test bench for dynamic loads must be developed, to allow testing under different circumstances. With that in mind, a multi-axial and hydraulic equipment was proposed. This document will describe the whole design process, witch was based on the methodology present in the VDI 2221 [12]. Therefore, it includes: understanding the requirements and concept creation; dynamics analysis of the multi-physics system through simulations; dimensioning and construction of the parts; implementation and testing.

2 Concept

For the concept design creation of this test bench, few alternatives were discussed, with the objective of selecting the best system. The challenge of this project was the generalization of the test bench, since it should not be constructed for a specific use case. With that in mind, the equipment should in future be able to test the large amount of different components, as well as the use of different configurations. That resulted in a very reduced requirements list.

Some of the defined requirements were: ability of applying static and dynamic loads; perform multi-axial force application; capability of testing components as big as vehicle parts; execution of long duration fatigue tests, with the possibility of low frequency vibration tests. These requirements were listed and quantified, and it was pre-determined that a hydraulic system should be implemented to achieve high forces.

There is a great variety of ways that the force of hydraulic cylinders can be arranged in order to obtain different degrees of freedom. However, after evaluating different configurations, a proposal was made based on the delta-robot concept. This is a parallel robot with a three translational degree of freedom system, projected for high speed robots, mostly used in low load applications [7] (Fig. 1a). Due to the use of parallel arms, the complexity of the system's control is decreased. The advantage

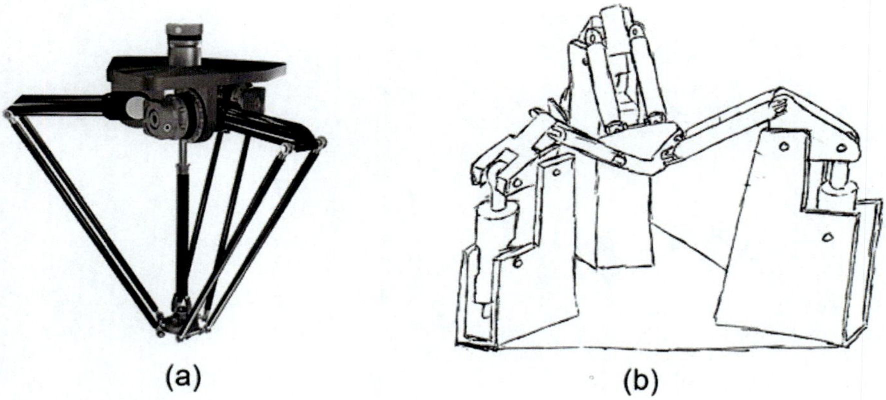

Fig. 1 (**a**) Delta-robot [13]; (**b**) Sketch of the final concept for the test bench

of this concept is also that the safety of the cylinders would be guaranteed, with no transverse force reaching the hub. However, it does not give much space for different configurations and it involves a high number of different components.

As seen in Fig. 1b, a concept that joins both of the proposed ideas was developed, in a way that the cylinders remain safe, a customizable arrangement is possible and a more simplistic construction can be performed. The final sketch shows the system composed of three independent sub-systems, called here *Towers*. Same as the delta-robot, the arms are connected in the centre, by an actuator, allowing three translational degrees-of-freedom.

3 System Simulations and Analysis

The second phase in the design of the test bench is the simulation of the dynamical behaviour for the proposed concept. The following steps will be described in this section: parametrization of the system, kinematics and inverse kinematics, hydraulics, rigid body dynamics and control. During the whole process, it was used computer supporting softwares. For the functions creation and modelling it was used Mathworks Matlab [4], including Simulink and one of it's toolboxes called Simscape. For the CAD drawing, it was used Autodesk Inventor [3].

3.1 Parametrization

In this section, all the important parameters are listed, where some of them were fixed by requirements, some were predicted, and some were found as function of

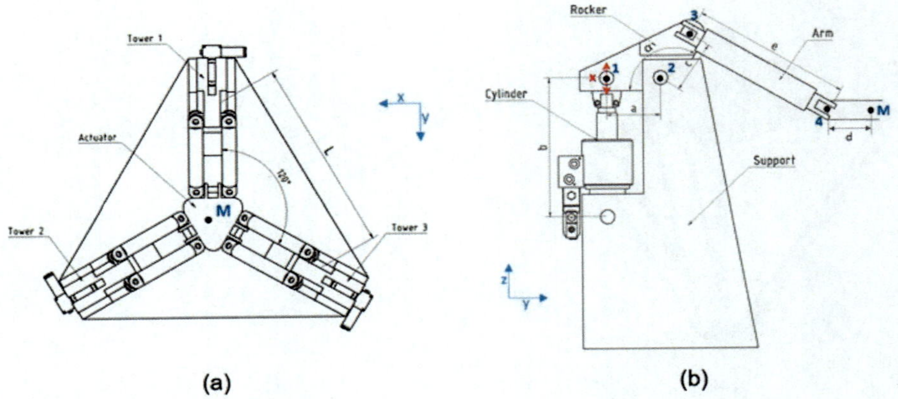

Fig. 2 (**a**) Upper view of the test bench; (**b**) Squematics for one of the three *Towers*

Table 1 Parameters specification

Par.	Value	Decision method
x	0 (initial)	Free controllable variable
a	150 mm	Chosen to provide space for the cylinder to move
b	395 mm	Set an angle of 90° between Cylinder and Rocker
c	150 mm	Chosen according to range/force requirements
d	145 mm	The smallest possible that can provide enough space
e	410 mm	Calculated to equalize force at different directions
L	1000 mm	Chosen to provide enough space for the components
α	116°	Calculated to have 90° between the Arm and Rocker

the others. Figure 2 shows the first sketch with the main paramentes, joints and components.

The points 1, 2, 3, 4 are coincident with the revolute joints. The point M is the center of the *Actuator*, where the resultant force is performed over the sample and where all other *Towers* are equally connected. The configuration of the three towers is made on an equilateral triangle, with L being the distance between the joint 2 of each *Tower*. Naturally, the parameters were chosen in an iterative process and Table 1 shows their final values and on what the decision was based.

3.2 *Kinematics and Inverse Dynamics*

Although the delta-robot configuration is a closed chain mechanism, its kinematics can be found through the intersection of the domain of each of the three *Arms*. Therefore, it was used here *Generalized Coordinates* instead of cartesian coordinates [1]. Thus, we choose the variables that we will control as our *Generalized*

Coordinates, that in our case is the vector **x** containing the displacement of each of the three cylinders. Based on that, the result will be the position of the *Actuator*, described with Cartesian coordinates.

The first step to be executed is a translation of the whole systems towards its center in a distance equivalent to d. The objective of this is to eliminate the space occupied by the actuator and place the origin O in the center, at joint 4 and in the same plane of the joint 2. This step eases the calculation of the kinematics, as described in detail in [6]. With the angle γ formed by the joints $O-2-3$ written as a function of x, we can write the position of point 3 as:

$$\mathbf{p}_3(x) = \mathbf{p}_2 + [0 \quad c\cos(\gamma(x)) \quad c\sin(\gamma(x))]^T \tag{1}$$

Now, it is needed to write the same position of the point 3 for the other *Towers*. For that purpose, a rotation transform Ω around z-axis in the angles 0°, 120° and 240° must be respectively applied Eq. (2).

$$\mathbf{p}_{31} = \Omega(0^\circ)\cdot\mathbf{p}_3(x_1), \quad \mathbf{p}_{32} = \Omega(120^\circ)\cdot\mathbf{p}_3(x_2), \quad \mathbf{p}_{33} = \Omega(240^\circ)\cdot\mathbf{p}_3(x_3) \tag{2}$$

After this description, the position of the point M in the center of the actuator can be found by the intersection of three spheres that have their centers in each point 3 of the *Towers* and a radius equivalent to the length e of the *Arm*. For that, the values must be substituted in the equation of a sphere:

$$(x - x_0)^2 + (y - y_0)^2 + (z - z_0)^2 = e^2 \tag{3}$$

Where x, y and z are the coordinates of the *Actuator*; r is the radius that will be equal to e; and x_0, y_0 and z_0 represent the coordinates of the center of the spheres, which will be replaced by each vector $\mathbf{p}_3$ at a time, forming a system of 3 equations. The intersection will happen at two different points, one upper and another lower, which gives a kinematic redundancy where the upper solution must be eliminated. With this function constructed, the positioning space of the test bench is fully defined.

For the use in a force control, where the force applied can be fully controlled, an inverse dynamic model must be also created. Differently from last section, here we want to know how much force the cylinders must make to achieve a certain resultant at the *Actuator*. Thereunto, a quasi-static model was constructed and simulated. The forces depend on the sample being tested and, for the simulations, the stiffness of this sample was approximated by a spring. Since the direction of the force actuation of each *Tower* will be the same of the *Arm*, the following normalized vector $\mathbf{p}_{\mathbf{Fi}}$ that describes the direction of the *Arm* can be written to represent the direction of the forces:

$$\mathbf{p}_{Fi} = \left[\frac{(\mathbf{p_M} - \mathbf{p_{3i}})}{\|\mathbf{p_M} - \mathbf{p_{3i}}\|} \right]; \qquad i = 1..3 \tag{4}$$

A subspace Δ can be written with these three normalized vectors as its columns:

$$\Delta = [\mathbf{p}_{F1}, \mathbf{p}_{F2}, \mathbf{p}_{F3}] \tag{5}$$

Than, a vector that describes the forces transmitted by the *Rocker* $\mathbf{F_T}$ as a function of the intensity of the force of the cylinder F_C is written. This force will be always perpendicular to the *Rocker* and can be written as:

$$\mathbf{F_{Ti}} = \Omega(\varepsilon_i) \cdot \frac{a\cos(\theta_i)}{c} \begin{bmatrix} 0 \\ \cos(90^\circ - \gamma_i) \\ \sin(90^\circ - \gamma_i) \end{bmatrix} F_{Ci}; \qquad i = 1..3 \tag{6}$$

Where θ is the very small angle of inclination between cylinder and *Rocker* ($<1.5^\circ$) and can be here neglected. Next, the force transmitted by the *Rocker* to the *Arm*, entitled $\mathbf{F}_{Arm}$, is written. This vector contains the intensity of the force of the three *Arms*, and can be calculated as the projection of the force $\mathbf{F}_T$ over the directional vector $\mathbf{p}_F$, as shown in Eq. (7).

$$\mathbf{F}_{Arm} = [< \mathbf{F}_{T1}, \mathbf{p}_{F1} >, \quad < \mathbf{F}_{T2}, \mathbf{p}_{F2} >, \quad < \mathbf{F}_{T3}, \mathbf{p}_{F3} >]^T \tag{7}$$

Finally, the resultant force $\mathbf{F_R}$ can be calculated by a liner system, in relation to the forces applied by the *Arms*, if the position is known:

$$\mathbf{F_R} = \Delta \cdot \mathbf{F}_{Arm} \tag{8}$$

With these equations, it is possible to find the required forces at the cylinders to apply a resultant force in any direction at the actuator.

3.3 Hydraulics

For the modelling of the hydraulic system, the software Simulink with the toolbox Simscape Hydraulics was used, as seen in Fig. 3.

The blocks give the possibility to implemented the exact real valve signal-flow behavior, as well as fluid and cylinder characteristics. The block entitled *Proportional and Servo-Valve Actuator* was equally calibrated, so that the same frequency response of the valve was achieved.

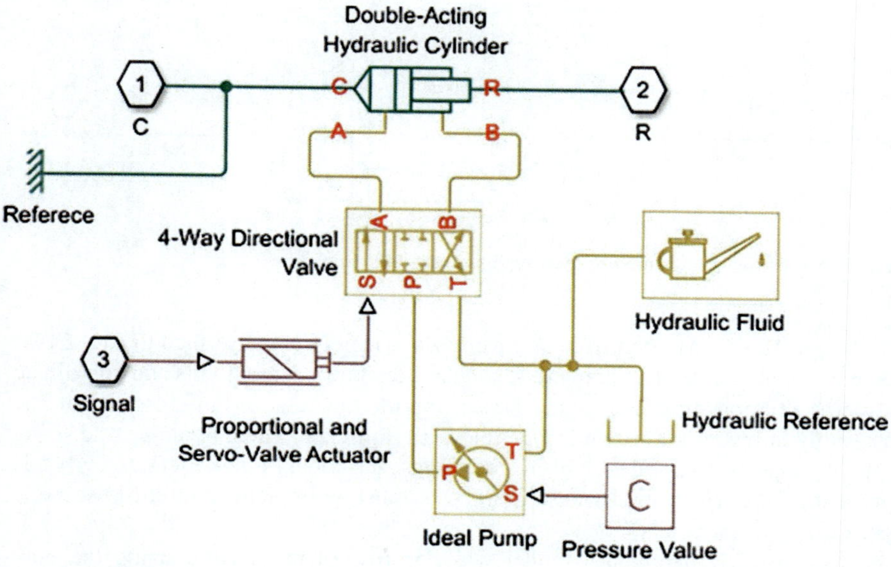

Fig. 3 Simscape hydraulics model

Fig. 4 Export-import process from Inventor to Simscape

3.4 *Rigid-body Dynamics*

The mechanical dynamics model was also constructed using the Simscape module, with the multi-body blocks. The advantages of this method is the very quick modeling, with a very clear visualization of the whole system. Despite that, it can also have interface with the other versions of Symscape, as well as the hydraulics one introduced before.

The objective now is to describe the test bench behavior, where the dynamics and kinematics of the cylinders are given as input and it is obtained the resultant dynamics and kinematics as the output. For the construction of the model, however, a method for the construction of the block diagram was performed using a module called *Simscape Multibody Link*. These module can be installed in some CAD platforms (such as Autodesk Inventor) and perform a data transfer between them through Matlab (Fig. 4).

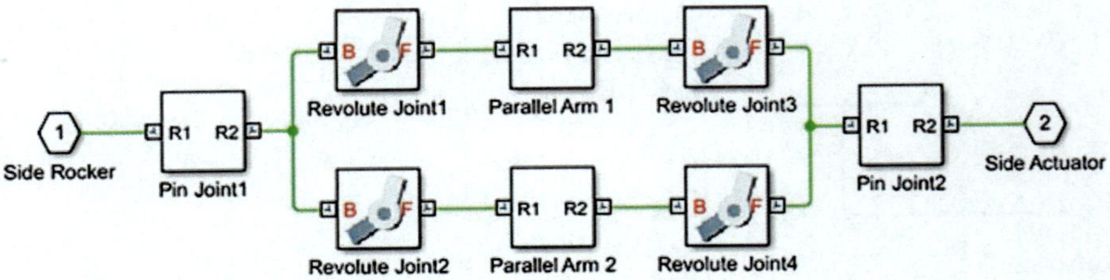

Fig. 5 Sub-system that composes the *Arm* block model in Simscape

The XML file exported from an Inventor's assembly is transcribed to a `m-file` containing all the design parameters, such as: geometry, mass, joints and coordinate systems relationships. For a successful export, the relationships created at the assembly in Inventor must be compatible with Simscape. Then, in the second stage, Matlab imports these files and create a Simscape block automatically. With this procedure, the whole model description that must be implemented in Simscape is performed in few clicks.

Thus, a CAD parametric model was constructed in Inventor using the core parameters of Sect. 3.1. At this moment, the other dimensions of the structure are estimated, since the mass and inertia of the system will have influence in the dynamic model. After the construction of the basic CAD model, the data transfer process is executed and a Simscape model was successfully generated by the toolbox. Figure 5 shows one of the subsystems created, for the *Arms* and its joints.

The resultant force of the hydraulic model presented in Fig. 3 is than connected to a prismatic joint in order to generate the forces of the cylinders. As a final result, a system with the following ports is created:

- Input: Signal for the valves in the hydraulic system.
- Output: Position and forces at the *Actuator*.

It is important to mention, that here, despite the assumption that the whole model is rigid, all the joints are set as ideal.

3.5 Sliding Mode Control

For this test bench, two types of control can be implemented: force or position control. In the first one, a chosen resultant force is desired at the *Actuator* and a certain differential pressure needed at each cylinder must be achieved. In the second, a path must be followed, where the displacement of the cylinders must be controlled. Since the main purpose of this test bench is for load application, only a force control will be developed, where a force sensor is positioned between the actuator and the component to be tested.

For that, Sliding Mode Control (SMC) was used, which is a control method with robust performance. By robust it is understood that this control is more independent from system's parameters and can provide good results even when uncertainties in modelling are present [11]. This theory was originally designed as a variation of the binary on/off type of control. In an on/off control, the signal assumes only two values, according to the signal of the error. In the standard SMC, the derivative of the error is also taken into account, as shown below:

$$s(t) = K_s \cdot sign(\sigma)\sigma = \lambda e + \dot{e} \tag{9}$$

where K_s and λ are parameters of control. The equation σ creates a straight line at the state space, where the signal assumes a positive or a negative value at each side of this line. Thus, the state of the system tends to approximate to this line and *slide* over it, which is the behavior that gave origin to the name of this method. This forces the non-linear dynamics of the system to behave in a simpler way. The stability and robustness of the control can be proved when $\lambda > 0$ [11].

However, with the presence of a proportional valve, an on/off control cannot work fast enough. Furthermore, the constant switch creates a undesirable vibration in the system. For that reason, it is introduced an improvement of SMC with a continuous function used instead of the signal one, which smooths the signal response when it gets close to zero. This can be implemented by the use of a hyperbolic tangent, as shown below [8]:

$$s(t) = K_s \cdot \tanh(k_t \sigma) \tag{10}$$

This model introduces a new control parameter k_t, which is related to how smooth the function is. High values of k_t is equivalent to a signal function. The calibration of the parameters is made using a step function. First it is attributed a very high value for k_t and λ, to approximate it to a simple on/off controller. The value of K_s is set to be the maximal value desired for the valve overture. This value may be calibrated according to the pressure used and here was set as 5% of the maximum command signal, where the command-pressure behaviour of the valve is still linear. Next, the parameter λ is decreased until a critical damping is achieved and k_t is decreased until the chattering behavior of the signal diminishes.

After the calibration of the parameters, the response for a desired force function was simulated. The force at the sample is $F_R = [300\sin(\pi t), 400\cos(\pi t), 0]^T$, describing a circle in the xy plane with 0.5 Hz frequency. The difference between the desired force and the real force measured is converted to cylinder forces, as seen in Sect. 3.2, where the force is really controlled. The results are presented in Fig. 6.

As a last step, the *Linear Analysis Tool* provided by Matlab was executed, to check the frequency response of the whole system (Fig. 7). Although this tool is just for an approximation and the values presented might not be accurate, it can serve as future reference. In a real application, this response will be dependent on the component to be tested.

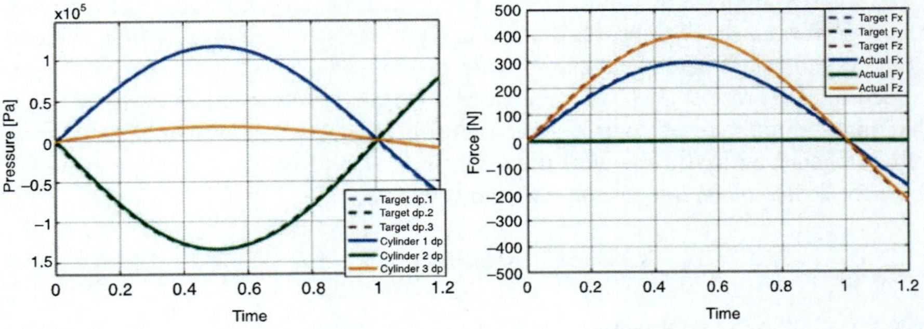

Fig. 6 Example of the force function controlled by SMC

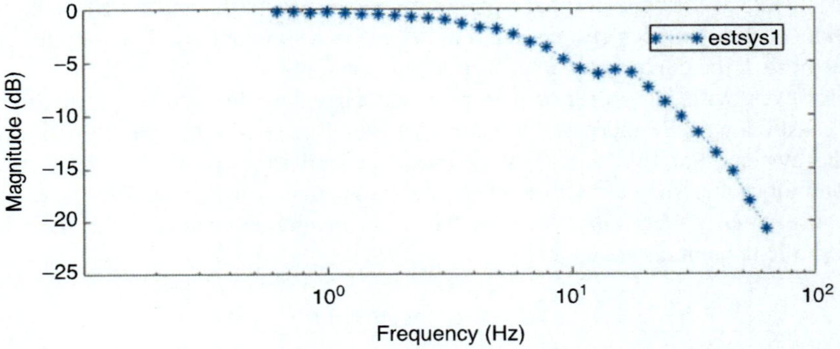

Fig. 7 Frequency response of the whole system with SMC

4 Construction and Implementation

The construction of the test bench was made in the Institute for Product Development of the Leibniz University Hannover. For that, a series of structural finite element simulations were also executed in order to guarantee the stiffness of the system. For some components of the structure, design alternatives were searched through template-based modeling [2]. After the conclusion of the design and construction, the signal acquisition, processing, control and user-interface were implemented through the software Labview. Figure 8 shows the constructed test bench, named *SAMBA*, with a suspension arm as component to be tested, as well as the real actuation forces being controlled.

In this test it is evaluated the capability of using strain gauges in the suspension arm in order to predict the forces being applied, which will characterize it as a smart component. For the first tests, forces in different directions were applied and controlled, while the signal of the strain gauges were recorded.

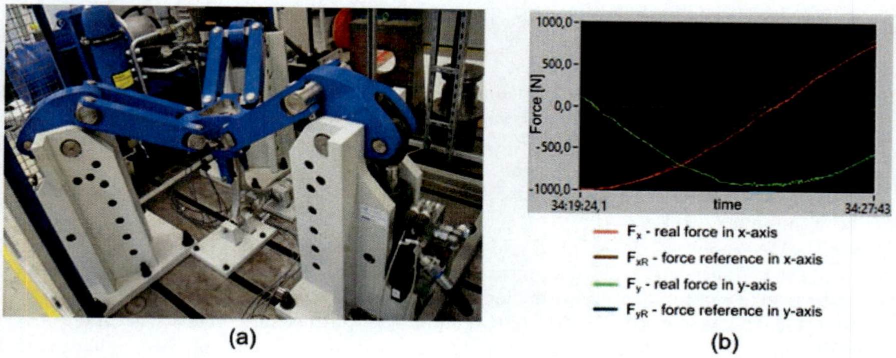

Fig. 8 (**a**) The constructed *SAMBA* test bench with a suspention arm as specimen; (**b**) Measured controlled force applied with 0.1 Hz

5 Conclusions

In this work a methodology for the development of a dynamic test bench was presented. The use of a delta-robot configuration for this kind of test bench was firstly introduced, which is a system not commonly seen. Through detailed parameter analysis, kinematic simulations and system modeling, the design of the machine was successfully performed. The computer-aided system showed to be a powerful tool to model the real system behavior in this process, generating detailed analysis in a relatively short time. Furthermore, the SMC was also suitable for use, since some parameters of the hydraulic system could not be well predicted in the computational model and some robustness in the control was needed.

The construction of the test bench *SAMBA* was also successfully executed. The control through Labview, however, is not fast enough in cases where a high frequency force must be implemented. This is due to the fact that no real-time controller was used and, although the mechanical system is much slower than the electrical one, the computational loop passing though the computer is not fast enough. Therefore, future implementation will require a real-time control system for high frequencies analysis. However, a great potential is seen for future tests with *SAMBA* for a variety of smart and multi-material components.

Acknowledgments The results presented in this paper were obtained under the umbrella of Collaborative Research Center 1153 "Process Chain for Manufacturing Hybrid High-Performance Components by Tailored Forming", preliminary inspection project C2. The authors would like to thank the German Research Foundation (DFG) and the CRC 1153 for its financial and organizational support.

References

1. Baruh, H.: Analytical dynamics. In: Engineering Mechanics Series. WCB/McGraw-Hill, New York (1999). https://books.google.de/books?id=oANZKAAACAAJ
2. Gembarski, P.C., Li, H., Lachmayer, R.: Template-based modelling of structural components. Int. J. Mech. Eng. Robotics Res. **6**, 336–342 (2017)
3. Inventor, A.: Autodesk Inventor Professional 2017, Build: 142, Release: 2017RTM. Autodesk Inc., California (2016)
4. Matlab, M.: version 9.2 (R2017a). The MathWorks Inc., Massachusetts (2017)
5. Mozgova, I., Yanchevskyi, I., Gerasymenko, M., Lachmayer, R.: Mobile automated diagnostics of stress state and residual life prediction for a component under intensive random dynamic loads. Procedia Manuf. **24**, 210–215 (2018). http://www.sciencedirect.com/science/article/pii/S2351978918305547. Proceedings of the 4th International Conference on System-Integrated Intelligence: Intelligent, Flexible and Connected Systems in Products and Production
6. Olsson, A.: Modeling and Control of a delta-3 Robot. Lund, Sweden (2009). ISSN 0280-5316
7. Pierrot, F., Reynaud, C., Fournier, A.: Delta: a simple and efficient parallel robot. Robotica **8**(2), 105–109 (1990)
8. Siqueira, R., Castro Pinto, F.: Model of an active magnetic bearing using sliding mode control. In: Proceedings of the 20th International Congress on Sound and Vibration 2013. Bangkok, Thailand (2013)
9. Siqueira, R., Bibani, M., Duran, D., Mozgova, I., Lachmayer, R., Behrens, B.A.: An adapted case-based reasoning system for design and manufacturing of tailored forming multi-material components. Int. J. Interact. Des. Manuf. (IJIDeM) **13**(3), 1175–1184 (2019). https://doi.org/10.1007/s12008-019-00566-7
10. Sonderforschungsbereich SFB 1153 (German) [Collaborative Research Center CRC 1153]: Process chain for the production of hybrid high-performance components through tailored forming. Leibniz Universität Hannover (2018). Accessed: 2018-05-23
11. Utkin, V.I.: Sliding mode control design principles and applications to electric drives. IEEE Trans. Ind. Electron. **40**(1), 23–36 (1993)
12. VDI-Richtlinie 2221: Methodik zum entwickeln und konstruieren technischer systeme und produkte. VDIVerlag, Düsseldorf (1993)
13. Wittenstein: Delta robots—gearboxes and actuators by wittenstein se. Wittenstein, Germany (2019). Accessed: 2019-07-31

Towards Online Transient Simulation of a Real Heat Pump

Mariusz Zamojski, Paul Sumerauer, Christoph Bacher, and Fadi Dohnal

Abstract Efficiency and flexibility are key aspects of modern heat pumps for the household. A nonlinear model of the refrigeration cycle is developed in the framework of Matlab/Simulink. This allows for simulation and control design of multiphase fluid dynamics of an existing heat pump. The complexity of the model is balanced against the calculation speed since the ultimate aim is to embed the model-predictive capability in existing products. A finite difference model of the evaporator and the condenser is tuned and benchmarked against real measurements at stationary operation. This capability is the basis for transient startup and shutdown dynamics which enables robust model-predictive control design.

Keywords Refrigeration cycle · FCV · Heat pump

1 Introduction

The future of our society is strongly related to energy consumption and climate protection is playing an increasingly important role in our lives. The reduction of energy consumption, the development of renewable energy and the better use of existing ones are constantly being driven forward. A significant part of the energy consumption can be assigned to the heating of buildings, especially private households in which increasingly a heat pump is used. Due to climate change towards warmer temperatures and the fact that a modern heat pump can also be used for cooling in summer, the purchasing tendency of a heat pump is steadily

M. Zamojski (✉) · F. Dohnal
UMIT—Private University for Health Sciences, Medical Informatics and Technology, Division for Mechatronics Lienz, Lienz, AT, Austria
e-mail: mariusz.zamojski@umit.at; fadi.dohnal@umit.at
https://www.umit.at/melz

P. Sumerauer · C. Bacher
IDM Energiesysteme GmbH, Matrei in Osttirol, AT, Austria
https://www.idm-energie.at

J. Awrejcewicz (ed.), *Perspectives in Dynamical Systems I: Mechatronics and Life Sciences*, Springer Proceedings in Mathematics & Statistics 362,
https://doi.org/10.1007/978-3-030-77306-9_7

increasing. According to the Austrian study [1] there were about 20,000 installed devices in 2004. In 2015 there were already 240,000 pumps. The forecast for 2030 is up to 600,000–900,000 units. Thus, the total energy consumption by heat pumps will increase significantly (relocation and new consumption). In order to maximize the efficiency of consumption, around 2.5 million EUR are already being invested by public authorities in research in Austria every year.

A heat pump is a thermal machine that exploits the Carnot cycle. As a working medium, refrigerants with special thermodynamic and chemical properties are used. The cycle can be divided into four sections, which are processed cyclically (see Figure 1) and Figure 2 shows the circular process in LogPH diagram:

- Evaporation:
  At a constant, low temperature and pressure level, heat is added to the refrigerant from the heat source and the refrigerant evaporates.

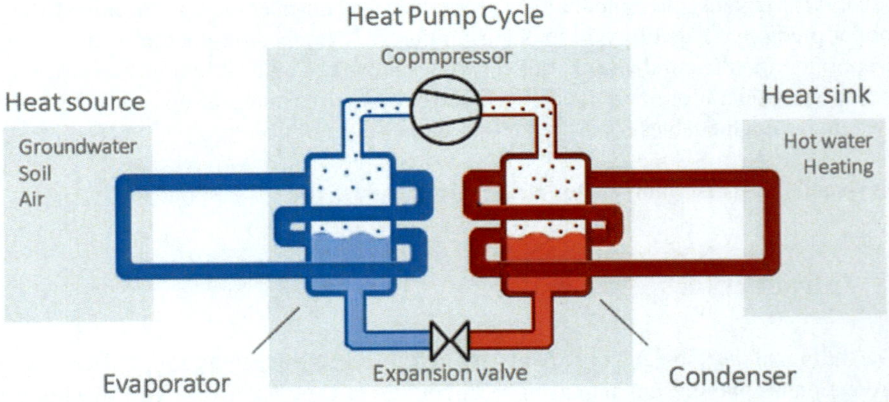

Fig. 1 Schematic representation of the heat pump. The refrigerant inside the heat pump exchanges heat between the heat source and the heat sink

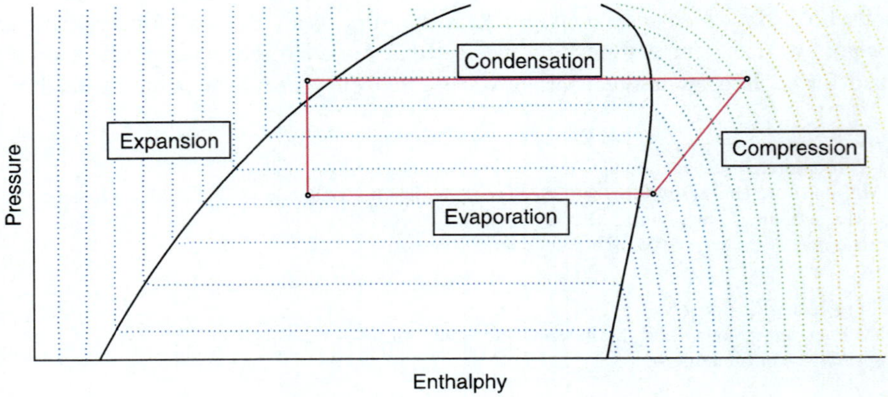

Fig. 2 Pressure and Enthalpy diagram of the heat pump

- Compression:
  By supplying energy from the outside, the refrigerant is compressed to a higher pressure level. Due to the Joule-Thomson effect, the temperature of the refrigerant gas rises.
- Condensation:
  At a high temperature and pressure level, the refrigerant transfers the heat to the heat sink, cools down and condenses into a liquid.
- Relaxation:
  The refrigerant is expanded in a valve to outlet pressure and evaporated. Now the Joule-Thomson effect leads to drop in the temperature of the gas back to the initial level.

For compression heat pumps, care must be taken to compress only superheated gaseous refrigerant to ensure permanent lubrication of the bearings and sliding surfaces of the compressor. Overheating is controlled by the electronic expansion valve, a stepper motor controlled injection valve located in front of the evaporator, and determines the amount of refrigerant to be evaporated. In addition, the heating or cooling capacity of the heat pump must be adapted to the prevailing environmental conditions and the changing needs of the user. This is done in modern heat pumps via a variable speed compressor.

The focus of the present contribution lies on the modelling of the components involved in the heat exchange of a heat pump.

2 Heat Exchanger Modeling

To meet the ever-increasing demand for modeling accuracy and product efficiency, the models of large-scale nonlinear system are of computational complexity, high-order, and complex nonlinearities. The modelling complexity is a challenging task in which the balance between complexity and accuracy must be considered. Two heat exchanger modeling approaches are commonly used: finite-volume distributed-parameter and moving-boundary lumped parameter methods [2]. In this contribution we present the development, simulation, and first experimental validation of a first principles modelling framework that captures the dynamics of the heat exchange components of an industrial heat pump based on [3]. Because of high nonlinear nature of refrigerant the simulation of heat exchanges is time consuming. The following assumptions are made for a compromise between accuracy and simplification:

1. The refrigerant flows through a long, thin, circular, and uniform horizontal tube.
2. The refrigerant flows only in the longitudinal direction.
3. Axial heat conduction in the refrigerant is negligible.
4. Momentum change and viscous friction in the refrigerant are negligible.

Table 1 Notation for governing partial differential equations

Variable	Description
ρ	Refrigerant density
A_{cs}	Tube inner cross-sectional area
$\dot{m}$	Refrigerant mass flow rate
h	Refrigerant enthalpy
P	Refrigerant pressure
α_i	Heat transfer coefficient between refrigerant and tube wall
α_o	Heat transfer coefficient between tube wall and secondary fluid
p_i	Inner surface area per unit length
p_o	Outer surface area per unit length
T_r	Refrigerant temperature
T_w	Tube wall temperature
T_a	Secondary fluid temperature
$(C_p\rho A)_w$	Tube wall thermal capacitance per unit length

Applying the above assumptions to laws of fluid dynamics and thermodynamics, mass and energy conservation equations for the refrigerant Eqs. (1a)–(1d) and tube wall energy conservation Eq. (1e) can be written as:

$$\frac{\partial \rho}{\partial t} + \nabla \cdot (\rho \vec{u}) = 0 \tag{1a}$$

$$\frac{\partial (\rho A_{cs})}{\partial t} + \frac{\partial (\dot{m})}{\partial z} = 0 \tag{1b}$$

$$\frac{\partial (\rho \vec{u})}{\partial t} + \nabla \cdot (\rho \vec{u} \vec{u}) = \rho \vec{f} + \nabla \cdot \sigma \tag{1c}$$

$$\frac{\partial (\rho A_{cs} h - A_{cs} h)}{\partial t} + \frac{\partial (\dot{m} h)}{\partial z} = p_i \alpha_i (T_w - T_r) \tag{1d}$$

$$(C_p \rho A)_w \frac{\partial (T_w)}{\partial t} = p_i \alpha_i (T_r - T_w) + p_o \alpha_o (T_a - T_w) \tag{1e}$$

The notation herein is listed in Table 1.

The model is written in the framework of Matlab/Simulink® and requires ordinary differential equations (ODE) containing the time derivatives of the states. Using the previously mentioned assumptions, as well as Leibniz integral rule Eq. (2) according to [4]

$$\int_{z_1(t)}^{z_2(t)} \frac{\partial f(z,t)}{\partial t} dz = \frac{d}{dz}\left[\int_{z_1(t)}^{z_2(t)} f(z,t)dz\right] - f(z_2(t),t)\frac{d(z_2(t))}{dt} + f(z_1(t),t)\frac{d(z_1(t))}{dt} \tag{2}$$

for removing the derivative with respect to length, it is possible to transform the partial differential equations (PDE) Eqs. (1b), (1d), and (1e)) into more efficiently solvable ordinary differential equations (ODE). In this article we will only show the derivation for refrigerant mass conservation. The full derivation for all conservation equations can be found in [3].

2.1 *Example Derivation for one Control Volume*

To derive ODEs for refrigerant mass conservation, we have to integrate the Eq. (1b) with integration limits from $z = 0$ to $z = L_{cv}$ for every term. Assuming the cross-sectional area of the tube as constant yields the Eq. (3a). By applying the Leibniz rule, we can write the result in Eq. (3b). Assuming the density in CV as an average and performing the integration results in Eq. (3c). Taking the time derivative results in Eq. (3d). Density derivative depends on Pressure and Enthalpy, by taking this into account gives Eq. (3e). Integrating the second term of the conservation PDE yields Eq. (3f). Combining both terms integration and reorganizing results in Eq. (4).

$$\int_0^{L_{cv}} \frac{\partial(\rho A_{cs})}{\partial t} dz = A_{cs,i}\left[\int_0^{L_{cv}} \frac{\partial(\rho_i)}{\partial t} dz\right] \tag{3a}$$

$$\int_0^{L_{cv}} \frac{\partial(\rho A_{cs})}{\partial t} dz = A_{cs,i}\left[\frac{d}{dt}\int_0^{L_{cv}} \rho_i dz\right] \tag{3b}$$

$$\int_0^{L_{cv}} \frac{\partial(\rho A_{cs})}{\partial t} dz = A_{cs,i}\left[\frac{d}{dt}(\rho_i L_{cv})\right] \tag{3c}$$

$$\int_0^{L_{cv}} \frac{\partial(\rho A_{cs})}{\partial t} dz = A_{cs,i} L_{cv}\left[\dot{\rho}_i\right] \tag{3d}$$

$$\int_0^{L_{cv}} \frac{\partial(\rho A_{cs})}{\partial t} dz = A_{cs,i} L_{cv}\left[\left(\frac{\partial \rho}{\partial P}\bigg|_h\right)_i \dot{h}_i + \left(\frac{\partial \rho}{\partial h}\bigg|_P\right)_i \dot{P}\right] \tag{3e}$$

$$\int_0^{L_{cv}} \frac{\partial(\dot{m})}{\partial z} dz = \dot{m}_{out} - \dot{m}_{in} \tag{3f}$$

$$\left[\left(\frac{\partial \rho}{\partial P}\bigg|_h\right)_i A_{cs,i} L_{cv}\right] \dot{h}_i + \left[\left(\frac{\partial \rho}{\partial h}\bigg|_P\right)_i A_{cs,i} L_{cv}\right] \dot{P} + \dot{m}_{out} - \dot{m}_{in} = 0 \tag{4}$$

Integrated equations for refrigerant energy conservation and Wall energy conservation can be written as in Eqs. (5) and (6)

$$\begin{aligned}&\left[\left(\frac{\partial \rho_{hx}}{\partial P_{hx}}\bigg|_{h_{hx}}\right) h_{hx} - 1\right] A_{cs,i} L_{cv} \dot{P}_{hx} + \left[\left(\frac{\partial \rho}{\partial h}\bigg|_P\right) h_i + \rho_i\right] A_{cs,i} L_{cv} \dot{h}_i \\ &+ \dot{m}_{out} \dot{h}_{out} - \dot{m}_{in} \dot{h}_{in} = \alpha_i A_i (T_w - T_r)\end{aligned} \tag{5}$$

$$(C_p \rho V_{cv})_w \frac{\partial (T_w)}{\partial t} = \alpha_i A_i (T_r - T_w) + \alpha_o A_o (T_a - T_w) \tag{6}$$

The ODEs can be combined in matrix form as follows

$$\begin{pmatrix} \left(\frac{\partial \rho}{\partial P}\Big|_h\right)_i A_{cs,i} L_{cv} & \left(\frac{\partial \rho}{\partial h}\Big|_P\right)_i A_{cs,i} L_{cv} & 0 \\ \left[\left(\frac{\partial \rho_{hx}}{\partial P_{hx}}\Big|_{h_{hx}}\right) h_{hx} - 1\right] A_{cs,i} L_{cv} & \left[\left(\frac{\partial \rho}{\partial h}\Big|_P\right) h_i + \rho_i\right] A_{cs,i} L_{cv} & 0 \\ 0 & 0 & (C_p \rho V)_{w,i} \end{pmatrix} \begin{pmatrix} \dot{P} \\ \dot{h}_i \\ T_{w,i} \end{pmatrix}$$
$$= \begin{pmatrix} \dot{m}_{in} - \dot{m}_{out} \\ \dot{m}_{in} \dot{h}_{in} - \dot{m}_{out} \dot{h}_{out} + \alpha_i A_i (T_{w,i} - T_{r,i}) \\ \alpha_i A_i (T_r - T_w) + \alpha_o A_o (T_a - T_w) \end{pmatrix} \tag{7}$$

2.2 *Derivation for Three CV*

The formulation for one CV, as shown in Fig. 3, can be extended easily to an arbitrary number of connected volumes [5]. To make it the intermediate mass flow rates between volumes have to be made a part of the state vector. Equations for refrigerant mass (4) and energy (5) conservation need to be solved together. Equation for wall tube energy (6) can be solved independent. Denoting $A_{cs} L_{cv}$ as V_{cv}, combined conservation equation of refrigerant mass and energy in matrix form for three control volumes can be expressed as

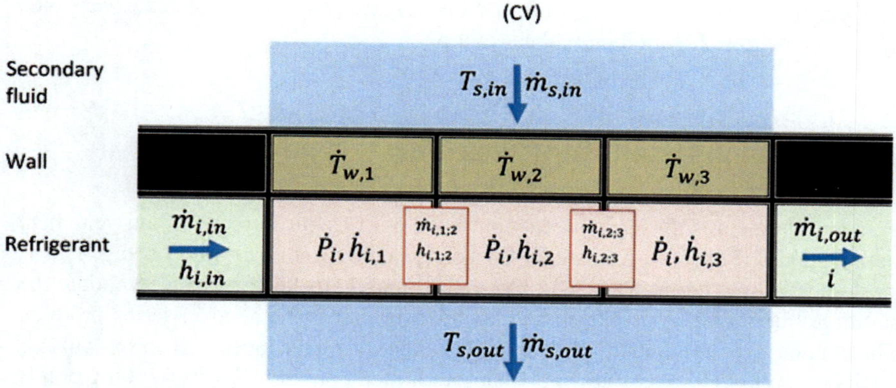

Fig. 3 Heat exchanger discretized with three control volumes

$$\begin{pmatrix} \left(\frac{\partial\rho}{\partial P}\Big|_h\right)_1 V_{cv} & & 1 & 0 \\ \left(\frac{\partial\rho}{\partial P}\Big|_h\right)_2 V_{cv} & I_{3x3}\times\left(\frac{\partial\rho}{\partial h}\Big|_P\right)_i V_{cv} & -1 & 1 \\ \left(\frac{\partial\rho}{\partial P}\Big|_h\right)_3 V_{cv} & & 0 & -1 \\ \left[\left(\frac{\partial\rho}{\partial P}\Big|_h\right)_1 h_1 - 1\right] V_{cv} & & h_{1,2} & 0 \\ \left[\left(\frac{\partial\rho}{\partial P}\Big|_h\right)_2 h_2 - 1\right] V_{cv} & I_{3x3}\times\left[\left(\frac{\partial\rho}{\partial h}\Big|_P\right) h_i + \rho_i\right]_i V_{cv} & -h_{1,2} & h_{2,3} \\ \left[\left(\frac{\partial\rho}{\partial P}\Big|_h\right)_3 h_3 - 1\right] V_{cv} & & 0 & -h_{2,3} \end{pmatrix} \begin{pmatrix} \dot{P} \\ \dot{h}_1 \\ \dot{h}_2 \\ \dot{h}_3 \\ \dot{m}_{1,2} \\ \dot{m}_{2,3} \end{pmatrix}$$
$$= \begin{pmatrix} \dot{m}_{in} \\ 0 \\ -\dot{m}_{out} \\ \dot{m}_{in}\dot{h}_{in} + \alpha_1 A_{cv}(T_{w,1} - T_{r,1}) \\ \alpha_2 A_{cv}(T_{w,2} - T_{r,2}) \\ -\dot{m}_{out}\dot{h}_{out} + \alpha_3 A_{cv}(T_{w,3} - T_{r,3}) \end{pmatrix} \tag{8}$$

Conservation equation of tube wall energy in matrix form for three CV

$$(C_p\rho V)_{w,cv}\begin{pmatrix} \dot{T}_{w,1} \\ \dot{T}_{w,2} \\ \dot{T}_{w,3} \end{pmatrix} = \begin{pmatrix} \alpha_{r,1}A_{cv,r}(T_{r,1} - T_{w,1}) + \alpha_{a,1}A_{cv,a}(T_{a,1} - T_{w,1}) \\ \alpha_{r,2}A_{cv,r}(T_{r,2} - T_{w,2}) + \alpha_{a,2}A_{cv,a}(T_{a,2} - T_{w,2}) \\ \alpha_{r,3}A_{cv,r}(T_{r,3} - T_{w,3}) + \alpha_{a,3}A_{cv,a}(T_{a,3} - T_{w,3}) \end{pmatrix} \tag{9}$$

For above equations an unique solution exists if appropriate initial conditions and boundary conditions are Specified. Unknown are the independent state variables $\dot{P}$, $\dot{h}$, $\dot{m}$ and $\dot{T}_w$. Another set of unknown variables are T_r, $\frac{\partial\rho}{\partial P}\Big|_h$ and $\frac{\partial\rho}{\partial h}\Big|_P$. Those can be determined based on refrigerant thermodynamic properties. All other variables have been acquired from the real heat pump.

3 Simulation Results for Condenser

The simulation results have been compared with measurement of an real heat pump, model iPump-A, from IDM Energiesysteme GmbH showed on Fig. 4. It is an domestic heat pump with 3–11 kW heat capacity. In this article we validate the stationary operation conditions of the condenser that is installed in series product. The condenser is an standard plate heat exchanger. We have measured and calculated refrigerant output temperature and water output temperature for five working points with different mass flow and condensing temperature.

The benchmark between simulation and experiment is shown in Figs. 5 and 6. As can be seen the measured and calculated temperatures match very well. The measured water output temperatures are marginally above the calculated for all five working points. The refrigerant output temperatures are marginally below the calculated once. The outlier at working point 3 corresponds to extreme working operation of the heat pump with very low mass flow. Similar results are expected for the evaporator. This validation process is currently ongoing.

4 Conclusions

We developed a robust universal heat exchanger model capable in simulation and control design of multi phase fluid dynamics of an existing heat pump. A finite

Fig. 4 Heat pump model iPump-A (3-11), inner and outer unit

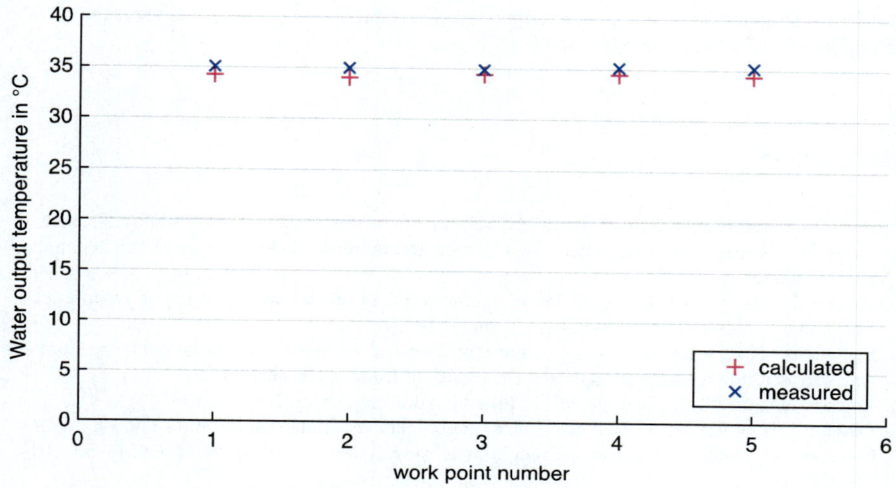

Fig. 5 Water output Temperature

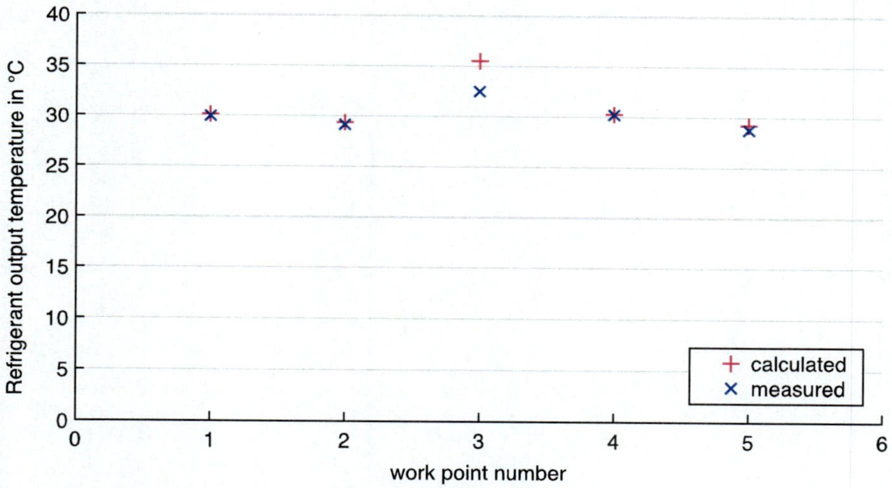

Fig. 6 Refrigerant output Temperature

difference model of the the condenser is tuned and bench marked against real measurements at stationary operation. The results demonstrate that the model is capable of predicting heat exchanger dynamics for condenser at normal operation level. Future work is needed to tune and benchmark the model to simulate the evaporator behavior. Also the capability to simulate the heat exchanger under extreme conditions need to be investigated.

Acknowledgments This work was supported by Tiroler Innovationsförderung and IDM Energiesysteme GmbH in Matrei in Osttirol, Austria.

References

1. Hartl, M., Biermayr, P., Schneeberger, A., Schöfmann, P.: Österreichische technologie-roadmap für wärmepumpen. Research project, Bundesministerium für Verkehr, Innovation und Technologie, Vienna (2016)
2. Satyam Bendapudi, J.E.B.: A review of literature on dynamic models of vapor compression equipment. Report ♯4036-5. ASHRAE, Atlanta (2002)
3. Rasmussen, B.P., Alleyne, A.G.: Dynamic Modeling and Advanced Control of Air Conditioning and Refrigeration Systems. Phd thesis, University of Illinois, Urbana, (2006)
4. Kaplan, W.: Advanced calculus, vol. 1, 4th edn. Addison-Wesley, Boston (1991)
5. Pangborn, H., Alleyne, A., Wu, N.: A comparison between finite volume and switched moving boundary approaches for dynamic vapor compression system modeling. Int. J. Refrig. **53**, 101–114 (2015)

Mathematical Approach to Assess a Human Gait

Wiktoria Wojnicz, Bartłomiej Zagrodny, Michał Ludwicki, and Jan Awrejcewicz

Abstract A purpose of the paper was to create a mathematical approach to assess a human gait. The scope of the study was to model a normal gait in the sagittal plane and frontal plane of the body. Applying the Newton-Euler formulation, three multibody biomechanical models were derived to describe single support phase and double support phase of the gait. To model a gait in the sagittal plane the open-close sagittal 6DOF model and the open-close sagittal 7DOF model can be used. To model a gait in the frontal plane the open-close frontal 7DOF model can be applied. Presented multibody models can be used to solve a forward dynamic task or an inverse dynamic task.

Keywords Gait · Modelling · Multibody model

1 Introduction

From the mechanical point of view a human gait is considered as repeated transition of the body weight from the stable state (double support phase) to the unstable state (single support phase) [12, 14]. This transition occurs due to the functioning of posture-stabilizing mechanisms controlled by the human nervous system. During gait the segments of the body moves in all anatomical planes. Qualifying the movements in each anatomical plane during a gait, some symmetries with respect to the anatomical axis can be considered [7]. Based on these observations, the gait is

W. Wojnicz (✉)
Gdansk University of Technology, Faculty of Mechanical Engineering and Ship Technology, Gdansk, Poland
e-mail: wiktoria.wojnicz@pg.edu.pl

B. Zagrodny · M. Ludwicki · J. Awrejcewicz
Department of Automation, Biomechanics and Mechatronics, Lodz University of Technology, Lodz, Poland
e-mail: bartlomiej.zagrodny@p.lodz.pl; michal.ludwicki@p.lodz.pl; jan.awrejcewicz@p.lodz.pl

J. Awrejcewicz (ed.), *Perspectives in Dynamical Systems I: Mechatronics and Life Sciences*, Springer Proceedings in Mathematics & Statistics 362,
https://doi.org/10.1007/978-3-030-77306-9_8

divided into two types: normal gait and pathological gait. Time dependent motions of body segments of normal gait have some symmetry with respect to the opposite sites of the body. On the other hand, during a pathological gait the movements of body segments are asymmetrical. From the point of view of mechanics these two types of gait can be analyzed by formulating different mathematical models. To assess a normal gait one has to derive planar dynamic models to predict motions in a sagittal and frontal plane of the body. To qualify a pathological gait one should formulate spatial dynamic models that can describe asymmetrical movements in space.

To formulate mathematical models to assess the gait a musculoskeletal system of the human body is treated as a multibody system with defined numbers of freedom (DOFs). Due to the fact that in practice during a gait examination one can measure kinematics of the body segments and the values of external loads (support reactions), an inverse dynamics approach is used to calculate joint moments (net join moments), joint intersegmental forces (net joint intersegmental forces) and joint powers (net joint powers) [14, 15]. This approach also demands to input biomechanical data of tested subjects: masses of the segments, lengths of the segments, segment radii of gyration and segment moments of inertia. Moreover, to estimate activity of the muscles during a gait testing one must to measure EMG signals.

Deriving a model to assess a human gait, one should take in mind that during each stable state (i.e. double support phase) a multibody system of the body becomes a closed system and problem of indeterminacy is occurred (i.e. distribution of supporting loads under each foot).

The aim of this work was to propose a mathematical approach that can be used to assess a human normal gait by calculating joint moments in the joints of the lower limbs during a single and double support phase. The scope of this work involved deriving three planar multibody dynamic models (open-close sagittal 6DOF model, open-close sagittal 7DOF model, open-close frontal 7DOF model) and their validations.

2 Materials and Methods

To derive planar multibody models a body was treated as a system composed of six segments (sagittal 6DOF model, see Fig. 1) and seven segments (sagittal 7DOF model and frontal 7DOF model, see Fig. 2) [9]. In the sagittal 6DOF model and the sagittal 7DOF model the following segments of the body were considered as a serially linked ones in the sagittal plane of the body: (1) right foot (from the right metatarsophalangeal joint to the right ankle joint); (2) right shin shank (from the right ankle to the right knee joint); (3) right thigh (from the right knee joint to the hip joint); (4) left thigh (from the hip joint to the left knee joint); (5) left shank (from the left knee joint to the left ankle joint); (6) left foot (from the left ankle joint to the left metatarsophalangeal joint). In the sagittal 7DOF model the influence of the upper part of the body was modelled as a seventh segment connected to the hip joints. In the sagittal 6DOF model an influence of the upper part of the body (the

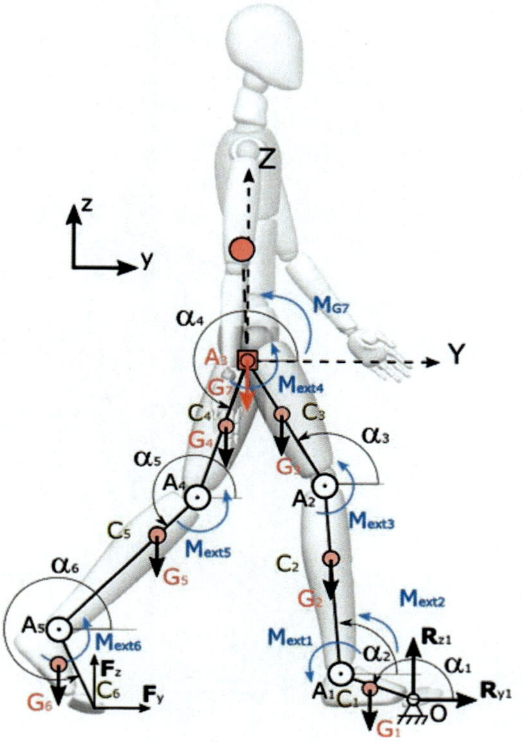

Fig. 1 The sagittal 6DOF model (O – the point between the support foot and the ground (the metatarsophalangeal joint); A_1 – the ankle joint of stance leg; A_2 – the knee joint of stance leg; A_3 – the hip joint; A_4 – the knee joint of swing leg; A_5 – the ankle joint of swing leg; α_i – the angle of the i-th segment (each angle is measured as an absolute coordinate); G_i – the gravity force of the i-th segment that acts at its center of gravity C_i; M_{exti} – the external moment loading the i-th segment; R_{y1} – the y-th component of stance leg reaction force (anterior-posterior component); R_{z1} – the z-th component of the stance leg reaction force (vertical component); F_y and F_z – the y-th and z-th component of reaction force of the swing leg during double supporting phase; y – the sagittal axis; z – the vertical axis) [11]

pelvis, torso, head, neck and upper limbs) was considered as one concentrated force applied at the centre of gravity of the upper part of the body and a moment of this force that both were applied to the hip joint of the stance leg.

In the frontal 7DOF model (Fig. 3) the seventh serially linked segments were considered in the frontal plane of the body: (1) right foot (from the right metatarsophalangeal joint to the right ankle joint); (2) right shank (from the right ankle to the right knee joint); (3) right thigh (from the right knee joint to the right hip joint); (4) pelvis (from the right hip joint to the left hip joint); (5) left thigh (from the left hip joint to the left knee joint); (6) left shank (from the left knee joint to the left ankle joint); (7) left foot (from the left ankle joint to the left metatarsophalangeal joint). An influence of the upper part of the body was modelled as one concentrated

Fig. 2 The sagittal 7DOF model (symbols are described in the Fig. 2) [11]

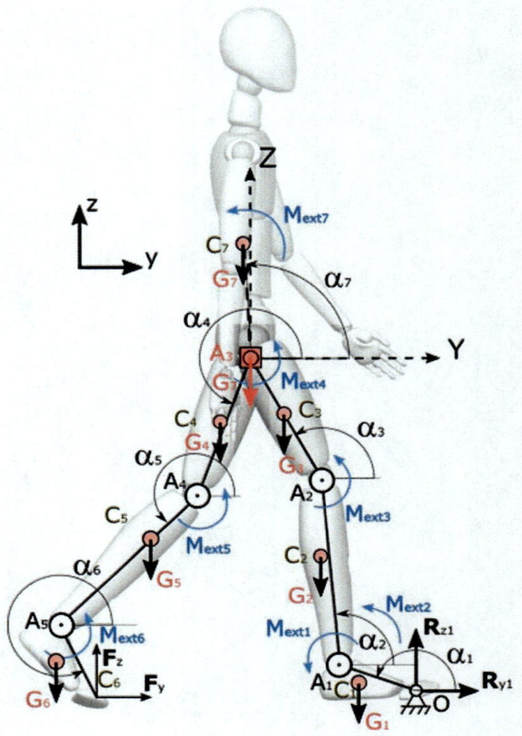

force applied at the centre of gravity of this part and a moment of this force that both were applied to the segment of the pelvis.

Three planar multibody dynamic models (open-close sagittal 6DOF model, open-close sagittal 7DOF model, open-close frontal 7DOF model) were formulated by using Newton-Euler formulation [1]. Each model can be used to solve an inverse dynamics task or forward dynamics task. It must be emphasized that proposed multibody models are more complex in comparison with ones presented in [5, 6, 8, 12].

Solving an inverse dynamics task one, an influence of passive soft tissues (tendons, ligaments, bursa) and active soft tissues (muscles) can be defined by assuming some optimization criterion to solve a redundancy problem.

Biomechanical parameters (parameters of the segments of the body) had been assessed on the base of method of Zatsiorsky and deLeva [2, 3]. Creating the author coding, numerical models of the proposed multibody models were implemented in MATLAB.

Considering any dissipation phenomena under the foot of the supporting leg [4], the metatarsophalangeal joint of the feet is treated as an attachment with the ground. This attachment is modelled as a pin joint (a joint O in Figs. 1, 2, and 3).

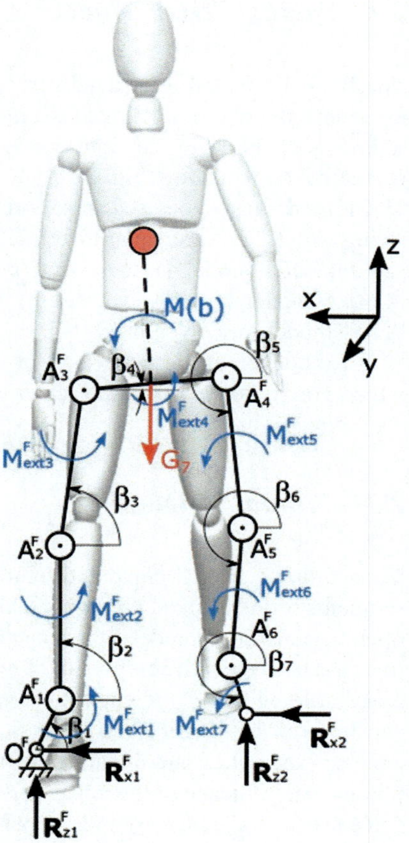

Fig. 3 The frontal 7DOF model: (O^F – the point between the support foot and the ground; $A^F{}_1$ – the ankle joint of stance leg; $A^F{}_2$ – the knee joint of stance leg; $A^F{}_3$ – the stance leg hip joint; $A^F{}_4$ – the swing leg hip joint; $A^F{}_5$ – the knee joint of swing leg; $A^F{}_6$ – the ankle joint of swing leg; β_i – the angle of the i-th segment in the frontal plane (each angle is measured as an absolute coordinate); G – gravity force of the upper part of the body; $M^F{}_{exti}$ – the external moment influenced the i-th segment in the frontal space; $R^F{}_{x1}$ – the x-th component of stance leg reaction force (medio-lateral component); $R^F{}_{z1}$ – the z-th component of the stance leg reaction force (vertical component); $R^F{}_{x2}$ – the x-th component of reaction force during double support phase; $R^F{}_{z2}$ – the z-th component of reaction force during double support phase; x – the transverse axis; y – the sagittal axis; z – the vertical axis) [9]

2.1 *Sagittal 6DOF Model*

The body is treated as a system composed of six segments serially linked in a sagittal plane of the body (Fig. 1). To model a single support phase the open sagittal 6DOF model has to be used (both y-th and z-th components of reaction force of the swing leg (F_y, F_z) are equal to zero). To model a double support phase the close sagittal 6DOF model has to be used (both y-th and z-th components of reaction force of the swing leg (F_y, F_z) are not zero). Mathematical models of these both models are published in [8–10].

An influence of the upper part of the body is considered as a load composed of one concentrate force G_7 (gravity force of this part) and a moment of this force M_{G7}.

2.2 Sagittal 7DOF Model

The body is treated as a dendritic structure, in which the hip joint spans three branches: three serially linked segments of the stance leg, three serially linked segments of the swing leg and one segment describing the influence of the upper part of the human body (Fig. 2). To describe a single support phase the open sagittal 7DOF model has to be used (both y-th and z-th components of reaction force of the swing leg (F_y, F_z) are equal to zero). To describe a double support phase the close sagittal 7DOF model has to be used (both y-th and z-th components of reaction force of the swing leg (F_y, F_z) are not zero). Detailed mathematical models of both sagittal 7DOF models are given in [8–10].

An influence of the upper part of the body is modeled as the seventh segment, which gravity force acts at the center of mass placed at the point C_7.

2.3 Frontal 7DOF Model

Considering a frontal plane and treating a body as a structure composed of seven segments serially linked through the hinge joints, there were created (Fig. 3): (1) the open frontal 7DOF model, which can be applied to model a single support phase (in this case both the x-th component of reaction force (${R^F}_{x2}$) and the z-th component of reaction force (${R^F}_{z2}$) are equal to zero); (2) the closed frontal 7DOF model, which can be used to describe a double support phase. Both models can be applied to analyse kinematics and dynamics of normal gait in a frontal plane during specific phases. An influence of the upper part of the body was modelled as one concentrate force G_7 (it is a gravity force of upper part of the body) and its moment $M(b)$.

A general mathematical description of the open frontal 7DOF model is a non-linear system of seven differential equations:

$$[B] \cdot \left\{ \frac{d^2 \beta_i}{dt^2} \right\} = \{M_i\}, i = 1, \ldots 7 \tag{1}$$

where:

$$[B] = \begin{bmatrix} B_{11} & B_{12}(\beta_1, \beta_2) & B_{13}(\beta_1, \beta_3) & B_{14}(\beta_1, \beta_4) & B_{15}(\beta_1, \beta_5) & B_{16}(\beta_1, \beta_6) & B_{17}(\beta_1, \beta_7) \\ B_{21}(\beta_1, \beta_2) & B_{22} & B_{23}(\beta_2, \beta_3) & B_{24}(\beta_2, \beta_4) & B_{25}(\beta_2, \beta_5) & B_{26}(\beta_2, \beta_6) & B_{27}(\beta_2, \beta_7) \\ B_{31}(\beta_1, \beta_3) & B_{32}(\beta_2, \beta_3) & B_{33} & B_{34}(\beta_3, \beta_4) & B_{35}(\beta_3, \beta_5) & B_{36}(\beta_3, \beta_6) & B_{37}(\beta_3, \beta_7) \\ B_{41}(\beta_1, \beta_4) & B_{42}(\beta_2, \beta_4) & B_{43}(\beta_3, \beta_4) & B_{44} & B_{45}(\beta_4, \beta_5) & B_{46}(\beta_4, \beta_6) & B_{47}(\beta_4, \beta_7) \\ B_{51}(\beta_1, \beta_5) & B_{52}(\beta_2, \beta_5) & B_{53}(\beta_3, \beta_5) & B_{54}(\beta_4, \beta_5) & B_{55} & B_{56}(\beta_5, \beta_6) & B_{57}(\beta_5, \beta_7) \\ B_{61}(\beta_1, \beta_6) & B_{62}(\beta_2, \beta_6) & B_{63}(\beta_3, \beta_6) & B_{64}(\beta_4, \beta_6) & B_{65}(\beta_5, \beta_6) & B_{66} & B_{67}(\beta_6, \beta_7) \\ B_{71}(\beta_1, \beta_7) & B_{72}(\beta_2, \beta_7) & B_{72}(\beta_2, \beta_7) & B_{74}(\beta_4, \beta_7) & B_{75}(\beta_5, \beta_7) & B_{76}(\beta_6, \beta_7) & B_{77} \end{bmatrix}$$

$$\left\{\frac{d^2\beta_i}{dt^2}\right\} = \frac{d^2}{dt^2}\begin{Bmatrix}\beta_1\\ \beta_2\\ \beta_3\\ \beta_4\\ \beta_5\\ \beta_6\\ \beta_7\end{Bmatrix}$$

$$\{M_i\} = \begin{Bmatrix}M_{1F}\\ M_{2F}\\ M_{3F}\\ M_{4F}\\ M_{5F}\\ M_{6F}\\ M_{7F}\end{Bmatrix}$$

β_i – the i-th angular displacement of the i-th segment (the i-th joint angle) in the frontal plane; $\frac{d^2\beta_i}{dt^2}$ – the i-th angular acceleration of the i-th segment in the frontal plane, $B_{ij}(\beta_i, \beta_j)$ – the ij-th coefficient depending on the mechanical characteristics; M_i – the i-th moment acting on the i-th segment depending on its weight and load caused by an acceleration of the neighbouring segments.

A general mathematical description of the closed sagittal 7DOF model has the same left part given in the Eq. (1) but the right part (the vector of moments acting on the segments) has the following form:

$$\{M_i\} = \begin{Bmatrix} M_{1F} - L_1 \cdot \sin(\beta_1) \cdot {R^F}_{x2} + L_1 \cdot \cos(\beta_1) \cdot {R^F}_{z2} \\ M_{2F} - L_2 \cdot \sin(\beta_2) \cdot {R^F}_{x2} + L_2 \cdot \cos(\beta_2) \cdot {R^F}_{z2} \\ M_{3F} - L_3 \cdot \sin(\beta_3) \cdot {R^F}_{x2} + L_3 \cdot \cos(\beta_3) \cdot {R^F}_{z2} \\ M_{4F} - L_4 \cdot \sin(\beta_4) \cdot {R^F}_{x2} + L_4 \cdot \cos(\beta_4) \cdot {R^F}_{z2} \\ M_{5F} - L_5 \cdot \sin(\beta_5) \cdot {R^F}_{x2} + L_5 \cdot \cos(\beta_5) \cdot {R^F}_{z2} \\ M_{6F} - L_6 \cdot \sin(\beta_6) \cdot {R^F}_{x2} + L_6 \cdot \cos(\beta_6) \cdot {R^F}_{z2} \\ M_{7F} - M\left({R^F}_{x2}\right) + M\left({R^F}_{z2}\right) \end{Bmatrix}, \quad (2)$$

where $M({R^F}_{x2})$ and $M({R^F}_{z2})$ – moments originating from supporting reactions applied to the seventh segment ${A^F}_6{A^F}_7$ (Fig. 3); L_i– length of the i-th segment that is inclined at the angle β_i to the horizontal plane.

2.4 Approach to Model an Interaction

Solving an inverse dynamics task for the double support phase, ground interaction components (the y-th component (F_y) and z-th component (F_z) in the close 6DOF sagittal model or the close 7DOF sagittal model; the x-th component ($R^F{}_{x2}$) and z-th component ($R^F{}_{z2}$) in the close 7DOF frontal model) have to be inputted. Their values can be assessed by using a second force plate (the first force plate is used to determine ground reaction components at the joint O (Figs. 1 and 2) or joint O^F (Fig. 3)).

In the case of lack of possibility to use the second force plate in practice one could assume an additional model of foot-ground interaction. In the paper [9] this model was derived by assuming that the reaction components have to have the values that allow a strike foot (of swing leg) to stay in the given narrow range of the ground level (vertical displacement).

3 Results

To validate proposed multibody models an inverse dynamics tasks were solved without using any optimization methods. To obtain kinematic data and load interaction data the group of health males were tested. None of the volunteer declared any kind of cardiovascular, neurological or pulmonary problems, none of them took cardiovascular medication and none had problems with motor system or postural stability. All volunteers provided written informed consent in accordance with procedures approved by the Committee of Research Ethics with Human participation at Gdansk University of Technology.

During study a marker setting protocol given on Fig. 4 was used. Kinematic data were defined by using the OPTITRACK motion capture system (six cameras working with 120 Hz frequency and MOTIVE software). Load interaction data (supporting reactions) were assess by using the Steinbichler force plate. During testing, the volunteer was given an oral instruction to perform a natural gait (each trial contained three full steps) by walking barefoot with open eyes. Single and double support phases were identified by using the motion capture system. In this study we present data of validations for one of the chosen volunteers (69.75 kg and 1.765 cm) that successfully performed trials.

Biomechanical parameters of segments and their centers of gravity (right and left foot, right and left calf, right and left thigh, upper body part) were assessed in each frame of the trial recorded by using method of Zatsiorsky and de Leva. Angular displacements of the tested segments were defined in: 1) a sagittal plane and the data are presented as relative angular displacements ($Hip = \alpha_3 - \alpha_7$, $Knee = \alpha_3 - \alpha_2$ and $Ankle = \alpha_1 - \alpha_2 - \pi/2$) in Fig. 5 [8]; b) a frontal plane (Fig. 6).

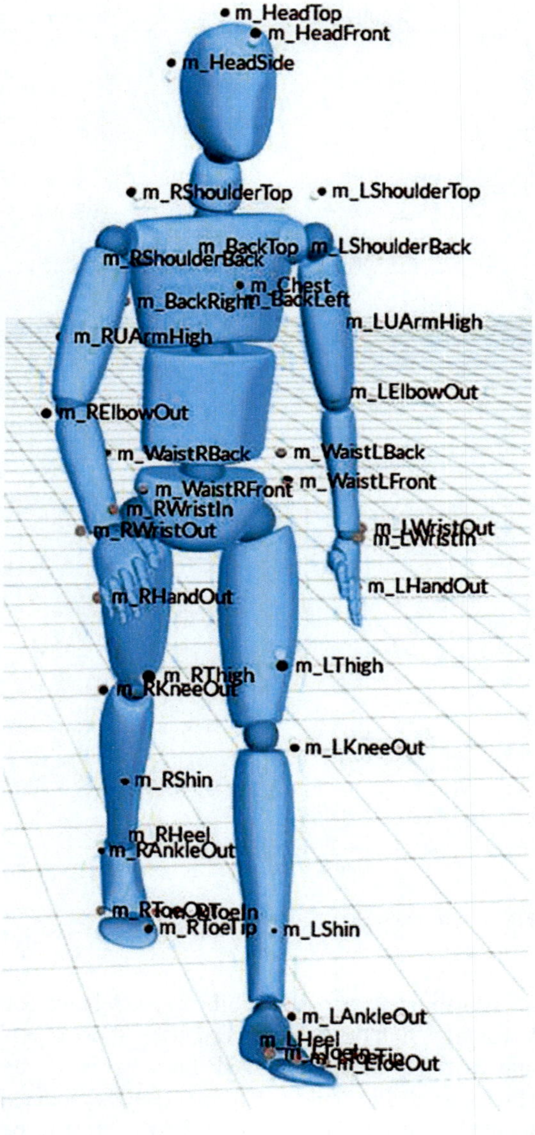

Fig. 4 Marker setting protocol

Data describing the segment angular velocities and segment angular acceleration were defined by applying postprocessing that included: (1) filtering (the 4th order Butterworth filter with 5 Hz cut-off frequency); (2) cubic spline interpolation; (3) smoothing implemented as three-point difference method.

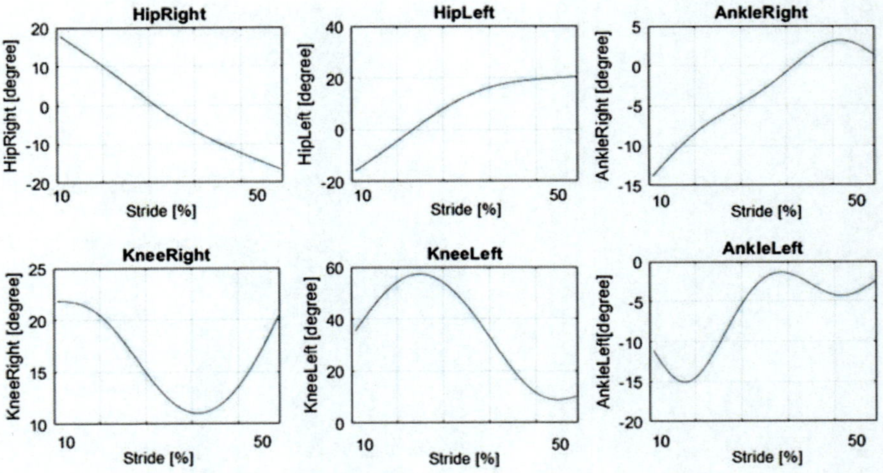

Fig. 5 Kinematic data (sagittal plane)

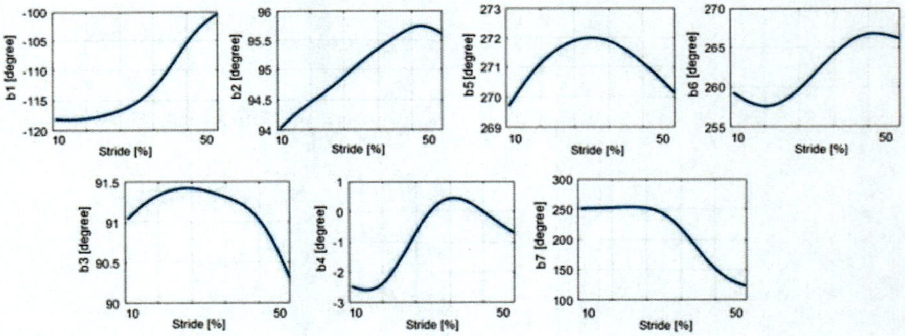

Fig. 6 Kinematic data (frontal plane)

To validate proposed multibody models we compared vertical support force and horizontal support force recorded by the force plate with the ones calculated by applying: (1) a sagittal 6DOF model (Figs. 7 and 10); (2) sagittal 7DOF model (Figs. 8 and Fig. 11); (3) frontal 7DOF model (Figs. 9 and 12). We limited our validation to the single support phase because only one force plate was available during experimental testing.

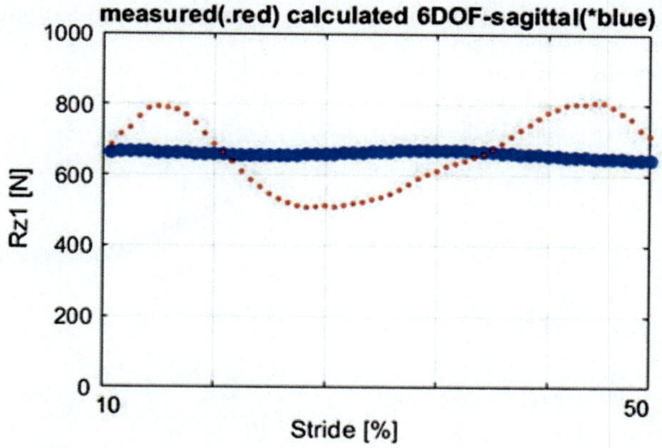

Fig. 7 Vertical support force: measured force and calculated one by using a sagittal 6DOF model

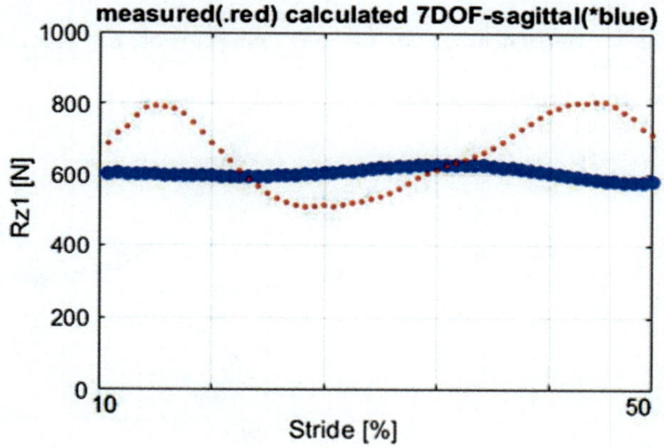

Fig. 8 Vertical support force: measured force and calculated one by using a sagittal 7DOF model

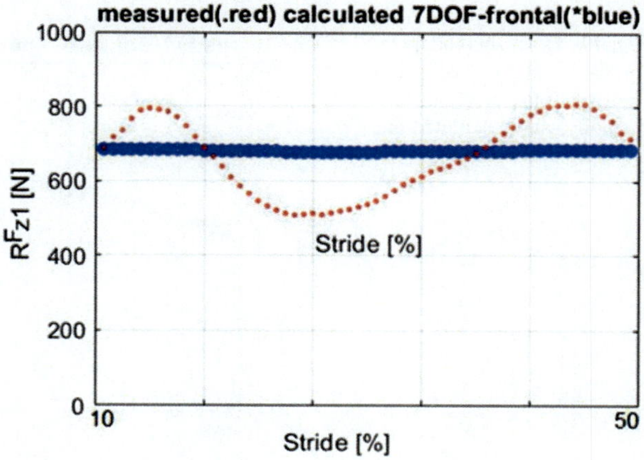

Fig. 9 Vertical support force: measured force and calculated one by using a frontal 7DOF model

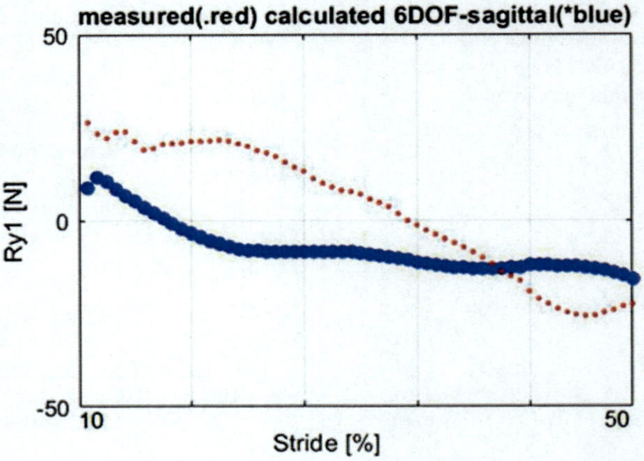

Fig. 10 Horizontal support force: measured force (towards sagittal axis of the body) and calculated one by using a sagittal 6DOF model

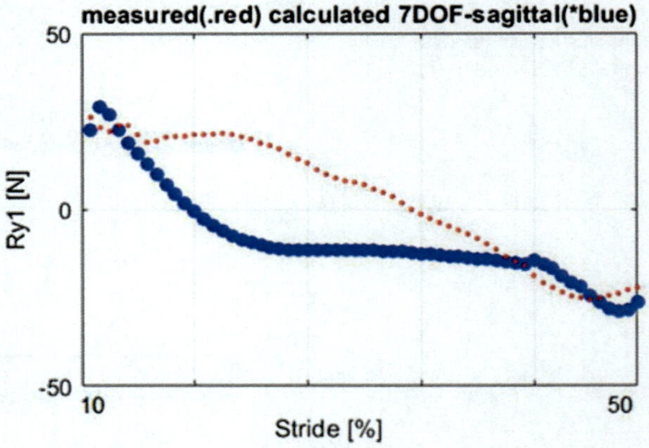

Fig. 11 Horizontal support force: measured force (towards sagittal axis of the body) and calculated one by using a sagittal 7DOF model

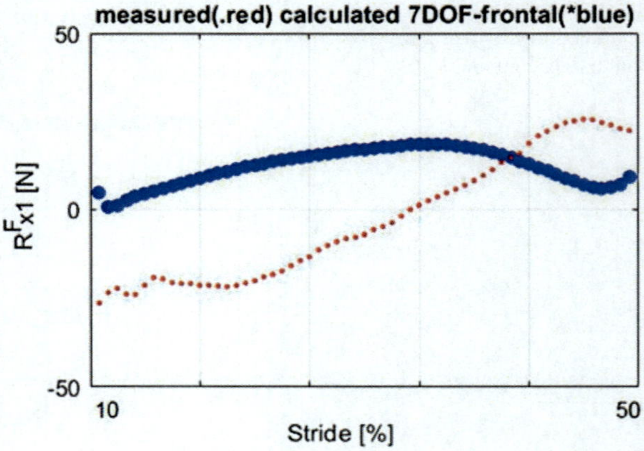

Fig. 12 Horizontal support force: measured force (towards transverse axis of the body) and calculated one by using a frontal 7DOF model

4 Discussion

Comparing experimental vertical results of with calculated ones, it was concluded that all three vertical forces calculated by using the sagittal 6DOF model (Fig. 7), the sagittal 7DOF model (Fig. 8) and the frontal 7DOF model (Fig. 9) have very similar shapes and their values are close to the measured one. Estimating an absolute maximum relative error of calculated vertical component with respect to the measured one, we obtain: 30% (sagittal 6DOF model (Fig. 7)), 28% (sagittal 7DOF model (Fig. 8)) and 34% (frontal 7DOF model (Figs. 9) and (Fig. 10).

On the other hand, comparing measured horizontal force with calculated one by using the sagittal 6DOF model (Fig. 11) and the sagittal 7DOF model (Fig. 11), we can notice that their values are very close at the beginning range and final range of the stride. However, calculated forces and measured one are a bit divergent in the middle range of the stride. Although, the value of this numerical divergence does not exceed 30 N.

Comparing measured horizontal force and calculated one by using the frontal 7DOF model (Fig. 12), we can notice that this calculated force looks like a mirror of the measured one in the middle range and final range of the stride. Although the value of numerical divergence in the beginning range does not exceed 25 N.

Analysing experimental results with calculated ones, one should take in mind that solutions of the inverse dynamics task were obtained by inputting only biomechanical parameters of the body and the kinematic data describing time motion characteristics of considered segments. An influence of upper part of the body was considered in the different way in all the presented multibody planar models. Calculated reaction forces occurred at the point of interaction during the single support phase were compared with the measured forces. Moreover, in the present approach any optimization method did not used to fit the calculated forces with the experimental ones.

5 Conclusions

A purpose of the paper was to create a mathematical approach to assess a human gait. The scope of the study covered modelling a normal gait in two anatomical planes of the body: sagittal and frontal. Using the Newton-Euler formulation, three multibody biomechanical models were derived that can be used to model a single support phase and double support phase of the gait. To model a gait in the sagittal plane the open-close sagittal 6DOF model or open-close sagittal 7DOF model can be used. To model a gait in the frontal plane the open-close frontal 7DOF model can be applied. Presented multibody models can be used to solve a forward dynamic task or an inverse dynamic task.

Using a sagittal 7DOF model, one can take into consideration an influence of the moment of inertia of the upper part of the body. On the other hand, according to the principles of the sagittal 6DOF model and the frontal 7DOF model, the influence of the upper part of the body is considered as a load, i.e. one force applied at the centre of gravity of the upper part of the body and one moment of this force.

It is worth paying attention that proposed biomechanical models describe motion in the sagittal or frontal anatomical plane of the human body. That is why they have to be used with caution to model an asymmetrical gait caused by the pelvis rotations with respect to the vertical axis and sagittal axis of the body.

Proposed multibody models can be used to design a control system of the exoskeleton that is used in gait rehabilitation. These models can be easily implemented in the software of the control system to work in the real-time to achieve the goals of rehabilitation. In this case one should consider that motions are performed in some range of variability [13, 15].

To improve the accuracy of the proposed multibody models the following crucial factors have to be considered. First, a method of segmentation has to be updated by considering the non-linear distribution of the segment masses and segment moments of inertia. Second, kinematic data should be properly postprocessed by reducing all non-physiological jerks. Third, an influence of the upper part of the body has to be considered by using proper biomechanical parameters.

Acknowledgments The work has been supported by the Polish National Science Centre under the grant OPUS 9 No. 2015/17/B/ST8/01700 for years 2016-2018. Calculations were carried out at the Academic Computer Centre in Gdansk, Poland.

References

1. Al-Shuka, H., Corves, B., Zhu, W.H.: Dynamics of biped robots during a complete gait cycle: Euler-Lagrange vs. Newton-Euler formulations. School of Control Science and Engineering. Shandong University, pp 1–39 (2019)
2. De Leva, P.: Adjustments to Zatsiorsky-Seluyanov's segment inertia parameters. J. Biomech. **29**(9), 1223–1230 (1996). https://doi.org/10.1016/0021-9290(95)00178-6
3. De Leva, P.: Joint center longitudinal positions computed from a selected subset of Chandlers' data. J. Biomech. **29**(9), 1231–1233 (1996). https://doi.org/10.1016/0021-9290(96)00021-8
4. Grzelczyk, D., Awrejcewicz, J.: Modeling and control of an eight-legged walking robot driven by different gait generators. Int. J. Struct. Stab. Dyn. **19**(5), 1–23 (2019). https://doi.org/10.1142/S0219455419410098
5. Onyshko, S., Winter, D.A.: A mathematical model for the dynamics of human locomotion. J. Biomech. **13**, 361–368 (1980). https://doi.org/10.1016/0021-9290(80)90016-0
6. Troy, J.J.: Dynamic balance and walking control of biped mechanisms. Retrospective theses and dissertations, 11095, Iowa State University (1995)
7. Vaughan, C.L., Davis, B.L., O'Connor, J.C.: Dynamics of Human Gait. Human Kinetics, Champaign (1992)
8. Winter, D.A.: Overall principle of lower limb support during stance phase of gait. J. Biomech. **13**, 923–927 (1980). https://doi.org/10.1016/0021-9290(80)90162-1
9. Wojnicz, W., Zagrodny, B., Ludwicki, M., Mrozowski, J., Awrejcewicz, J., Wittbrodt, E.: Multibody models for gait analysis. In: Awrejcewicz, J., et al. (eds.) Applicable Solutions in Non-Linear Dynamical Systems, pp. 523–538. Lodz University of Technology, Lodz (2019)

10. Wojnicz, W., Zagrodny, B., Ludwicki, M., Syczewska, M., Mrozowski, J., Awrejcewicz, J.: Approach for determination of functioning of lower limb muscles. In: Awrejcewicz, J. (ed.) Dynamical Systems in Applications Springer Proceedings in Mathematics & Statistics, vol. 249, pp. 423–438 (2018)
11. Wojnicz, W.: Biomechaniczne modele układu mięśniowo-szkieletowego człowieka (Biomechanical Models of the Human Musculoskeletal System). Wydawnictwo Politechniki Gdańskiej, Gdańsk, Poland, pp 1–209, ISBN 978–83–7348-727-7 (2018)
12. Yamaguchi, G.T., Zajac, F.E.: Restoring unassisted natural gait to paraplegics via functional neuromuscular stimulation: a computer simulation study. IEEE Trans. Biomed. Eng. **37**(9), 886–902 (1990). https://doi.org/10.1109/10.58599
13. Zagrodny, B., Ludwicki, M., Wojnicz, W., Mrozowski, J., Awrejcewicz, J.: Cooperation of mono- and bi-articular muscles: human lower limb. J. Musculoskelet. Nueronal Interact., 1–7 (2018)
14. Zajac, F.E., Neptune, R.R., Kautz, S.A.: Biomechanics and muscle coordination of human walking. Part I: introduction to concepts, power transfer, dynamics and simulations. Gait Posture. **16**, 215–232 (2002). https://doi.org/10.1016/s0966-6362(02)00068-1
15. Zhang, X., Riemer, R., Hsiao-Wecksler, E.T.: Uncertainties in inverse dynamics solutions: a comprehensive analysis and an application to gait. Gait Posture. **27**, 578–588 (2008). https://doi.org/10.1016/j.gaitpost.2007.07.012

Role of the Immune System in AIDS-defining Malignancies

João P. S. Maurício de Carvalho and Carla M. A. Pinto

Abstract The Center for Disease Control and Prevention considers AIDS-defining illnesses Kaposi's sarcoma, non-Hodgkin's lymphoma and cervical cancer. These cancers have higher incidence in HIV-infected individuals than in the general population. Additionally, cancers' clinical courses in HIV-positive individuals are increasingly aggressive when compared to those in HIV-negative patients. It is thus compelling to further understand the dynamics of AIDS-related cancer growth. We propose a non-integer order model to describe the role of the immune system in cancer cells' growth in a HIV-infected individual. The model incorporates anti-retroviral therapy and chemotherapy. We simulate the model for different proliferation functions of the cytotoxic T lymphocytes (CTLs), and other parameters, namely the HIV-infection rate, the elimination rate of infected T cells by CTLs, and the elimination rate of cancer cells by the immune system and discuss the results from a physiological perspective. The order of the fractional derivative completes the discussion of the results.

Keywords Immune system · AIDS-defining malignancies · Non-integer order system

J. P. S. Maurício de Carvalho (✉)
Faculty of Sciences, University of Porto, Porto, Portugal
e-mail: up200902671@fc.up.pt

C. M. A. Pinto
School of Engineering, Polytechnic of Porto, Porto, Portugal

Centre for Mathematics of the University of Porto, Porto, Portugal
e-mail: cpinto@fc.up.pt

J. Awrejcewicz (ed.), *Perspectives in Dynamical Systems I: Mechatronics and Life Sciences*, Springer Proceedings in Mathematics & Statistics 362,
https://doi.org/10.1007/978-3-030-77306-9_9

1 Introduction

The human immunodeficiency virus (HIV) is a retrovirus that attacks the T-helper lymphocytes ($CD4^+$ T cells) and is responsible for the development of the acquired immune deficiency syndrome (AIDS). $CD4^+$ T cells are key in protecting our immune system. The role is to signal $CD8^+$ T cells to destroy bacteria and viruses, such as HIV [1]. Unprotected sexual intercourse with potentially infected partners, sharing of contaminated syringes, mother-to-child transmission, and also blood transfusions are some of the ways to contract HIV. Currently there is no cure for AIDS [2].

According to the latest data available 36.9 million people globally were living with HIV in 2017. Thirty five point one million are adults and 1.8 million are children.

People infected with HIV/AIDS are more likely to develop some types of cancer, such as Kaposi's sarcoma, non-Hodgkin's lymphoma and cervical cancer. Kaposi's sarcoma is caused by Kaposi sarcoma-associated herpesvirus (KSHV), also known as human herpesvirus 8 (HHV-8). Non-Hodgkin's lymphoma is caused by Epstein-Barr virus (EBV) and cervical cancer is caused by human papillomaviruses (HPV). These cancers develop largely as a result of HIV-related immunosuppression, which impairs the control of oncogenic viral infections [3, 4]. In 1996, with the introduction of antiretroviral therapy (ART), the risk of AIDS and AIDS-related death declined considerably in patients with HIV. Nevertheless, although the incidence of Kaposi's sarcoma and non-Hodgkin's lymphomas also decreased, they remained higher in HIV patients than in the general population [5]. There are several types of cancer treatment, namely chemotherapy, some types of surgeries, blood transfusions and bone marrow transplant, among others [3].

Mathematical models of infectious diseases are extremely useful to predict future behaviour, assessing, and controlling potential outbreaks. Within-host models can be used for example to measure the impact of different therapies on infected patients [3, 6, 7]. Chávez et al. [3] studied the effect of chemo in a mathematical model for HIV/AIDS-cancer dynamics. They discovered the existence of a bifurcation of limit-cycles that has influence on HIV control. The numerical simulations showed that for values of the drug application period below a bifurcation point the presence of HIV in the body is kept to a minimum. For values of the drug application period above a bifurcation point the viral load significantly increases, which can become dangerous to patient's health. Carvalho et al. [8] proposed a model for the dynamics of HIV/AIDS-related cancer cells with the aim of studying the effects of delay and treatment on their growth. The results showed that the chemotherapy efficacy is highly reliable in the values of drug growth rates and lower decay rates promote a greater elimination of cancer cells.

Fractional calculus (FC) is applied in various areas of science, biology, engineering, epidemiology, among others [9]. With the help of FC we can get a deeper insight into some biological interactions, since FC uses information from past events to better predict future ones. With this, we can help health professionals adapt and administer certain health care, according to the needs of each patient [10, 11]. In 2015, Pinto et al. [12] study a non-integer order model for the three-stages of HIV infection, where drug-resistance is considered. The model provides a wide range of dynamics for three states, the disease-free equilibrium, the rapid progressors and the long-term non-progressors. This variety may be useful to devise treatment protocols, according to HIV-infected individuals' specificities, reducing treatment burden (i.e., toxicity), for example. In 2018, Pinto et al. [13] proposed a fractional model for T cells and HIV interactions. The model encompasses medication administered periodically. For example, in the case of sinusoidal drug efficacies, a bifurcation from disease-free to periodic endemic balance can be observed. They concluded that this bifurcation occurs for any value of the order of the fractional derivative α.

Inspired by the work described above, we propose a mathematical model of FO for the dynamics of cancer cells' growth in HIV-infected patients. The goal is to analyze how immune functions, as well as the rates of elimination of cancer cells by healthy T cells, HIV infection and the order of the fractional derivative, can influence the response of our immune system. In Sect. 2 we introduce the model and prove the positiveness of solutions. In Sect. 3 we present and comment the numerical results of the system. We finalize our work in Sect. 4.

2 The Model

Five cell and virus populations and chemo dose are considered in the model: cancer cells, $C(t)$; healthy CD4$^+$ T cells, $T(t)$; infected CD4+ T cells, $I(t)$; HIV, $V(t)$; cytotoxic T lymphocytes (CTLs), $E(t)$; and chemotherapeutic dynamic, $D(t)$. The number of cancer cells increases with the term $r_1{}^\alpha \left[1 - \left(\frac{C}{C_0}\right)^{1-\gamma}\right] C^\gamma$, where $r_1{}^\alpha$ is the growth rate and C_0 is the maximum size of the cancer cells. Louzoun et al. [14] showed that the growth of the cancerous tissue can be described mathematically by a function with an exponent γ that varies between 2/3 and 1. This variation depends on the growth conditions and on the neoplastic vascular system's topology. We assume $\gamma = 3/4$. Cancer cells are eliminated by healthy CD4$^+$ T cells at a rate $k_1{}^\alpha$. The healthy CD4$^+$ T cells population grow at a rate λ^α. These cells die by apoptosis at a rate $\mu_T{}^\alpha$, and by the cancer cells' action at a rate p. They can also be infected by HIV at a rate $k_2{}^\alpha$. RTI-based treatment efficacy is included in healthy and infected CD4$^+$ T cells equations using the parameter $0 \leq \epsilon_{RT} \leq 1$, where 1 represents an efficacy of 100%. The infected cells die at a rate $\mu_I{}^\alpha$ and are killed by CTLs at a rate $k_0{}^\alpha$. HIV is produced by the infected CD4$^+$ T cells, with bursting size N. Parameter $0 \leq \epsilon_P \leq 1$ is the drug efficacy of the protease inhibitors (PIs). A value of 1 means 100% drug efficacy. HIV is removed at a rate $\mu_V{}^\alpha$. CTLs are produced by the $f_i(I, E)$ functions and die at a $\mu_E{}^\alpha$ rate. The immune functions are $f_1(I, E) = q^\alpha IE$ and $f_2(I, E) = \frac{q^\alpha IE}{\epsilon E+1}$. The proliferation rate of these cells is given by q. For f_1 the CTLs production depends on the number of cytotoxic T cells and infected T cells. The function f_2 includes the saturation level of the CTLs expansion, ϵ. We also consider the effect of chemotherapy on the variation of cancer cells. This is done by inserting the term $(1 - e^{-D})$, representing the fraction of dead cells per chemo dose, since the drugs are effective only at certain stages of the cell cycle and their effectiveness is limited. The terms for cellular dose response are introduced into the equations of cancer cells, healthy and infected T cells, in particular the terms ${P_C}^\alpha(1 - e^{-D})$ and ${P_T}^\alpha(1 - e^{-D})$. $u(t)$ describes the amount of drug administered (assumed to be intravenous) and the injection time. It is assumed an instantaneous drug distribution in all body parts. Constant ${d_D}^\alpha$ represents the drug elimination rate. The description of the model variables and all parameter values are given in Tables 1 and 2, respectively.

Table 1 Description of the variables of model (1)

Variable	Symbol
Cancer cells	$C(t)$
Healthy CD4$^+$ T cells	$T(t)$
Infected CD4$^+$ T cells	$I(t)$
Virus (HIV)	$V(t)$
Cytotoxic T cells	$E(t)$
Chemotherapeutic dose	$D(t)$

Table 2 Parameter values used in numerical simulations

Parameter	Symbol	Value	Unit
Proliferation rate of cancer cells	r_1	0.18	(cells/mL)$^{1/4}$ day$^{-\alpha}$
Maximum density of cancer cells	C_0	1.00×10^6	cells/mL
Elimination rate of T cells by CTLs	k_0	4.50×10^{-7}	mL day$^{-\alpha}$
Elimination rate of cancer cells by T cells	k_1	1.00×10^{-8}	mL^{-1} day$^{-\alpha}$
HIV infection rate	k_2	1.00×10^{-5}	mL day$^{-\alpha}$
Death rate of cancer cells by drug	P_C	0.90	day$^{-\alpha}$
Death rate of T cells by drug	P_T	0.60	day$^{-\alpha}$
Proliferation rate of T cells	λ	1.00×10^4	cells/mL day$^{\alpha}$
Death rate of T cells	μ_T	0.02	day$^{-\alpha}$
Death rate of infected T cells	μ_I	0.30	day$^{-\alpha}$
Death rate of virus	μ_V	23	day$^{-\alpha}$
Death rate of CTL	μ_E	4.12×10^{-2}	day$^{-\alpha}$
Drug elimination rate	d_D	0.90	day$^{-\alpha}$
Proportion of immune cells loss due to killing of cancer cells	p	0.10	–
RTI-based treatment efficacy	ϵ_{RT}	0.75	–
PI-based treatment efficacy	ϵ_P	0.70	–
Bursting size for virus growth	N	1.00×10^3	mL^{-1}
Proliferation rate of CTL	q	3.30×10^{-6}	mL day$^{-\alpha}$
CTL's half-saturation constant	ϵ	1.00×10^{-3}	–

The nonlinear system of FO equations is given by

$$\begin{aligned}
\frac{d^\alpha C}{dt^\alpha} &= {r_1}^\alpha\left[1-\left(\frac{C}{C_0}\right)^{\frac{1}{4}}\right]C^{\frac{3}{4}} - {P_C}^\alpha\left(1-e^{-D}\right)C - {k_1}^\alpha TC,\\
\frac{d^\alpha T}{dt^\alpha} &= \lambda^\alpha + T\left[-p{k_1}^\alpha C - {k_2}^\alpha(1-\epsilon_{RT})V - {P_T}^\alpha\left(1-e^{-D}\right) - {\mu_T}^\alpha\right],\\
\frac{d^\alpha I}{dt^\alpha} &= {k_2}^\alpha(1-\epsilon_{RT})VT - {P_T}^\alpha\left(1-e^{-D}\right)I - {\mu_I}^\alpha I - {k_0}^\alpha IE,\\
\frac{d^\alpha V}{dt^\alpha} &= N{\mu_I}^\alpha(1-\epsilon_P)I - {\mu_V}^\alpha V,\\
\frac{d^\alpha E}{dt^\alpha} &= f_i(I,E) - {\mu_E}^\alpha E,\\
\frac{d^\alpha D}{dt^\alpha} &= u(t) - {d_D}^\alpha D,
\end{aligned} \tag{1}$$

where $i \in \{1, 2\}$ and $u(t) = 2.3869$ for $t = 21n$ ($n \in \mathbb{N}$) and $u(t) = 0$ otherwise. The order of the fractional derivative is given by α, where $\alpha \in (0, 1]$. We consider the definition of a FO derivative proposed by Caputo:

$$\frac{d^\alpha y(t)}{dt^\alpha} = I^{p-\alpha}y^{(p)}(t), \quad t > 0, \tag{2}$$

in which $p = [\alpha]$ is the integer part of α, $y^{(p)}$ is the p-th derivative of $y(r)$, and I^{p_1} is the Riemann–Liouville fractional integral

$$I^{p_1}z(t) = \frac{1}{\Gamma(p_1)}\int_0^t (t-t')^{p_1-1}z(t')dt'. \tag{3}$$

2.1 Model Properties

All solutions of the system (1) with non-negative initial conditions will remain non-negative for all $t > 0$. Let $R_+^6 = \{x \in R^6 \mid x \geq 0\}$ and $x(t) = (C(t), T(t), I(t), V(t), E(t), D(t))^T$. We begin by quoting the following Generalized Mean Value Theorem [15] and corollary.

Lemma 1 ([15]) *Suppose that $f(x) \in C[a,b]$ and $D_a^\alpha f(x) \in C(a,b]$, for $0 < \alpha \leq 1$, then we have*

$$f(x) = f(a) + \frac{1}{\Gamma(\alpha)}(D_a^\alpha f)(\xi) \cdot (x-a)^\alpha \tag{4}$$

with $a \leq \xi \leq x, \forall x \in (a,b]$ and $\Gamma(\cdot)$ is the gamma function.

Corollary 1 *Suppose that $f(x) \in C[a,b]$ and $D_a^\alpha f(x) \in C(a,b]$, for $0 < \alpha \leq 1$.*

1. *If $D_a^\alpha f(x) \geq 0$, $\forall x \in (a,b)$, then $f(x)$ is non-decreasing for each $x \in [a,b]$;*
2. *If $D_a^\alpha f(x) \leq 0$, $\forall x \in (a,b)$, then $f(x)$ is non-increasing for each $x \in [a,b]$.*

This proves the main theorem.

Theorem 1 *There is a unique solution $x(t) = (C(t), T(t), I(t), V(t), E(t), D(t))^T$ of the system (1) throughout the domain $(t \geq 0)$ and it remains in R_+^6.*

Proof As we can see from Theorem 3.1 and Remark 3.2 of [16], the solution for $t \geq 0$ of the initial value problem exists and is unique. For this, it is enough to prove that the non-negative orthant R_+^6 is positively invariant. To show that this happens we have to demonstrate that the vector field points to R_+^6 in each hyperplane, limiting the non-negative orthant. So, for system (1), we get:

$$\begin{aligned}
D^\alpha C\,|_{C=0} &= 0 \geq 0 \\
D^\alpha T\,|_{T=0} &= \lambda^\alpha \geq 0 \\
D^\alpha I\,|_{I=0} &= {k_2}^\alpha(1-\epsilon_{RT})VT \geq 0 \\
D^\alpha V\,|_{V=0} &= N{\mu_I}^\alpha(1-\epsilon_P)I \geq 0 \\
D^\alpha E\,|_{E=0} &= f_i(I,E) \geq 0 \\
D^\alpha D\,|_{D=0} &= u(t) \geq 0.
\end{aligned} \tag{5}$$

By Corollary 1 we conclude that the solution will remain in R_+^6.

3 Numerical Simulations and Discussion

In this section we simulate the model (1) for different values of the (i) elimination rate of cancer cells by T cells, k_1, (ii) HIV infection rate, k_2. Moreover, we also consider two distinct values of the order of the fractional derivative, α, and the two immune functions, f_1 and f_2. The numerical solutions of the model were obtained by a subroutine provided by Diethelm and Freed [17].

In Fig. 1 we can verify that the number of cancer cells is lower when the proliferation rate of CTLs only depends on the number of infected T cells and the population of CTLs (f_1). The same is true for HIV. This happens regardless of the value of α. We also conclude that the lower the value of the order of the fractional derivative, the lower the number of cancer cells and the viral load, despite of the rate of proliferation of CTLs. In Fig. 2 we vary the elimination rate of cancer cells by healthy T cells, k_1, for $\alpha = 1$ and $\alpha = 0.8$. It is observable that when the rate of elimination of cancer cells by healthy T cells is higher, more cancer cells are removed and consequently less will be the number of these cells in the body, for both values of the order of the fractional derivative. We can also see that a lower order of the fractional derivative suggests a lower number of infected T cells in the body and a decrease in the immune response, number of CTLs, regardless of the rate of elimination of cancer cells by T cells. In Fig. 3 we simulated the variation of the HIV infection rate, k_2, also for $\alpha = 1$ and $\alpha = 0.8$. We can see that an HIV infection rate of $k_2 = 4.5 \times 10^{-7}$, and for $\alpha = 1$, suggests that the model approaches the HIV-free equilibrium. For the other values of k_2 the viral load tends

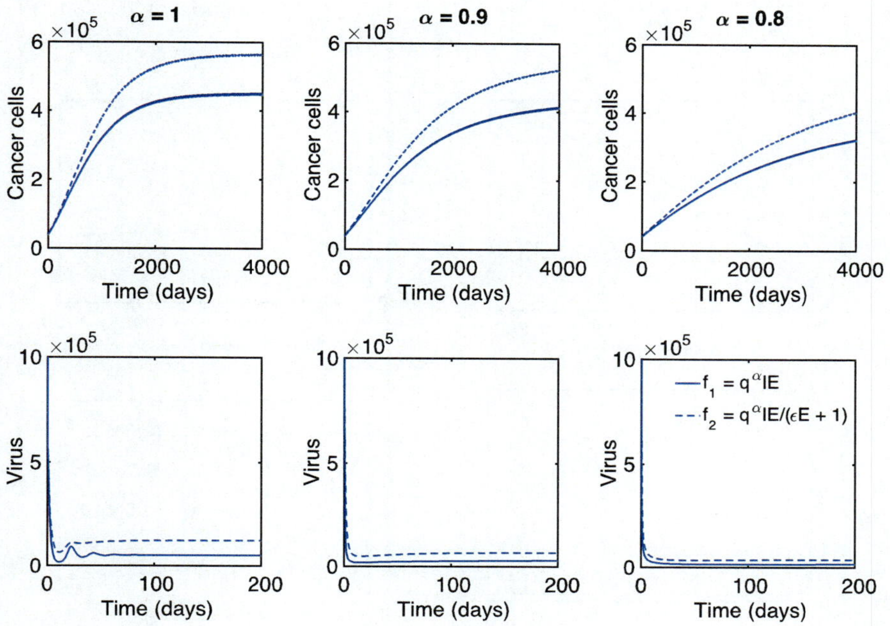

Fig. 1 Variation of cancer cells number and HIV for immune functions $f_1 = q^{\alpha} I E$ and $f_2(I, E) = \frac{q^{\alpha} I E}{\epsilon E + 1}$, and for $\alpha = \{1, 0.9, 0.8\}$. All parameter values are given in Table 2. Initial conditions: $C(0) = 4.0 \times 10^4$, $T(0) = 1.0 \times 10^4$, $I(0) = 1.7 \times 10^5$, $V(0) = 5.0 \times 10^6$, $E(0) = 333 \times 10^3$ and $D(0) = 3.9 \times 10^{-2}$

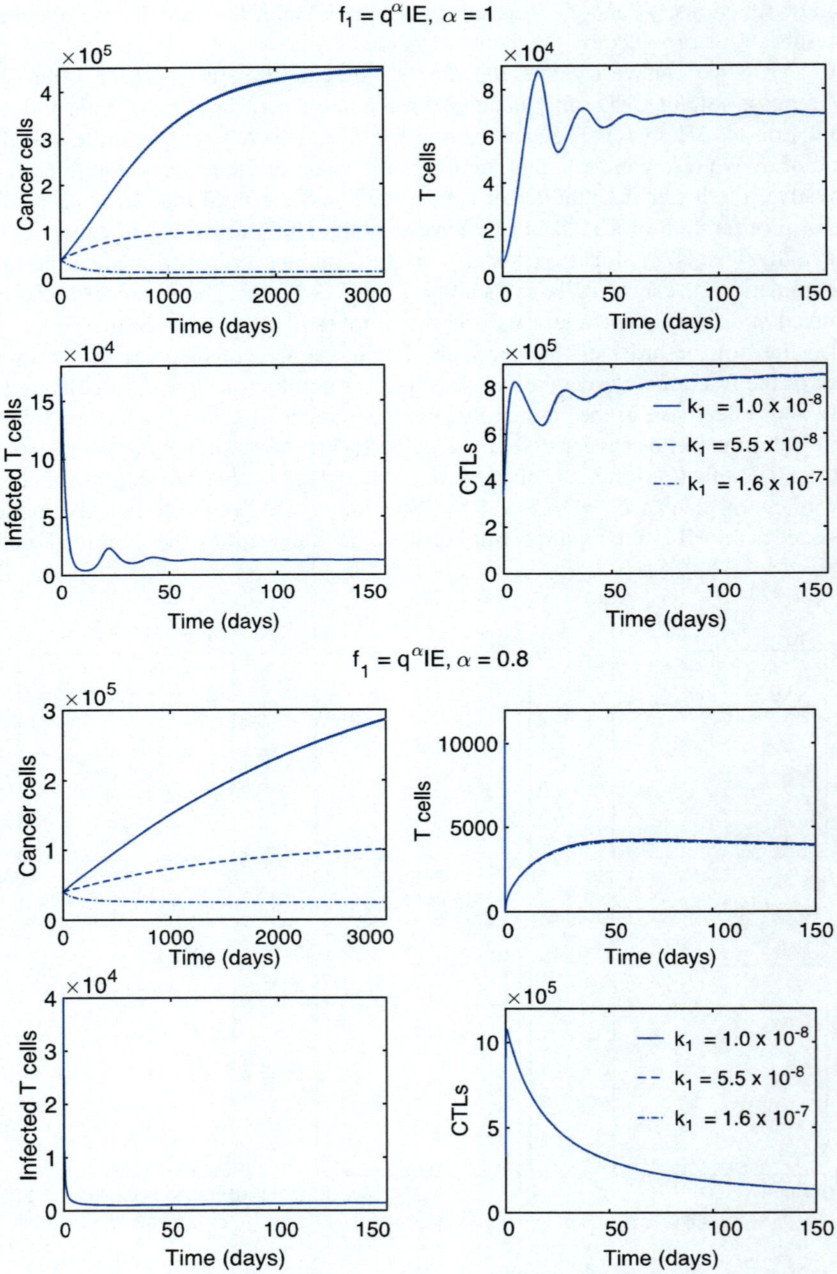

Fig. 2 Dynamics of model cancer cells, T cells, infected T cells e CTLs for $\alpha = \{1, 0.8\}$ and $k_1 = \{1.0 \times 10^{-8}, 5.5 \times 10^{-8}, 1.6 \times 10^{-7}\}$. We use the immune function $f_1 = q^{\alpha} I E$. We consider the parameter values given in Table 2. Initial conditions: $C(0) = 4.0 \times 10^4$, $T(0) = 1.0 \times 10^4$, $I(0) = 1.7 \times 10^5$, $V(0) = 5.0 \times 10^6$, $E(0) = 333 \times 10^3$ and $D(0) = 3.9 \times 10^{-2}$

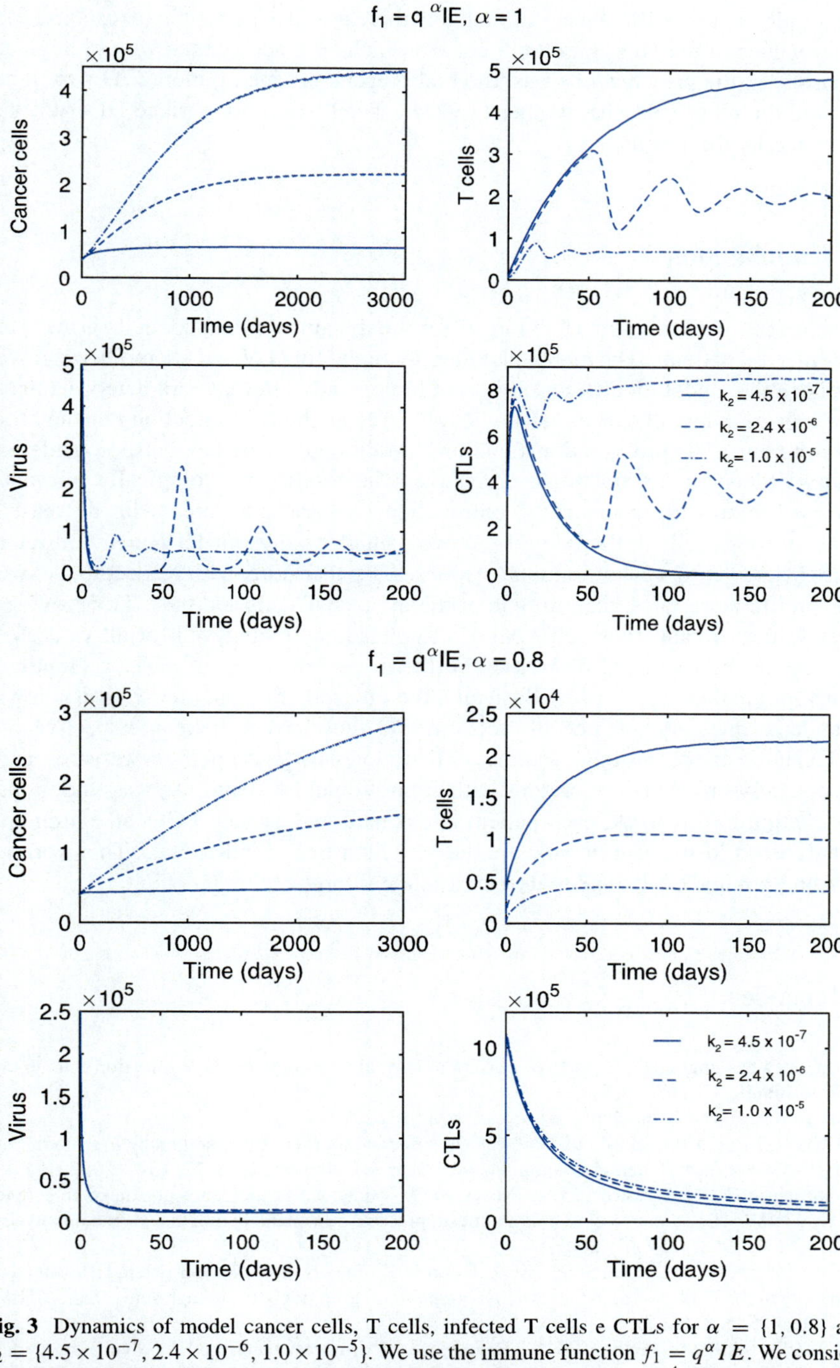

Fig. 3 Dynamics of model cancer cells, T cells, infected T cells e CTLs for $\alpha = \{1, 0.8\}$ and $k_2 = \{4.5 \times 10^{-7}, 2.4 \times 10^{-6}, 1.0 \times 10^{-5}\}$. We use the immune function $f_1 = q^\alpha IE$. We consider the parameter values given in Table 2. Initial conditions: $C(0) = 4.0 \times 10^4$, $T(0) = 1.0 \times 10^4$, $I(0) = 1.7 \times 10^5$, $V(0) = 5.0 \times 10^6$, $E(0) = 333 \times 10^3$ and $D(0) = 3.9 \times 10^{-2}$

asymptotically to 5×10^4. We also found that there is a stronger immune response for higher values of the HIV infection rate, k_2, which is in accordance with biological scenarios. In the presence of virus, the body reacts, producing more CTLs to fight HIV. On the other side, the number of $CD4^+$ T cells decreases, since HIV attacks preferentially these cells.

4 Conclusions

In this paper, we proposed a FO model for the dynamics of cancer cells growth in HIV-infected patients. The model includes treatment for HIV and chemotherapy. We simulated the model for different values of biologically relevant parameters, namely the elimination rate of cancer cells by $CD4^+$ T cells, the HIV infection rate, and for two different CTL proliferation functions. Additionally, we have also considered various orders for the fractional derivative. The results are biologically relevant. Increased values of cancer cells' elimination rate are associated with decreased values of these cells in the body. Moreover, smaller values of HIV infection rates are associated with weaker immune responses. Furthermore, with respect to the two CTL proliferation rates, the multiple of the infected T cells and the CTLs provides lower values of cancerous cells. All of these results are observed for all values of the order of the fractional derivative. Furthermore, are observed lower asymptotic values for smaller values of α. Basically, the order of fractional derivatives allows us to have one more degree of freedom and thus have a better perspective on the behavior of the system dynamics. This allows us to deepen our research and consequently broaden our horizons with regard to public health. As was said in the introduction to this work, each patient is a patient and as such FC is an extremely advantageous instrument in what are the particularities of individuals. This work is a starting point and we hope to deepen it in the future.

References

1. Lifson, J.D., Engleman, E.G.: Role of CD4 in normal immunity and HIV infection. Immunol. Rev. **109**(1), 93–117 (1989)
2. https://www.who.int/hiv/en/. Last accessed 28 May 2019
3. Chávez, J.P., Gürbüz, B., Pinto, C.M.A.: The effect of aggressive chemotherapy in a model for HIV/AIDS-cancer dynamics. Commun. Nonlinear Sci. Numer. Simul. **75**, 109–120 (2019)
4. Grulich, A.E., van Leeuwen, M.T., Falster, M.O., Vajdic, C.M.: Incidence of cancers in people with HIV/AIDS compared with immunosuppressed transplant recipients: a meta-analysis. Lancet **370**(9581), 59–67 (2007)
5. Hernández-Ramírez, R.U., Shiels, M.S., Dubrow, R., Engels, E.A.: Cancer risk in HIV-infected people in the USA from 1996 to 2012: a population-based, registry-linkage study. Lancet HIV **4**(11), 475–528 (2017)
6. Carvalho, A.R.M., Pinto, C.M.A.: Emergence of drug-resistance in HIV dynamics under distinct HAART regimes. Commun. Nonlinear Sci. Numer. Simul. **30**(1–3), 207–226 (2016)

7. The HIV Modelling Consortium Treatment as Prevention Editorial Writing Group. HIV Treatment as Prevention: Models, Data, and Questions—Towards Evidence-Based Decision-Making. PLoS Med. **9**(7), 8 (2012)
8. Carvalho, A.R.M., Pinto, C.M.A.: New developments on AIDS-related cancers: the role of the delay and treatment options. Math. Methods Appl. Sci. **41**(18), 8915–8928 (2018)
9. Pinto, C.M.A., Carvalho, A.R.M.: Time-varying pharmacodynamics in a simple non-integer HIV infection model. Math. Biosci. **307**, 1–12 (2019)
10. Carvalho, A.R.M., Pinto, C.M.A., Baleanu, D.: HIV/HCV coinfection model: a fractional-order perspective for the effect of the HIV viral load. Adv. Difference Equ. **2018**(2), 22 (2018)
11. Pinto, C.M.A., Carvalho, A.R.M.: Diabetes mellitus and TB co-existence: clinical implications from a fractional order modelling. Appl. Math. Model. **68**, 219–243 (2019)
12. Pinto, C.M.A., Carvalho, A.R.M.: Fractional Modeling of Typical Stages in HIV Epidemics with Drug-Resistance. Prog. Fract. Differ. Appl. **1**(2), 111–122 (2015)
13. Pinto, C.M.A., Carvalho, A.R.M.: Fractional dynamics of an infection model with time-varying drug exposure. Prog. J. Comput. Nonlinear Dynam. **13**(9), 16 (2018)
14. Louzoun, Y., Xue, C., Lesinski, G.B., Friedman, A.: A mathematical model for pancreatic cancer growth and treatments. J. Theor. Biol. **351**, 74–82 (2014)
15. Odibat, Z.M., Shawagfeh, N.T.: Generalized Taylor's formula. Appl. Math. Comput. **186**, 286–293 (2007)
16. Lin, W.: Global existence theory and chaos control of fractional differential equations. J. Math. Anal. Appl. **332**(1), 709–726 (2007)
17. Diethelm, K., Freed, A.D.: The FracPECE subroutine for the numerical solution of differential equations of fractional order. In: Heinzel, S., Plesser, T. (eds.) Orschung und Wissenschaftliches Rechnen: Beiträge zum Heinz-Billing-Preis 1998, pp. 57–71 (1999)

Bio-Inspired Tactile Sensing: Distinction of the Overall Object Contour and Macroscopic Surface Features

Moritz Scharff

Abstract Vibrissae of rats are part of the somatosensory system. A tactile stimulus along the hair shaft is transmitted to the Follicle-Sinus complex and transduced into an action potential by mechanoreceptors. The signal contains information about the texture of the contacted object including the overall contour, macroscopic features, and microscopic features. Here, the overall contour and the macroscopic features of an object are analyzed using an artificial vibrissa-like sensor which is dynamically swept along the object. The natural vibrissa is replaced by a cylindrical steel wire and the Follicle-Sinus complex by a force/torque sensor, respectively. In Experiment, the overall object contour is designed as a sine wave (long wavelength) and is superimposed by a second sine wave with a shorter wavelength in order to represent the macroscopic features. A procedure to distinguish both components is developed and successfully applied. The combination of the sensor shape and the scanning conditions—for example, the large, nonlinear deformation of the sensor shaft—operate like a morphological filter and consequently influences the detected profile features.

Keywords Vibrissa · Tactile sensing · Surface texture · Object contour

1 Introduction

The hairs located on the sides of a muzzle of e.g. a rat are named vibrissae, see Fig. 1. Those are powerful tactile sensors. A vibrissa can be divided into two different parts. First, there is the hair shaft that transmits a mechanical stimulus caused by e.g.

M. Scharff (✉)
Technical Mechanics Group, Department of Mechanical Engineering, Technische Universität Ilmenau, Ilmenau, Germany

Section of Mechanical Engineering, Department of Engineering, Pontificial Catholic University of Peru, Lima, Peru
e-mail: moritz.scharff@tu-ilmenau.de

J. Awrejcewicz (ed.), *Perspectives in Dynamical Systems I: Mechatronics and Life Sciences*, Springer Proceedings in Mathematics & Statistics 362,
https://doi.org/10.1007/978-3-030-77306-9_10

Fig. 1 Rat in interaction with the surrounding environment

contact with an object to the hair follicle, Follicle-Sinus complex respectively. The hair shaft is a slender, tapered (decreasing diameter from base to tip) and inherent curved structure that is very flexible. The Follicle-Sinus complex supports this hair shaft and includes mechanoreceptors to transduce the mechanical stimulus into an action potential [2]. There are two groups of mechanoreceptors, slow and rapid adapting ones.

Slow adapting mechanoreceptors correspond to signals with low frequencies and rapid ones to signals with high frequencies [4]. The mechanoreceptors are activated by the forces and moments at the base of the hair shaft in consequence of a mechanical stimulus [3]. Using this arrangement consisting of transmitter and sensors, animals can recognize the contour of an object and identify properties of the surface texture as well [1].

Since the term surface texture is not standardized, it can be interpreted in different ways. In the present work, the term surface texture is defined to include two types of features: macroscopic and microscopic ones. The macroscopic ones are related to surface features that are larger than the actual hair shaft diameter. In contrast, microscopic ones are of less size than the hair shaft diameter and can be summarized in the frictional contact properties. All three types of information principle object contour, macroscopic and microscopic features are combined in the signals at the base of the hair shaft recorded by the mechanoreceptors. The relations between signal components and the form of signal decomposition to extract the desired signal component only are still not exactly understood. But it is demonstrated that animals can extract and divided between these three types of information. Furthermore, artificial sensors inspired by natural vibrissae demonstrated those functionalities too with the restriction that normally only one of the three types of information is in focus [7, 9, 11, 12].

In this work the interactions of principle object contour and superimposed macroscopic features are analyzed. In a first step, microscopic features are neglected. The goal is to identify how principle object contour and macroscopic features can be extracted out of the signals measured at the base of an artificial vibrissa-like sensor. Therefore, a theoretical model and an experiment are designed, see Sect. 2. The results are presented and discussed in Sect. 3. Finally, four hypotheses are formulated and summarized in Sect. 4.

2 Material and Methods

The present problem is analyzed for two different objects in multiple configurations, e.g. spatial position. First, the relation between object distance/contour and macroscopic features is evaluated by simulation, analyzing a straight contour with superimposed sinus undulation, see Fig. 2b. The corresponding shape function is given by (1).

$$g_s(x, \eta) = \eta + 0.001\,\mathrm{m} + 0.001\,\mathrm{m}\,\sin\left(\frac{2\pi}{0.004\,\mathrm{m}}x\right) \tag{1}$$

Based on this preliminary analysis, a more general shape function (2) is investigated in experiments, see Figs. 2c and 3.

$$g_w(x, \eta) = \eta + 0.001\,\mathrm{m} + 0.001\,\mathrm{m}\,\sin\left(\frac{2\pi}{0.004\,\mathrm{m}}x\right) + 0.008\,\mathrm{m}\,\sin\left(\frac{2\pi}{0.1\,\mathrm{m}}x\right) \tag{2}$$

The natural vibrissa is replaced by a slender steel filament characterized by the following properties: length 0.1 m, diameter $0.5e^{-3}$ m, density $7850\,\mathrm{kg\,m^{-3}}$, Young's modulus $2.06e^{11}\,\mathrm{N\,m^{-2}}$, Poisson's ratio 0.3, Rayleigh damping $\alpha = 0\,\mathrm{s}$, $\beta = 0.75\,\mathrm{s^{-1}}$. Further information about the sensor shaft in simulation and experiment are given in Sects. 2.1 and 2.2. In all scenarios, the sensor shaft gets

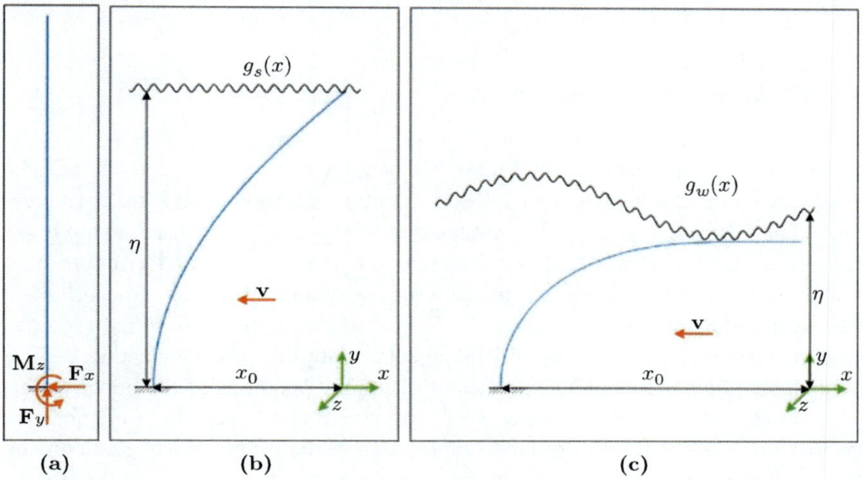

Fig. 2 **(a)** Initial state with support reactions $\vec{F}_x$, $\vec{F}_y$ and $\vec{M}_z$. **(b)** The undulated straight object is described by the contour function (1), and **(c)** the undulated wavy object by the contour function (2) whereby η corresponds to the distance between clamping and the first local minima of the undulation

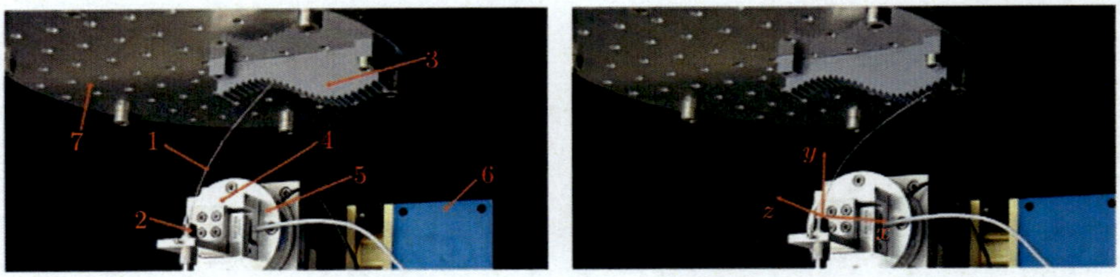

(a) (b)

Fig. 3 Scanning of the undulated wavy object in experiment: Contacting the **(a)** convex and **(b)** concave part of the overall shape while displacing the sensor along a straight trajectory. Components: 1—sensor shaft; 2—jaw chuck; 3—object; 4—force sensor; 5—torque sensor; 6—linear guide; 7—hexapod robot

sweep across the object by displacing its clamping along the straight trajectory x_0 with a constant velocity $v = 1e^{-3}\,\mathrm{m\,s^{-1}}$. In accordance with the natural example, the recorded support reactions are decomposed in a low frequent and high frequent part by a low pass filter. The reconstructions of the object contours using the simulated or experimental recorded support reactions only is done by applying the algorithm presented in [10]. This algorithm is based on the Euler-Bernoulli beam theory considering large deflections but limited to quasi-static displacement, single point contacts between sensor, and object and excluded friction. These restrictions are partly broken by the simulation as well as the experiment. The quantities of peaks/macroscopic features are determined using the peak count algorithm `findpeaks()` included in MATLAB2015b.

2.1 Finite Element Analysis

The problem is modeled using Finite Element Analysis incorporated into the software ANSYS Mechanical v.19.1. The simulation is limited to 2D- space ($x-y-$ plane). The vibrissa/sensor shaft is modeled as straight and cylindrical using beam elements of type `beam188`. It is discretized by 200 nodes and consists of a homogenous, isotropic, linear elastic material. A clamping replaces the FSC and the corresponding support reactions $\vec{F}_x$, $\vec{F}_y$, $\vec{M}_z$ are interpreted as signal recorded by the mechanoreceptors, see Fig. 2. The object is assumed to be rigid and its surface is represented by a series of 1500 nodes whereby the positions of the nodes follow (1). As contact pair, contact elements of type `conta175` and target elements of type `targe169` are used in order to represent a force based line to line contact. The contact elements correspond to the beam and the target ones to the object. A pure Lagrange multiplier contact algorithm describes the relation of the contact pair. Friction between beam and object is neglected but global Rayleigh Damping is considered. The clamping is dynamically displaced along a straight trajectory in order to sweep the beam across the object shape. A minimal timestep of $1e^{-3}$ s is used to satisfy Nyquist–Shannon sampling theorem with respect to the first natural frequency of the beam $\approx$ 32 Hz, smaller timesteps are possible since an adaptive timestep control is activated. The support reactions $\vec{F}_x$, $\vec{F}_y$, $\vec{M}_z$ are recorded for every timestep.

2.2 *Experiment*

In experiment, the straight sensor shaft and the connected sensors are displaced horizontal ($x-$ direction) with a linear guide of type AMTEC Power Cube PLB 090 (position repeatability ± 0.005 mm). The sensor shaft consists of spring steel according to DIN EN 10270-1:2017-09 and is clamped by a miniature jaw chuck, see Fig. 3. It is cut from a larger piece and the tip/cutting edge is ground by sandpaper. The miniature jaw chuck is attached to a 3D force sensor of type K3D40 (ME-Meßsysteme), accuracy class 0.5, nominal load ± 2 N and a 1D torque sensor of type TD70 (ME-Meßsysteme), accuracy class 0.1, nominal load ± 50 mNm. The torque sensor measures signals with respect to the $z-$ direction. All signals are recorded using a GSV-1A4 M12/2 (ME-Meßsysteme) amplifier, a NI PXI 6221 M-Series multifunction data acquisition device, and the software LabVIEW 2017 with a sampling rate of 1000 Hz. Additionally, to the linear guides, a hexapod of type PI M-850.50 is used to position and align the object to the scanning trajectory. The object was 3D-printed using an Ultimaker S5 3D-printer and ABS filament. The length in x- direction is 0.1 m. The object was scanned five times for $\eta \in \{0.05\,\text{m}; 0.07\,\text{m}\}$ and the support reactions recorded. The mean and the standard deviation were calculated and used to reconstruct the object contour.

3 Results and Discussion

The simulations are analyzed for $\eta = 0.03$ m (0.01 m) 0.09 m with exception of $\eta = 0.06$ m, see Fig. 2b. The support reactions for chosen η are illustrated in Fig. 4.

Remark 1 Using the parameters mentioned in Sect. 2, it was not possible to simulate the scanning process for $\eta = 0.06$ m as consequence of conditionally convergent solution.

The force component F_x oscillates for all η around zero. For large η, e.g. 0.09 m, there are distinct, separated peaks in a regular low frequency with large amplitude while for smaller η the signals shows smaller amplitudes and higher frequency, see Fig. 4a. An alike behavior is indicated by the signals for F_y and M_z in Fig. 4b,c. But, in contrast to F_x, F_y, and M_z increase with decreasing η. The simulated support reactions are used to reconstruct the scanned object contour (see (1)) by applying the algorithm presented in [10], see Fig. 5. The reconstructed contours for large η are inaccurate but e.g. in the case of $\eta = 0.07$ m there a notable peaks. On the

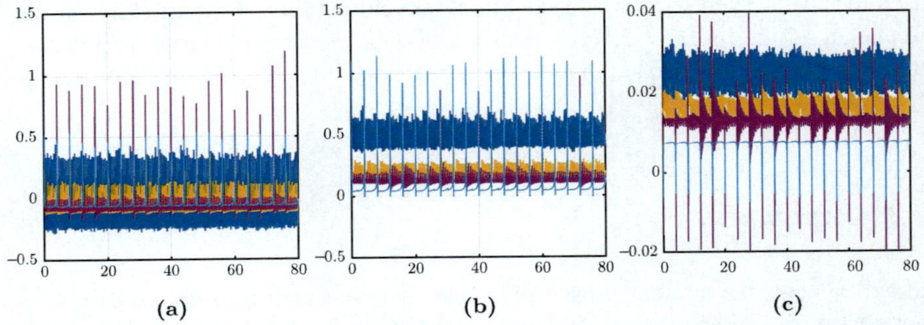

Fig. 4 Scanning the object corresponding to (1) in simulation. Support reactions for: $\eta = 0.03$ m (blue); $\eta = 0.05$ m (yellow); $\eta = 0.07$ m (magenta); $\eta = 0.09$ m (cyan). (**a**) F_x[N]vs.t[s] (**b**) F_y[N]vs.t[s] (**c**) M_z[Nm]vs.t[s]

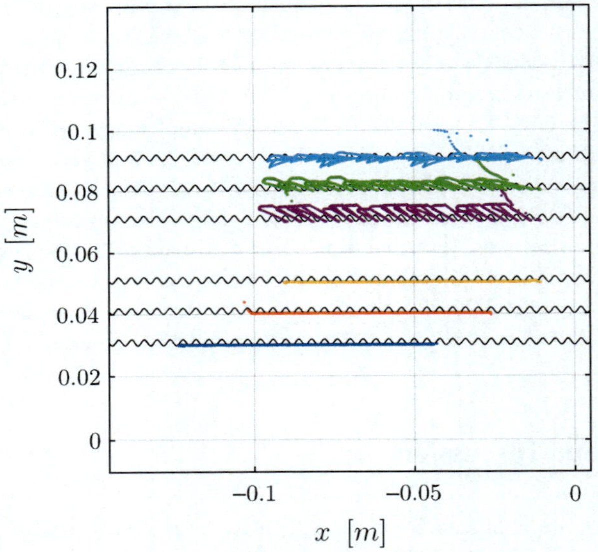

Fig. 5 Object contour reconstruction for: $\eta = 0.03$ m (blue); $\eta = 0.04$ m (red); $\eta = 0.05$ m (yellow); $\eta = 0.07$ m (magenta); $\eta = 0.08$ m (green); $\eta = 0.09$ m (cyan) and given object contour by (1) (black)

contrary, the results for small η show almost a straight line without any peaks. In all cases, the principal object distance is determined correctly. Figure 6 summarizes the results of the simulations. Alike to the observations in Fig. 4, the mean of the norm $\overline{F}$ of the signals F_x and F_y reduces for an increasing η as well as the mean of the bending moment $\overline{M}_z$ does, see Fig. 6a,b. The quantity macroscopic features are evaluated by determining the number of peaks included in the input signal. As input signals the signal of F_y and the reconstructed object contours are used, see Fig. 6c.

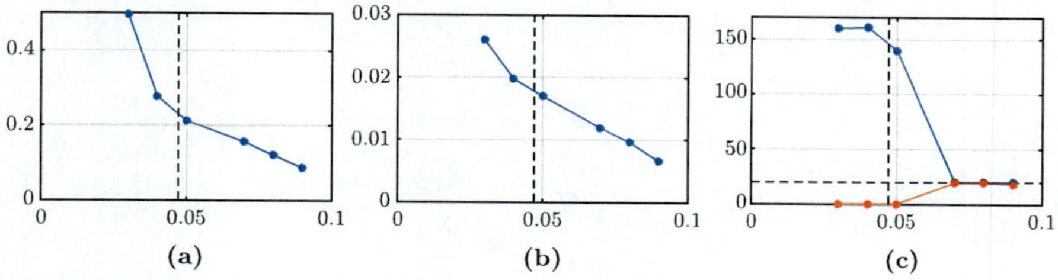

Fig. 6 In **(a)** the mean of the norm $\overline{F}$ of the signals F_x and F_y and in **(b)** the mean of the bending moment $\overline{M}_z$ are evaluated for different η. **(c)** The quantities of detected peaks for different η are shown for the case of F_y (blue) and reconstructed object shape (red) as input signal. In **(a)**, **(b)** and **(c)** the vertical black dashed line marks the transition from a contact occurring at the sensor tip to a contact occurring along the shaft in the case of a flat, straight object contour according to [10]. In **(c)** the horizontal black dashed line corresponds to the true quantity of existing macroscopic features/local maxima of (1) respectively for a scanned section of 0.08 m **(a)** $\overline{F}$[N]vs.η[m] **(b)** $\overline{M}_z$[N]vs.η[m] **(c)** $\sum$ Peaksvs.η[m]

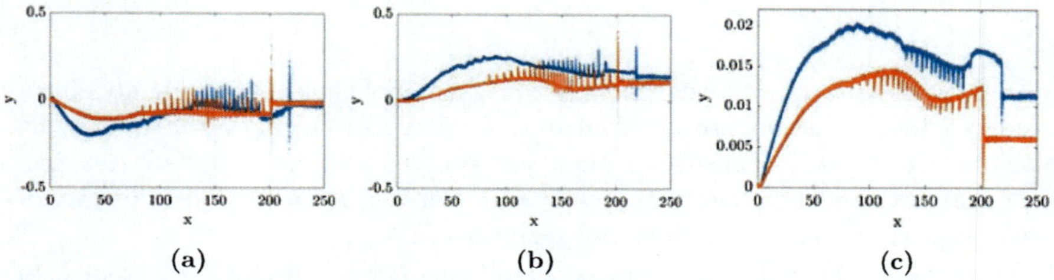

Fig. 7 Scanning the object corresponding to (2) in experiment. Support reactions for: $\eta = 0.05$ m (blue); $\eta = 0.07$ m (red). The grey shade corresponds to the three times the standard deviation. **(a)** F_x[N]vs.t[s] **(b)** F_y[N]vs.t[s] **(c)** M_z[Nm]vs.t[s]

Scanning the object contour (1) for 80 s there are 20 detectable features. Analyzing F_y, the quantity of detected peaks reduces with decreasing η. First, the quantity of peaks exceeds the true quantity but for larger η it is determined correctly. In the case of the object contour as input signal, for large η the quantity of peaks is determined correctly too, but there are nearly no detected features for small η.

In Experiment the object contour (2) is analyzed for $\eta \in \{0.05\,\text{m}; 0.07\,\text{m}\}$. The recorded support reactions are shown in Fig. 7. In every case, there is a major trend superimposed by peaks. The peaks appear for larger t. At the end of the scanning process, there is always a large peak that corresponds to the release of the sensor shaft from the object.

The reconstructed object contours match the principle given contours according to (2), see Fig. 8a. In the beginning, the convex part of the contour is scanned. Therefore, for both distances, the reconstructed contours are almost smooth and do not show any macroscopic features/peaks. Scanning the concave part of the contour, the reconstructed contours get wavy and show macroscopic features corresponding to the macroscopic features of the given object contour. The quantity of detected

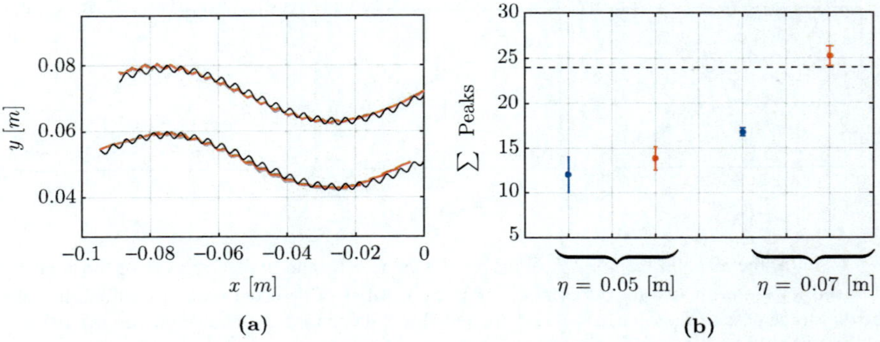

Fig. 8 **(a)** The reconstructed object contour is plotted in red and the reference one in black according to (2) for $\eta = 0.05$ m and $\eta = 0.07$ m. **(b)** Quantity of detected peaks with respect to the reconstructed object shape (blue) as input signal and F_y (red) as input signal. In **(b)** the horizontal black dashed line corresponds to the true quantity of existing macroscopic features/local maxima of (2) respectively

macroscopic features depends on η, see Fig. 8b. For small clamping to object distances fewer features are detected than for large distances. Furthermore, the quantities of detected peaks using the object contour as input signal are less than the quantities corresponding to F_y. For $\eta = 0.07$ m and F_y as input the quantity of detected peaks matches almost the true quantity of 24.

The results indicate that the principle clamping to object distance corresponds to the major trend of the recorded support reactions, see Fig. 6a,b. This is confirmed by the findings of [5]. Consequently, the object contours can be calculated using the low frequency part of the support reactions. Since a low pass filter is used to decompose the original signal in a low and a high frequent part, the chosen cut-off frequency determines the relation between major object contour and superimposed macroscopic feature. If the cut-off frequency is low only the principle object contour will be reconstructed or if it is large enough macroscopic features will be reconstructed as well. In nature, this decomposition of the signals is may be implemented by slow and rapid adapting mechanoreceptors included in the FSC, see Sect. 1. But, Figs. 5 and 8 indicates that there is a morphological filtering as well. For small distances between clamping and object the contact point occurs along the sensor shaft and not at the tip, compare Fig. 2b,c. If the contact occurs far from the tip the macroscopic features will be not completely contacted by the sensor shaft consequently, they will be not reconstructed. Here, the sensor shaft sweeps over the tops of the macroscopic features and by doing so the deflection induced curvature of the sensor shaft suppresses the macroscopic features. So, there is a relation between the spatial distribution of macroscopic features and the position of the contact point along the sensor shaft. Obviously, the diameter of the sensor shaft is a morphological filter as well. If the difference in size between the macroscopic features is less than the sensor shaft diameter it cannot be recognized. Taking Figs. 6c and 8b into account, the hypothesis of a deflection induced morphological filter is forced because the accuracy of the number of peaks reduces with decreasing clamping to object distance. But it shows also that counting the peaks included in the support reactions can give information about the present macroscopic features. Here, a small clamping to object distance increases the number of detected peaks in order of an increasing number of contact points. Both types of signals are strongly dependent on the clamping to object distance whereby for analyzing the surface macroscopic features there is an important difference. The support reactions, as well as the reconstructed object contour, contain information about the number of macroscopic features. But the reconstructed object contour contains information about the size and shape of the macroscopic features, too. If the size of a macroscopic feature corresponds to the amplitude of the peak in the signal of the support reactions is very questionable since there are dynamical impacts between the sensor shaft and object. Furthermore, a peak in the signals of the support reactions does not correspond necessarily to a macroscopic feature. Peaks in the signals of the support reactions can be caused by e.g. Stick-Slip events or other effects. An advantage of using F_y as input signal is that there is no loss of information by any transformation of the signal like in the case of using the reconstructed object contour as input signal. Here, the algorithm of [10] causes a loss of information due to the restrictions of the underlying theoretical model.

Considering the fact that the magnitudes of the support reactions increase with decreasing clamping to object distance in combination with the hypothesis about the morphological filter functionality, it can be summarized that by reducing the clamping to object distance the signal components of the principle object contour get pronounced/amplified and for an increasing distance increases the ability to recognize macroscopic features. This hypothesis is supported by the findings in [8] which report that rats try to limit the amount of bending of their vibrissae while contacting an object. Furthermore, for rats it is known that there is a connection between sensor and motor control in the cerebrum [6]. In context to the present findings, it can be supposed that active controlling of the distance between clamping and object by observing the low frequent part of the support reactions can be used to optimize the detection of macroscopic features. In the case of animals, the relation between increasing signal strength and decreasing ability of detecting macroscopic features is further promoted by Weber' Law which indicates that the measurable change of a neuronal signal is proportional to the initial magnitude of the signal. Adapted to the present problem this means that a close clamping to object distance yields a large initial signal and small changes due to macroscopic features are difficult to detect.

4 Conclusions

The present work describes how an artificial vibrissa-like tactile sensor interacts with an object contour that is superimposed with macroscopic features. First, the scenario was analyzed in simulation for a straight horizontal contour with superimposed sinus undulation, macroscopic features respectively. The results were confirmed by the second step. Here, a wavy contour with superimposed macroscopic features was analyzed in experiment. All findings can be summarized by the following hypotheses:

- The deformation of the sensor shaft caused by contact with an object represents a morphological filter for surface features.
- A closer clamping to object distance pronounces/amplifies the signal components related to the principal object contour.
- A larger clamping to object distance improves the ability to detect macroscopic features.
- The process of detecting macroscopic features can be enhanced by controlling the distance between clamping and object in a way that there is tip contact only.

In future investigations, the analyses must be extended to surface micro features, too. Here, the investigation of a frictional contact and its effect on the signals of the support reactions will be in focus. Another interesting point will be to include more properties of the natural paragon. For example, the natural vibrissa shaft is inherently curved and has a conical body. Both properties will affect the described morphological filter properties. Considering these properties, the sensor shaft maybe builds up an inherent, adaptive, morphological filter bank consisting of: the sensor tip (tapered body!); the inherent curvature; the bending induced curvature (adaptive due to controlling the clamping object distance), and finally the slow and rapid adapting mechanoreceptors. This idea can be even more advanced by including more details of the natural example like properties of the Follicle-Sinus complex.

Acknowledgments Thanks to Philipp Schorr from Technische Universität Ilmenau for valuable discussion and support.

References

1. Ahl, A.S.: The role of vibrissae in behavior: a status review. Vet. Res. Commun. **10**(1), 245–268 (1986)
2. Brecht, M., Preilowski, B., Merzenich, M.M.: Functional architecture of the mystacial vibrissae. Behav. Brain Res. **84**(1–2), 1792–1799 (1997)
3. Campagner, D., Evans, M.H., Loft, M.S.E., Petersen, R.S.: What the whiskers tell the brain. Neuroscience **368**(1), 95–108 (2018)
4. Ebara, S., Furuta, T., Kumamoto, K.: Vibrissal mechanoreceptors. Scholarpedia **12**(3), 32372 (2017)
5. Evans, M.H., Fox, C.W., Lepora, N.F., Pearson, M.J., Sullivan, J.C., Prescott, T.J.: The effect of whisker movement on radial distance estimation: a case study in comparative robotics. Front. Neurorobot. **6**, 12 (2012)
6. Kleinfeld, D., Ahissar, E., Diamond, M.E.: Active sensation: insights from the rodent vibrissa sensorimotor system. Curr. Opin. Neurobiol. **16**(4), 435–444 (2006)
7. Lucianna, F.A., Farfán, F.D., Pizá, G.A., Albarracín, A.L., Felice, C.J.: Functional specificity of rat vibrissal primary afferents. Physiol. Rep. **4**(11), e12810 (2016)

8. Mitchinson, B., Martin, C.J., Grant, R.A., Prescott, T.J.: Feedback control in active sensing: rat exploratory whisking is modulated by environmental contact. Proc. R. Soc. B Biol. Sci. **274**(1613), 1035–1041 (2007)
9. Schwarz, C.: The slip hypothesis: tactile perception and its neuronal bases. Trends Neurosci. **39**(7), 449–462 (2016)
10. Steigenberger, J.: A continuum model of passive vibrissae. Preprint **M 13/03**, 17 (2013)
11. Will, C., Behn, C., Steigenberger, J.: Object contour scanning using elastically supported technical vibrissae. J. Appl. Math. Mech. **79**(1), 11 (2017)
12. Wolfe, J., Hill, D.N., Pahlavan, S., Drew, P.J., Kleinfeld, D., Feldman, D.E.: Texture coding in the rat whisker system: slip-stick versus differential resonance. PLoS Biol. **6**(8), e215 (2008)

Modelling and Control of a Lower Limb Exoskeleton Driven by Linear Actuators

Dariusz Grzelczyk, Olga Jarzyna, and Jan Awrejcewicz

Abstract In this paper, a design of lower limb exoskeleton driven by linear electric actuators was proposed and investigated. A general, three-dimensional simulation model of a lower limb exoskeleton was developed to study crucial kinematic parameters of the proposed device. Also, biocompatibility aspects of the simulated walking machine were considered both in the presented design and the simulation model. Time histories of human joint angles in normal gait, captured with the use of a motion capture system in our previous study, were employed as articulation variables of individual joints of the investigated device. Moreover, a new gait generator, which can be used to produce rhythmic movements in hip and knee joints of both limbs, was developed and tested. Finally, the possibility of using the proposed control method was verified by using the constructed prototype of a single limb of an exoskeleton controlled by a popular Arduino Uno microcontroller. Experimental tests gave a promising outcome regarding the applied control approach. As a result, a relatively simple, inexpensive and efficient mechanical design and control system are expected, which can provide better access to lower limb exoskeletons for the public and reduce the workload of physiotherapists.

Keywords Exoskeleton · Lower limb · Linear actuators

1 Introduction

In highly developed societies, the impairment of human locomotion is a phenomenon commonly observed among both young people and the elderly. Regardless of the reasons (such as, for instance, sports injuries and traffic accidents among young people, and osteoarthritis or osteoporosis in elderly people), such dysfunc-

D. Grzelczyk (✉) · O. Jarzyna · J. Awrejcewicz
Department of Automation, Biomechanics and Mechatronics, Lodz University of Technology, Lodz, Poland
e-mail: dariusz.grzelczyk@p.lodz.pl

J. Awrejcewicz (ed.), *Perspectives in Dynamical Systems I: Mechatronics and Life Sciences*, Springer Proceedings in Mathematics & Statistics 362,
https://doi.org/10.1007/978-3-030-77306-9_11

tions of the locomotor system have a negative influence on medical and social spheres. Different reports show that hundreds of million people in the world live with some difficulties in locomotion, and, unfortunately, this number is constantly growing [1]. One of the most popular and often used methods restoring human mobility is physiotherapy [2]. However, in general, it requires a large number of physiotherapists and their hard work, which often leads to occupational conditions such as lower-back problems [3, 4]. Fortunately, it has been shown that lower limb rehabilitation can be improved by external stimulation of muscular and nervous systems realized by so-called Lower Limb Exoskeletons (LLEs), see papers [5–10]. Recent developments and challenges in LLEs can be found in one of the review papers [2]. LLEs should meet numerous requirements regarding mechanical strength, stability, kinematic and dynamic biocompatibility with the human lower limb, and control possibility [11–14]. The abovementioned aspects are still subjects of the study of many academic and commercial research centers, although several decades have passed since the first lower limb exoskeleton was made. Unfortunately, detailed information about advanced commercial solutions is not available to the public. Therefore, our study was inspired by similar devices found in academic literature. As a result, relatively simple, efficient and inexpensive device to assist restoration of motor functions of the disabled has been proposed. The developed LLE can increase power in hip, knee and ankle joints of a human, causing an increase in the efficiency of movement of the operator. The literature overview indicates that DC motors with reduction gears are one of the most popular actuators used to drive active joints of LLEs, for instance, see papers [15–19]. That is why this type of electric actuators has been used in our design as well. In this paper, we proposed CAD model of lower limb exoskeleton made of easily accessible aluminum profiles and actuated by electric linear actuators (DC motors equipped with reduction gears and screw-nut systems). To perform some virtual experiments of the locomotion process, we developed a general full parametric 3D simulation model in Mathematica software. The model is useful for obtaining crucial kinematic parameters of the proposed device. Experimental data obtained in our previous study by capturing the motion of real human gait were applied as kinematic excitations (i.e. articulated variables in the LLE joints corresponding to human hip, knee and ankle joints) [20]. Besides, smooth analytical approximations of experimental articulated variables, in both hip and knee joints, were proposed. Although the actuation of rotation in the ankle has a positive influence on the movement of the exoskeleton, it is usually overlooked to reduce the total mass and power consumption of the device. As a result, only two degrees of freedom per each limb (i.e. hip and knee joints) are actuated in many devices found in the literature, for instance, see papers [16, 18]. A similar approach was used in the present study as well. To verify the proposed control method, experimental investigations were carried out by using a constructed prototype of a single lower limb of the exoskeleton. To control motion in hip and knee joints, a popular and inexpensive microcontroller Arduino Uno with a digital PID controller was used. The outcomes of the study can be used as guidelines for further improvement of the proposed method and its practical applications in real LLE control systems.

2 CAD Model of the Lower Limb Exoskeleton

The efficiency of machine operation can be improved with the use of computer modelling through studying simulation results obtained for different modifications of mechanical construction. Engineers can investigate numerous virtual models of different machines before creating the final construction. Currently, there are a lot of computer programs supporting researches in virtual prototyping of different constructions, including lower limb exoskeletons. In this study, Inventor Professional 2019 was used for this purpose. CAD model of the proposed human lower limb exoskeleton is presented in Fig. 1. Mechanical design of the presented model consists of the main static frame which corresponds to the human pelvis, back support for improving patient's posture, two lower limbs and three linear actuators per each robot's limb. When modelling lower limbs of the exoskeleton, we were inspired by the morphology of human lower limbs and similar solutions met in the literature. As a result, the limb is reduced to three main segments actuated by three joints, which correspond to human hip, knee and ankle joints. By changing the lengths of particular segments of limbs, the presented model can be adapted to people of different heights. Moreover, the application of linear actuators with

Fig. 1 CAD model of the designed lower limb exoskeleton driven by linear actuators equipped with DC motor, gear ratio and screw-nut system

limited strokes guarantees mechanical limitation of motion in the patient's joints, which increases safety during operation.

As an actuation element in each active joint, a linear actuator with DC motor, gear ratio and screw-nut system was used. It should be noted that DC motors were often used in numerous previous LLEs because of their low cost and easy control, for instance, see papers [21–25].

3 Simulation Model

To investigate the proposed CAD model of lower limb exoskeleton, a 3D parametric simulation model of the LLE was developed (see Fig. 2). The model has all parts of the exoskeleton discussed above and presented in Fig. 1. It is fully parametric, therefore arbitrary values of all parameters determining the kinematic model of the exoskeleton can be used. The created simulation model can be used to visualize the investigated mechanical design and control the correctness of the simulated results. Especially, it allows one to control spatial positions of individual elements of the device as well as configurations of its limbs. As a result, it can help understand crucial kinematic and dynamic parameters of the investigated device in the further, more advanced analysis and virtual experiments.

The presented simulation model is embedded in a global coordinate system *Oxyz* that is fixed to the ground. Moreover, a local coordinate system *O'x'y'z'* is fixed to

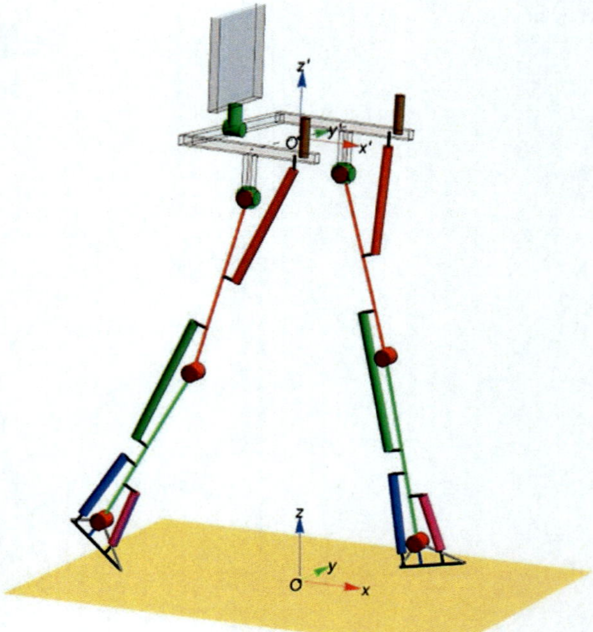

Fig. 2 A general, three-dimensional full parametric simulation model of a lower limb exoskeleton, created in Mathematica software

the pelvic part of the exoskeleton. When a patient wearing the LLE is walking, the distance between the centres of these coordinate systems changes, and is equal

$$\Delta \mathbf{r}(t) = [x(t), y(t), z(t)]^T \tag{1}$$

where $x(t)$, $y(t)$, and $z(t)$ denote the distances between the coordinate systems in the forward, lateral and vertical directions, respectively. To ensure biocompatibility, it is possible to rotate the pelvic frame of the exoskeleton around all three main axes of the coordinate system. Moreover, back support is connected to the pelvic part via two passive revolute joints (marked by two green cylinders). Rotation in the frontal plane $y'z'$ (around x'-axis) in described by the angle $\alpha(t)$. Rotation in the sagittal plane $x'z'$ (around rotated axis y') in described by the angle $\beta(t)$. In turn, rotation in the transverse plane $x'y'$ (around rotated axis z') is given by the angle $\gamma(t)$. As a result, vectors $\mathbf{r}(t)$ of positions of points of the pelvic part in the coordinate system $Oxyz$ are given as follows

$$\mathbf{r}(t) = \mathbf{R}\ (\alpha(t), \beta(t), \gamma(t)) \cdot \mathbf{r}' + \Delta \mathbf{r}\ (t) \tag{2}$$

where $\mathbf{r}'$ are vectors describing positions of points of the pelvis in the $O'x'y'z'$ system,

$$\mathbf{R}\ (\alpha(t), \beta(t), \gamma(t)) = \mathbf{R}_z\ (\gamma(t)) \cdot \mathbf{R}_y\ (\beta(t)) \cdot \mathbf{R}_x\ (\alpha(t)) \tag{3}$$

is the rotation matrix between two abovementioned coordinate systems, whereas

$$\mathbf{R}_x\ (\alpha(t)) = \begin{bmatrix} 1 & 0 & 0 \\ 0 & \cos\alpha(t) & -\sin\alpha\ (t) \\ 0 & \sin\alpha\ (t) & \cos\alpha(t) \end{bmatrix} \tag{4}$$

$$\mathbf{R}_y\ (\beta(t)) = \begin{bmatrix} \cos\beta(t) & 0 & \sin\beta\ (t) \\ 0 & 1 & 0 \\ -\sin\alpha\ (t) & 0 & \cos\beta(t) \end{bmatrix} \tag{5}$$

$$\mathbf{R}_x\ (\gamma(t)) = \begin{bmatrix} \cos\gamma(t) & -\sin\gamma\ (t) & 0 \\ \sin\gamma\ (t) & \cos\gamma(t) & 0 \\ 0 & 0 & 1 \end{bmatrix} \tag{6}$$

are rotation matrices for x-, y- and z-axes, respectively. The abovementioned rotations are visualized in Fig. 3.

As one can see in Fig. 2, each lower limb is attached to the main pelvic part of the exoskeleton via revolute hip joints (two passive joints marked by green cylinders

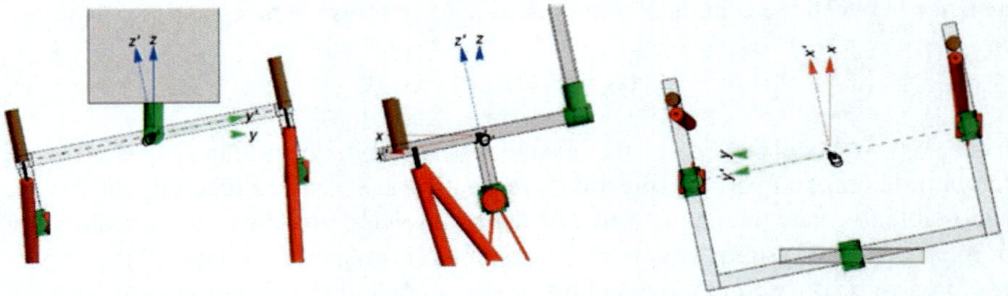

rotation of the pelvic in the frontal plane ($\alpha(t)$ is the angle around x' axis)

rotation of the pelvic in the sagittal plane ($\beta(t)$ is the angle around y' axis)

rotation of the pelvic in the transverse plane ($\gamma(t)$ is the angle around z' axis)

Fig. 3 Rotations of the exoskeleton pelvic frame: (**a**) in the frontal plane (angle $\alpha(t)$); (**b**) in the sagittal plane (angle $\beta(t)$); (**c**) in the transverse plane (the angle $\gamma(t)$)

and one active hip joint marked by a red cylinder). Three main segments of each limb are connected by active knee and ankle joints, also marked by red cylinders. As a result, regardless of the rotation of the pelvic part, each limb can perform its movement in two dimensions, i.e. in the sagittal plane. The exact positions of all characteristic points of both lower extremities of the exoskeleton can be determined with the use of relatively simple mathematical relations. In this paper, we do not focus on the formulation of a detailed mathematical model of a single limb since a similar model can be found in one of our previous studies [20] as well as in other papers found in the literature. Here, we presented only examples of some numerical simulations and focused on the experimental results obtained with the help of the constructed experimental stand.

4 Numerical Results

In our previous study [20], we conducted a kinematic analysis of the normal human gait. To obtain articulated variables corresponding to particular human joints we used an Optitrack motion capture system, which has been also successfully used to similar biomechanical problems [26–29]. Mean angular positions of individual lower limb joints in a single gait cycle are shown in Fig. 4. According to the literature, we assumed that the gait cycle begins when the heel of one lower limb (in our study – the left limb) touches the ground. A literature review indicates that the presented results are usually similar regardless of time T of a single gait cycle. Snapshots of simulations of the exoskeleton in different gait phases (captured at regular time intervals) are shown in Fig. 5. Articulated variables in particular joints were taken from Fig. 4, whereas time histories of the rotation angles $\alpha(t)$, $\beta(t)$ and $\gamma(t)$ were estimated based on our previous study.

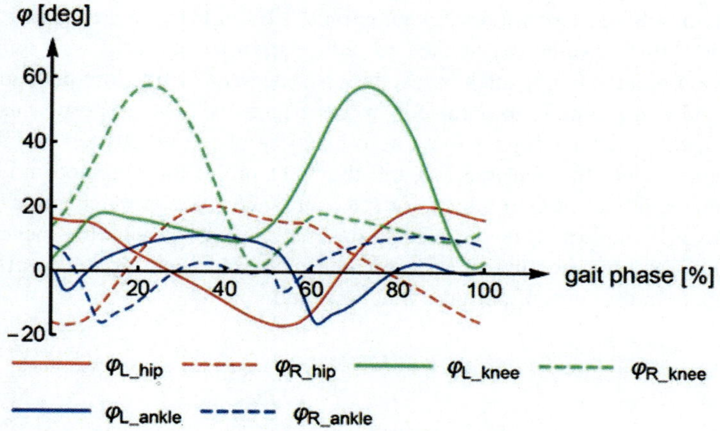

Fig. 4 Mean angles in human hip, knee and ankle joints of both legs (left L and right R) in a single gait cycle

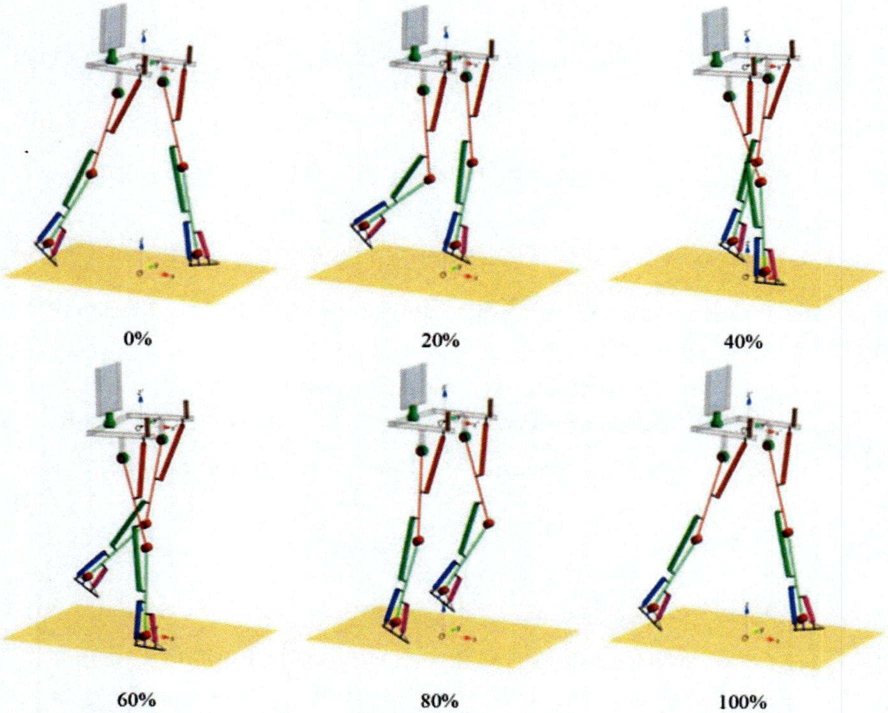

Fig. 5 Snapshots of simulations of the exoskeleton in different phases (in %) of a single gait cycle, captured at regular time intervals

As has already been mentioned, in numerous LLE found in the literature, rotation in the ankle joint is often overlooked to reduce the total mass as well as energy consumption of a device. In such cases, only two degrees of freedom per each limb (i.e. hip and knee joints) are actuated. In this paper, we also proposed analytical smooth functions which approximate time histories of the articulated variables in hip and knee joints for both the left and the right limb. The obtained model can be treated as a model of Gait Generator (GG) or Central Pattern Generator (CPG). After analysing the experimental results presented in Fig. 4, including locations of local minima and maxima in different moments of a single gait cycle, we proposed a CPG model in the form of periodic functions, namely

$$\varphi_{\text{L\_hip}}(t) = \varphi_1(t) \tag{7}$$

$$\varphi_{\text{R\_hip}}(t) = \varphi_1(t - 0.5T) \tag{8}$$

$$\varphi_{\text{L\_knee}}(t) = \varphi_2(t) \tag{9}$$

$$\varphi_{\text{R\_knee}}(t) = \varphi_2(t - 0.5T) \tag{10}$$

where

$$\varphi_1(t) = \phi_1(\text{mod}\,[t - T_1, T]) \tag{11}$$

$$\varphi_2(t) = \phi_2(\text{mod}\,[t - T_2, T]) \tag{12}$$

mod states for the modulo operation that returns the remainder of a division of t - T_1 and t - T_2 by T, respectively, and

$$\phi_1(t) = \left\{ \begin{array}{ll} \phi_{11\,\min} + (\phi_{11\,\max} - \phi_{11\,\min}) \cdot \sin^2\left(\frac{\pi}{2t_{11}}t\right) & \text{if } t \in \left< 0, t_{11} \right), \\ \phi_{11\,\max} - (\phi_{11\,\max} - \phi_{11\,\min}) \cdot \sin^2\left(\frac{\pi}{2(T - t_{11})}\left(t - t_{11}\right)\right) & \text{if } t \in \left< t_{11}, T \right), \end{array} \right\} \tag{13}$$

$$\phi_2(t) = \left\{ \begin{array}{ll} \phi_{21\,\min} + (\phi_{21\,\max} - \phi_{21\,\min}) \cdot \sin^2\left(\frac{\pi}{2t_{21}}t\right) & \text{if } t \in \left< 0, t_{21} \right), \\ \phi_{21\,\max} - (\phi_{21\,\max} - \phi_{21\,\min}) \cdot \sin^2\left(\frac{\pi}{2(t_{22} - t_{21})}(t - t_{21})\right) & \text{if } t \in \left< t_{21}, t_{22} \right), \\ \phi_{22\,\min} + (\phi_{22\,\max} - \phi_{22\,\min}) \cdot \sin^2\left(\frac{\pi}{2(t_{23} - t_{22})}(t - t_{22})\right) & \text{if } t \in \left< t_{22}, t_{23} \right), \\ \phi_{22\,\max} - (\phi_{22\,\max} - \phi_{21\,\min}) \cdot \sin^2\left(\frac{\pi}{2(T - t_{23})}(t - t_{23})\right) & \text{if } t \in \left< t_{23}, T \right), \end{array} \right\} \tag{14}$$

The best fit of the proposed approximations to the experimental results was obtained for the following values of the parameters: $T_1 = 0.545\ T$, $T_2 = 0.435\ T$, $t_{11} = 0.33\ T$, $t_{21} = 0.30\ T$, $t_{22} = 0.55\ T$, $t_{23} = 0.70\ T$, $\phi_{11\min} = -16.9$ deg.,

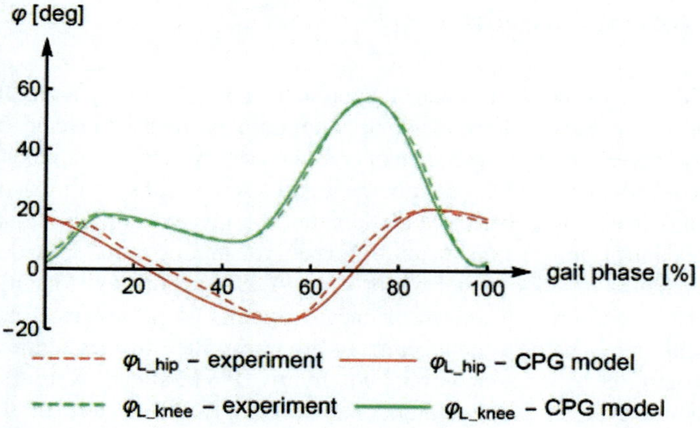

Fig. 6 Comparison of the proposed gait generator with experimental data obtained for the left limb

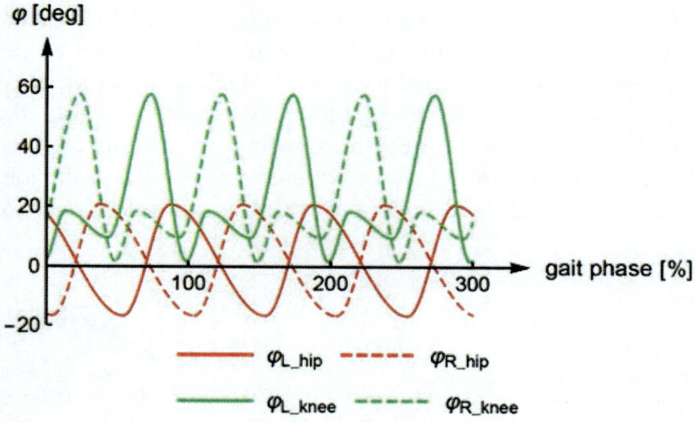

Fig. 7 Angles φ_{L_hip}, φ_{R_hip}, φ_{L_knee} and φ_{R_knee} produced by the proposed gait generator for 3 full gait cycles

ϕ_{11max} = 20.5 deg., ϕ_{21min} = 9.4 deg., ϕ_{21max} = 57.6 deg., ϕ_{22min} = 1.7 deg., ϕ_{22max} = 18.2 deg. A comparison of the obtained time histories of articulated variables φ_{L_hip} and φ_{L_knee} with experimental results for left limb is presented in Fig. 6. Based on the proposed gait model (formulas (7)–(14)), periodic time histories of both hip and knee angles can be generated for any number of full walking cycles, both for the left and right limb of the exoskeleton (for instance, see Fig. 7). It should be noted that, depending on the values of the parameters T, T_1, T_2, t_{11}, t_{21}, t_{22}, t_{23}, ϕ_{11min}, ϕ_{11max}, ϕ_{21min}, ϕ_{21max}, ϕ_{22min}, ϕ_{22max}, the proposed gait generator can produce different gait types.

5 Experimental Results

To verify the proposed LLE control approach, a single exoskeleton limb was constructed – see Fig. 8. It is made of aluminum profiles connected by rotary aluminum joints. Each of the two segments of the prototype is driven independently by two linear actuators with a reduction gear and a screw-nut system. Also, each of the two joints is equipped with a potentiometer that gives a signal in the feedback loop. As a control unit, a popular and inexpensive microcontroller Arduino Uno was used. Thanks to a 2-channel motor driver L298N dedicated for DC motors, it was possible to control the speed of linear actuators by pulse-width-modulation (PWM) technique. Analog voltage signals were transmitted from potentiometers to the microcontroller by using analog inputs. In turn, PWM signals were transmitted to the motor driver by using digital outputs. Finally, fill factors of the PWM signals were calculated by two PID controllers implemented in the memory of the microcontroller.

Figure 9 shows the results of control of the constructed exoskeleton leg. Figure 9a was obtained for experimental hip and knee angles (see Fig. 4) used as desired trajectories. In turn, in the case of Fig. 9b, articulated variables calculated from Eqs. (7)–(14) were used as desired trajectories. Two independent trajectory tracking controllers were used to track the desired trajectories using feedback digital PID controllers. PID controllers produce control signals by comparing the desired input angles with actual output angles from potentiometers installed in the joints of the prototype. One can see that in all presented cases, actual trajectories (solid lines)

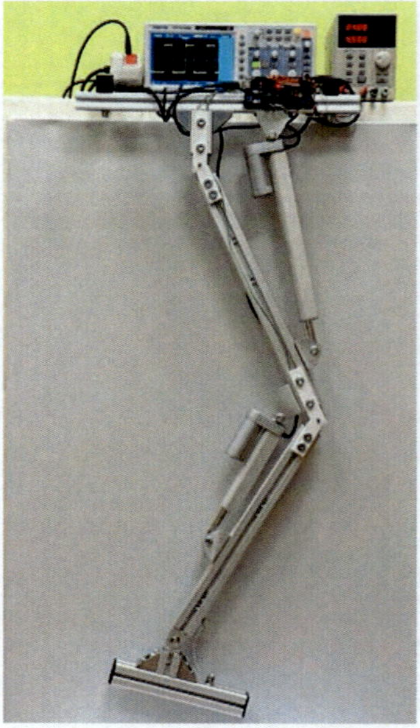

Fig. 8 The prototype of a single limb of the exoskeleton made of aluminum profiles, actuated by linear actuators and controlled by Arduino Uno microcontroller

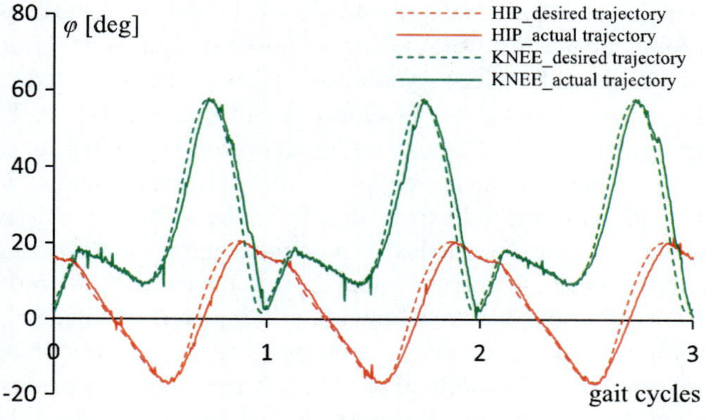

(a)

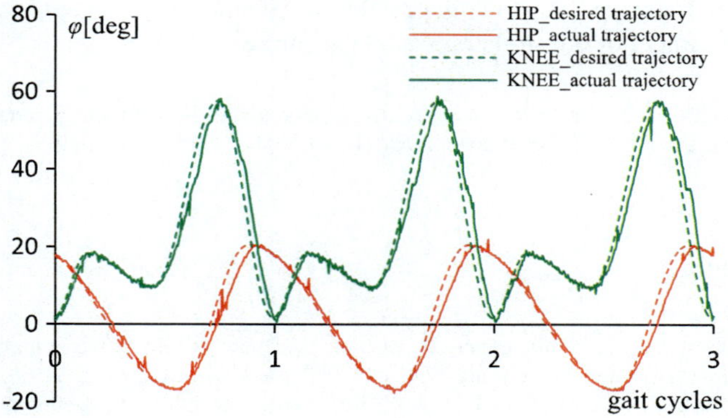

(b)

Fig. 9 Experimental results of control of the prototype of a single exoskeleton limb: (**a**) experimental articulated variables as the desired trajectories; (**b**) articulated variables produced by the proposed CPG model as the desired trajectories

of the hip and the knee follow the ideal trajectories (dashed lines) quite well, and control errors are acceptable. The digital PID controllers implemented in Arduino Uno gave promising results indicating that this approach can be successfully applied in real devices.

6 Conclusions

In this paper, we proposed a design of a lower limb exoskeleton made of aluminum profiles driven by linear electric actuators. To better investigate kinematic parameters of the designed device as well as human gait with the exoskeleton, a general three-dimensional and fully parametric simulation model of the exoskeleton was developed in Mathematica software. In the proposed design, and especially in the developed simulation model, biocompatibility aspects of the simulated device were considered. In a further study, the developed simulation model will be used for more accurate virtual studies of a walking process and determination of the most important gait parameters. In the present paper, time histories of human joint angles during normal walking (obtained experimentally in the authors' previous research) were used as the articulated variables in individual joints of the exoskeleton. In addition, a new gait generator was developed. It produces rhythmic movements of both limbs, in two joints that are most important during gait, i.e. hip and knee joints. To verify the proposed control method, a prototype of a single exoskeleton limb was constructed. The carried out experimental studies gave promising results regarding control of the device. To conclude, it is possible to develop a relatively inexpensive and efficient design of an LLE as well as a relatively simple and inexpensive control system for such devices. Also, it is possible to provide better access to LLE and reduce the workload of physiotherapists in the future.

Acknowledgments The work has been partially supported by the National Science Centre of Poland under the grant OPUS 9 no. 82015/17/B/ST8/01700 for years 2016-2019.

References

1. Bach, J.P., Deuschl, G., Doblhammer, G., Ziegler, U., Dodel, R.: Projected numbers of people with movement disorders in the years 2030 and 2050. Mov. Disord. **26**(12), 2286–2290 (2011)
2. Chen, B., Ma, H., Qin, L.-Y., Gao, F., Chan, K.-M., et al.: Recent developments and challenges of lower extremity exoskeletons. J. Orthop. Transl. **5**, 26–37 (2016)
3. Glover, W.: Work-related strain injuries in physiotherapists: prevalence and prevention of musculoskeletal disorders. Physiotherapy. **88**(6), 364–372 (2002)
4. Rugelj, D.: Low back pain and other work-related musculoskeletal problems among physiotherapists. Appl. Ergon. **34**(6), 635–639 (2003)
5. Agrawal, S., Banala, S., Fattah, A., Sangwan, V., Krishnamoorthy, V., et al.: Assessment of motion of a swing leg and gait rehabilitation with a gravity balancing exoskeleton. IEEE Trans. Neural Syst. Rehabil. Eng. **15**(3), 410–420 (2007)
6. Banala, S., Agrawal, S., Scholz, S.: Active leg exoskeleton (ALEX) for gait rehabilitation of motor-impaired patients. In: Proceedings of the 10th International Conference on Rehabilitation Robotics, June 12-15, 2007, Noordwijk, The Netherlands, 401–407 (2007)
7. Veneman, J.F., Kruidhof, R., Hekman, E.E.G., Ekkelenkamp, R., Van Asseldonk, E.H.F., et al.: Design and evaluation of the LOPES exoskeleton robot for interactive gait rehabilitation. IEEE Trans. Neural Syst. Rehabil. Eng. **15**(3), 379–386 (2007)
8. Gregorczyk, K., Adams, A., O'Donovan, P., Schiffman, J., Bensel, C.: Biomechanical and metabolic implications of wearing a powered exoskeleton to carry a backpack load. American Society of Biomechanics 2012 Proceedings, August 15–18, 2012, Gainesville (Florida), 2 pages (2012)
9. Guo, Z., Yu, H., Yin, Y.H.: Developing a mobile lower limb robotic exoskeleton for gait rehabilitation. J. Med. Dev. **8**, 044503-1–044503-6 (2014)
10. Petrarca, M., Patane, F., Rossi, S., Carniel, S., Cappa, P., et al.: A new robotic exoskeleton for gait recovery. Gait Posture. **40**, 26–27 (2014)

11. Ferrari, A., Benedetti, M., Pavan, E., Frigo, C., Bettinelli, D., et al.: Quantitative comparison of five current protocols in gait analysis. Gait Posture. **28**, 207–216 (2008)
12. Clark, T., Hawkins, D.: Are fixed limb inertial models valid for dynamic simulations of human movement? J. Biomech. **43**(14), 2695–2701 (2010)
13. Zach, L., Konvickova, S., Ruzicka, P.: Investigation of in-vivo hinge knee behaviour using a dynamic finite element model of the lower limb. Comput. Methods Biomech. Biomed. Engin. **15**, 326–327 (2012)
14. Dong, L., Zhu, F., Jin, X., Suresh, M., Jianga, B., et al.: Blast effect on the lower extremities and its mitigation: a computational study. J. Mech. Behav. Biomed. Mater. **28**, 111–124 (2013)
15. Kwa, H.K., Noorden, J.H., Missel, M., Craig, T., Pratt, J.E., et al.: Development of the IHMC mobility assist exoskeleton. In: Proceedings – IEEE International Conference on Robotics and Automation 2009, Kobe, Japan, 2556–2562 (2009)
16. Quintero, H.A., Farris, R.J., Goldfarb, M.: A method for the autonomous control of lower limb exoskeletons for persons with paraplegia. J. Med. Dev. **6**(4), 041003-1–041003-6 (2012)
17. Wang, L., Wang, S., van Asseldonk, E.H.F., van der Kooij, H.: Actively controlled lateral gait assistance in a lower limb exoskeleton. In: 2013 IEEE/RSJ International Conference on Intelligent Robots and Systems (IROS), pp. 965–970 (2013)
18. Farris, R.J., Quintero, H.A., Murray, S.A., Ha, K.H., Hartigan, C., et al.: A preliminary assessment of legged mobility provided by a lower limb exoskeleton for persons with paraplegia. IEEE Trans. Neural Syst. Rehabil. Eng. **22**(3), 482–490 (2014)
19. Wang, S., Wang, L., Meijneke, C., van Asseldonk, E., Hoellinger, T., et al.: Design and control of the Mindwalker exoskeleton. IEEE Trans. Neural Syst. Rehabil. Eng. **23**(2), 277–286 (2015)
20. Grzelczyk, D., Szymanowska, O., Awrejcewicz, J.: Gait pattern generator for control of a lower limb exoskeleton. Vib. Phys. Syst. 29, 2018007, 10 pages (2018)
21. Lu, R., Li, Z., Su, C.-Y., Xue, A.: Development and learning control of a human limb with a rehabilitation exoskeleton. IEEE Trans. Ind. Electron. **61**(7), 3776–3785 (2014)
22. Aguilar-Sierra, H., Yu, W., Salazar, S., Lopez, R.: Design and control of hybrid actuation lower limb exoskeleton. Adv. Mech. Eng. **7**(6), 1–13 (2015)
23. Li, L., Hoon, K.H., Tow, A., Lim, P.H., Low, K.H.: Design and control of robotic exoskeleton with balance stabilizer mechanism. IEEE/RSJ International Conference on Intelligent Robots and Systems (IROS), Congress Center Hamburg, September 28–October 2, 2015, Hamburg, Germany (2015)
24. Murray, S.A., Ha, K.H., Hartigan, C., Goldfarb, M.: An assistive control approach for a lower-limb exoskeleton to facilitate recovery of walking following stroke. IEEE Trans. Neural Syst. Rehabil. Eng. **23**(3), 441–449 (2015)
25. Long, Y., Du, Z., Wang, W., Dong, W.: Development of a wearable exoskeleton rehabilitation system based on hybrid control mode. Int. J. Adv. Robot. Syst., 1–10 (2016)
26. Grzelczyk, D., Biesiacki, P., Mrozowski, J., Awrejcewicz, J.: Dynamic simulation of a novel "broomstick" human forward fall model and finite element analysis of the radius under the impact force during fall. J. Theor. Appl. Mech. **56**(1), 239–253 (2018)
27. Grzelczyk, D., Biesiacki, P., Mrozowski, J., Awrejcewicz, P.: A 3-link model of a human for simulating a fall in forward direction. Springer Proc. Math. Stat. **249**, 135–146 (2018)
28. Grzelczyk, D., Biesiacki, P., Mrozowski, J., Awrejcewicz, J.: A 4-link model of a human for simulating a forward fall. Vib. Phys. Syst. 29, 2018008, 10 pages (2018)
29. Nigmatullin, R.R., Morozov, A.L., Awrejcewicz, J., Ludwicki, M.: Modeling and experimental validation of walking processes. Biocybernet. Biomed. Eng. **40**(1), 200–210 (2020)

Uncertainties in the Movement and Measurement of a Hexapod Robot

István Kecskés, Ákos Odry, and Péter Odry

Abstract Model uncertainties can be defined using the simulation model and real measurements, thereby the model accuracy is practically represented. The differences between the simulation and reality create both inaccuracy and uncertainty in control system development. Our previous researches presented these inaccuracies numerically and pointed out some structure imperfections of the Szabad(ka)-II hexapod robot. The performed sequential and parallel measurements on the Szabad(ka)-II robot highlighted notable uncertainties at (i) the left and right mechanical sides, (ii) in front and rear legs, (iii) current and voltage sensors and (iv) in case of repetitive walking scenarios. The presented analysis takes into account the 6-axis accelerometer measurements as well. The measurement errors and uncertainties should be estimated before the optimization of robot control or robot structure. It is also necessary to define the expected quality optimum and correctly interpret the simulation results and imperfections.

Keywords Hexapod robot · Uncertainty analysis · Simulation model

1 Introduction

The robustness is a key property in robot control, and thus, the robust optimization is an important step in control system design of complex mechanical systems. The sensitivity and uncertainty analysis are the two main tools in the robust control design, since the robust solution shows less model uncertainties compared to other solutions [1]. The minimization of the uncertainty is a general optimization goal in robot navigation problems [2] and in robot motion planning [3].

The construction of the dynamic simulation model is indispensable for the optimization of robot control, because such a model can estimate adequately

I. Kecskés (✉) · Á. Odry · P. Odry
University of Dunaújváros, Dunaújváros, Hungary
e-mail: kecskes.istvan@gmail.com; odrya@uniduna.hu; podry@uniduna.hu

J. Awrejcewicz (ed.), *Perspectives in Dynamical Systems I: Mechatronics and Life Sciences*, Springer Proceedings in Mathematics & Statistics 362,
https://doi.org/10.1007/978-3-030-77306-9_12

the robot's behavior. The Szabad(ka)-II hexapod robot with 18 DOF embedded mechatronic device system is suitable for complex drive control research [4], moreover, it requires a robust control solution [5]. Its kinematical and dynamical simulation model was built in Simulink environment, and published in detail in [6]. In the case of Szabad(ka)-II, the focus was on the dynamic modeling in order to be able to optimize both the motor controllers and walking algorithms, and the robot structure as well [6]. In real systems, like this robot, a single definitive optimal solution has been sought (not like a pareto set), and generally a robust solution is required, because there are many uncertainties, i.e., deviations between the simulation and measurements results, moreover, the real device is equipped with non-perfect measuring instruments and there are time-varying parameters [7].

The uncertain parameters can be divided into three groups in an optimization problem: physical, design, and scenario uncertainties [8]. The physical uncertainties occur in the parts of the model where estimated model or approximate solutions are used. The scenario uncertainties are related to the scenario parameters in a multi-scenario approach, whereas the optimized (design) variables carry the design uncertainties. These uncertainties belong to the simulation model, which was analyzed previously – we denoted as model uncertainty.

The uncertainties are analyzed in two approaches in case of Szabad(ka)-II robot:

- Model uncertainty obtainment, i.e., the deviation analysis between the real robot and simulation model by taking into account the measurement errors. This task was performed in our earlier work [6].
- Robot uncertainty obtainment, i.e., the deviation and repeatability analysis between two real situations or two parts of the robot. This task is elaborated in this paper.

Figure 1 illustrates both the aforementioned approaches and their role in the robust control design endeavors.

1.1 Accuracy, Repeatability and Uncertainty

Accuracy and repeatability are two important properties of the robot motion, moreover, these properties complement each other [9] in robot applications. Repeatability is a measure of the ability of the robot to consistently reach a specified point, while accuracy is a measure of the distance error associated with the desired point and achieved point [10].

In industrial robots the gear backlash is the most influencing factor in repeatability [11], which has been confirmed by our previous research from model accuracy point of view [6].

In case of model uncertainty issues, the model accuracy estimation is the key problem, however in case of robot uncertainty the repeatability analysis is required. Therefore, the uncertainty estimation is practically a repeatability analysis in this

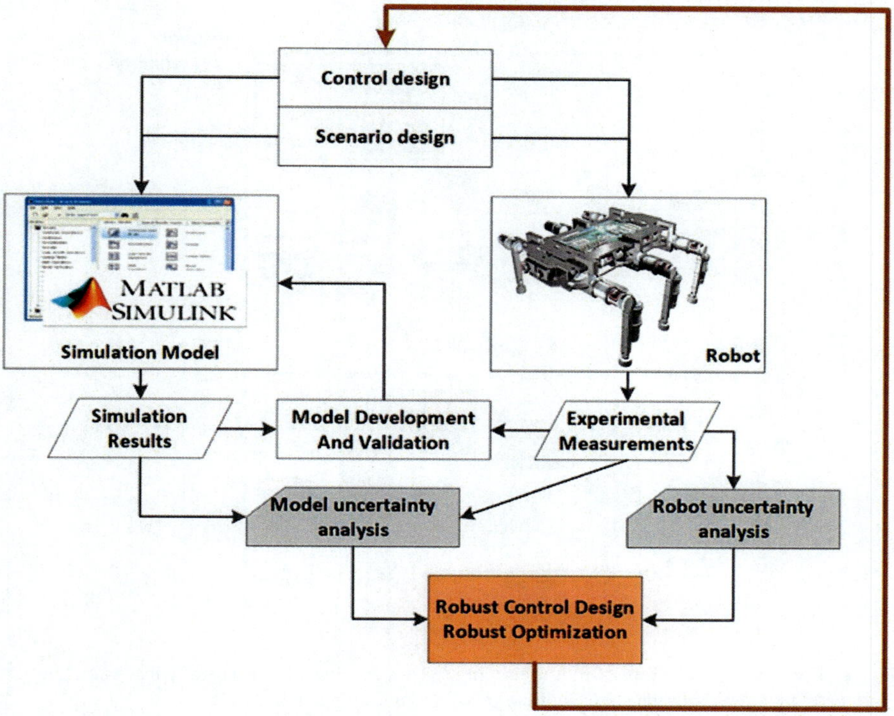

Fig. 1 Two kinds of uncertainty analysis in the robust control research of a hexapod robot

paper. The simple standard deviation and relative standard deviation are used as in most of such studies [4, 10]. Details are discussed in chapter "Nonlinear Dynamics of the Hierarchic System of Oscillators".

1.2 *Model Uncertainty*

There are many robot simulators available, each emphasizes different aspects of robot behavior simulation [12]. Simulation models have been published for several hexapod robots, but the quantification of the model validation mostly does not exist, i.e., the comparison between the simulation results and reality is rather descriptive [6].

Similarly to the results of model validation of Szabad(ka)-II robot, the model uncertainties were numerically expressed and classified by taking into account both the expectations and measurements errors. The measurements errors are defined by statistical evaluation of repeated measurements on the real robot, which are described and detailed in this paper.

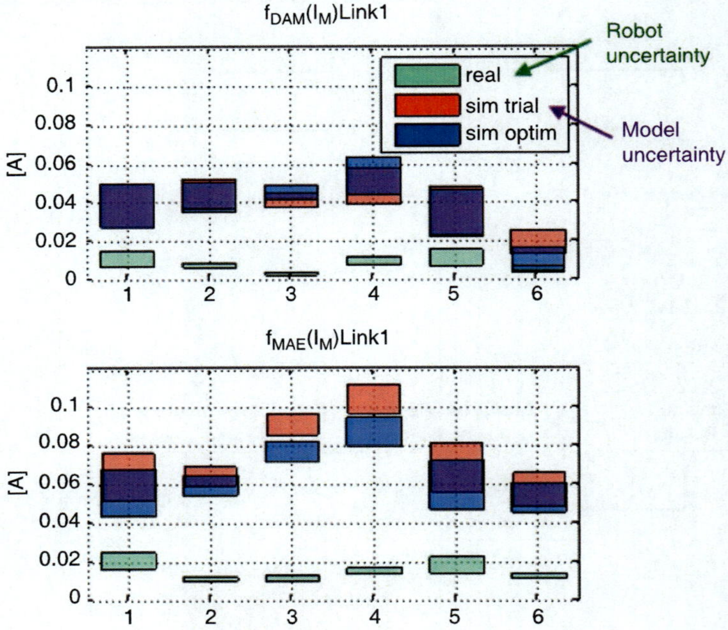

Fig. 2 The model uncertainty of motor current (I_M) was evaluated in sense of robot uncertainty. Horizontal bars represent the six legs of the robot. Originally published in [6]

Figure 2 shows the simulation error of motor current (I_M) expressed with Difference of Absolute Mean (f_{DAM}) and Mean Absolute Error (f_{MAE}) functions. The green (*real*) color represents the measurement errors gained by repeatability and referred as robot uncertainty in this paper. The red (*sim trial*) and blue (*sim optim*) results represent the simulation error calculated between real and simulation variables; these are considered as model uncertainties in the paper. The robot uncertainty is considered as both reference error and theoretical minimum of model error. In this example, the model error has approximately 5 times higher error compared to this reference error.

All the measurements errors and uncertainties are independent from any simulation model, therefore we distinguished and called them as robot uncertainties.

1.3 Robot Uncertainty

The Szabad(ka)-II robot is able to move based on a predefined trajectory curve, which allows it to walk on flat terrain with different gait. This version of Szabad(ka) robot series was not equipped with ground contact sensors, but it was sufficient to develop DC motor controller and simulation model. We have chosen to perform dedicated scenarios according to the aforementioned possibilities and limitations [5]. The analysis includes different robot walks on even terrain with variable load and direction. All the measurements used in this research are derived from these walking scenarios.

2 Analysis Method

2.1 Measurement Variables

The measurement unit utilized on the Szabad(ka)-II robot provides various electrical and navigation variables. Most of these variables are sampled with 500 Hz, however some variables are sampled only on 100 Hz or 200 Hz. Table IV in article [6] has already described these quantities, but Table 1 below provides the important information for this research.

Theoretically desired angles (*Drad*) are the same quantities between walk cycles or between legs, but due to the measurement system timing error this can also

Table 1 Measurement variables of Szabad(ka)-II robot

Name	Description	Symbol	Meas. unit	Dimensions
Desired angles of links	Calculated by the inverse kinematics	*Drad*	[rad]	time × 3 link × 6legs
Angles of links	The angles of robot leg links, derived from the encoders signal	*Arad*	[rad]	time × 3 link × 6legs
Control voltage of PWM amplifier	Calculated by the control algorithm	*Uvlt*	[V]	time × 3 link × 6legs
Motor currents	The absolute value of motor current measured by 10bit ADC	*Im*	[A]	time × 3 link × 6legs
Power voltage	The power voltage measured at each leg (nominal value is 12 V)	*Upow*	[V]	time × 6legs
Robot body 3D angular velocity	Measured by gyroscope mounted to the center of the robot body	*Gyro*	[dps]	time × 3 axis (X,Y,Z)
Robot body 3D acceleration	Measured by accelerometer mounted to the center of the robot body	*Accel*	[mg]	time × 3 axis (X,Y,Z)
Cumulative motor currents	Calculated from the motor currents and summarized all the 18 links. $Isum = \sum Im$	*Isum*	[A]	time × 1
Cumulative system resistance	Calculated from average power voltage and cumulative motor currents. The Ohm's law is used $Rcalc = Upow/Isum$	*Rcalc*	[ohm]	time × 1

show smaller deviations. The angle (*Arad*), current (*Im*) and control voltage (*Uvlt*) depend on the torque load on robot links, therefore these show higher deviations. The cumulative motor current (*Isum*) represents the average consumption of all the six legs and it is insensitive for the load imbalance between the legs. The 3D gyroscope (*Gyro*) and accelerometer (*Accel*) capture the 3D motion of robot body; however, their noise is relatively high.

2.2 Repeatability Analyses

The following repeatability analyses have been performed:

(a) Scenario-repetition analysis, which compares the measurement variables between the same scenarios repeated after each other.
(b) Walk-cycle repeatability analysis, which compares the measurement variables between consecutive walk cycles within one measurement. The cycle-deviations of several scenarios are averaged on and this average was analyzed.
(c) Different leg comparison analysis, which compares measurement variables between both left and right legs (c1) and front and rear legs (c2) on various scenarios.
(d) Scenario variability analysis, which compares measurement variables between various scenarios. This analysis is used as a reference value for the other analyses.
(e) Power variable analysis, which calculates the internal cumulate resistance from power voltage and summary current, moreover, scenario repeatability is analyzed for these three variables.

2.3 Statistics

In the statistical analysis at least two or more time-series variables $X \in \mathbb{R}^M$ are compared, where M denotes the number of digital samples (typically for one walk cycle $M = 620$) and N is the number of variables ($N \geq 2$). First, the synchronicity is ensured between these variables and the resampling to $Fs = 500Hz$ is performed (if it is required), using some special cycle cut and signal preparation algorithms.

The absolute mean value (AM), standard deviation (SD) and relative standard deviation (RSD) are calculated to express the variability between time-series variables, see Eqs. 1, 2 and 3. These equations are averaged on M samples to get an overall score value.

$$AM = \frac{1}{M}\sum\nolimits_{j=1}^{M}\left(\left|\frac{1}{N}\sum\nolimits_{i=1}^{N}X_i(j)\right|\right) \tag{1}$$

$$SD = \frac{1}{M}\sum\nolimits_{j=1}^{M}\left(\sqrt{\frac{1}{N-1}\sum\nolimits_{i=1}^{N}\left(X_i(j)-\overline{X}(j)\right)}\right) \tag{2}$$

$$RSD = \frac{SD}{AM} \tag{3}$$

The RSD can be expressed in percentage. This normalized score is the best appropriate marker to both compare and evaluate the aforementioned repeatability analysis.

3 Experimental Results

Altogether 16 measurements have been used from walking scenarios where the length of walk cycle is *1.26 sec*. Moreover, 10 different walking speeds or forward/backward motion scenarios were used for the scenario variability analysis.

Table 2 lists the statistical results from a), b), c) and d) analyses. Figures 2, 3, 4 and 5 illustrate one of the variables as examples belong to a), b), c) and e) analyses. Figure 7 summarizes and illustrates the statistical results.

Figure 3 shows the first robot leg angles for link variables related to one walk cycle. The curves highlight the mean value (MEAN), the minimum value (MIN), the maximum value (MAX), and the standard deviation (SD) calculated for each time points. The deviation is relatively small (less than 2%); this was expected based on the model uncertainty analysis made previously in [6].

Figure 4 shows the 3D acceleration of robot body in the b) analysis, the same way as Fig. 3. The deviation is significantly higher than the angles (50–250%). The uncertainty in direction Y and Z are notably higher, because the robot moves in X direction, and its movement in Y and Z axis is small, i.e., the effect of vibration is shown on these axes. It can be assumed that the magnitude of the acceleration sensor noise is close to these vibration magnitudes, and together the movement variability and noise produce such high uncertainty. The angular velocity of the robot body measured by gyroscope shows similar uncertainty rate as the acceleration.

Figure 5 shows the motor current in c2) analysis, where the front and real legs are compared. The plot shows the two motor current curves, and the standard deviation calculated between these samples. There are specific sections where the difference is small, and there are other parts where it is high in Link1 and Link2. This behavior

Table 2 Measurement variables of Szabad(ka)-II robot

Analysis		**a)**		**b)**		**c1)**		**c2)**		**d)**	
Variable	Channel	SD	RSD	SD	RSD	SD	RSD	SD	RSD	SD	RSD
Drad	Link1	0.001	0.010	0.009	0.069	0.001	0.011	0.000	0.004	0.157	4.281
Drad	Link2	0.001	0.011	0.008	0.064	0.001	0.011	0.001	0.005	0.144	1.075
Drad	Link3	0.001	0.001	0.007	0.004	0.001	0.001	0.001	0.000	0.125	0.074
Arad	Link1	0.002	0.019	0.009	0.065	0.003	0.021	0.002	0.018	0.150	4.161
Arad	Link2	0.002	0.016	0.007	0.056	0.002	0.014	0.005	0.041	0.129	0.952
Arad	Link3	0.002	0.001	0.007	0.004	0.003	0.002	0.002	0.001	0.115	0.068
Uvlt	Link1	0.203	0.057	0.369	0.104	0.222	0.063	0.162	0.046	2.834	5.727
Uvlt	Link2	0.194	0.032	0.512	0.092	0.158	0.028	0.484	0.084	5.561	2.424
Uvlt	Link3	0.159	0.062	0.423	0.163	0.237	0.090	0.126	0.049	2.736	2.477
Im	Link1	0.012	0.107	0.015	0.129	0.015	0.125	0.013	0.111	0.036	0.462
Im	Link2	0.020	0.088	0.028	0.172	0.018	0.114	0.058	0.293	0.116	0.606
Im	Link3	0.009	0.108	0.016	0.160	0.016	0.162	0.013	0.145	0.054	0.791
Isum		0.121	0.045	0.308	0.116					0.668	0.315
Upow		0.163	0.015	0.053	0.005					0.145	0.014
Gyro	X	2.372	2.975	2.007	1.227					1.733	1.972
Gyro	Y	3.380	0.640	4.268	0.781					2.430	1.593
Gyro	Z	4.294	3.728	2.413	0.905					2.333	1.867
Accel	X	26.258	0.259	47.352	0.481					44.048	1.584
Accel	Y	24.846	1.627	21.133	1.132					19.764	2.298

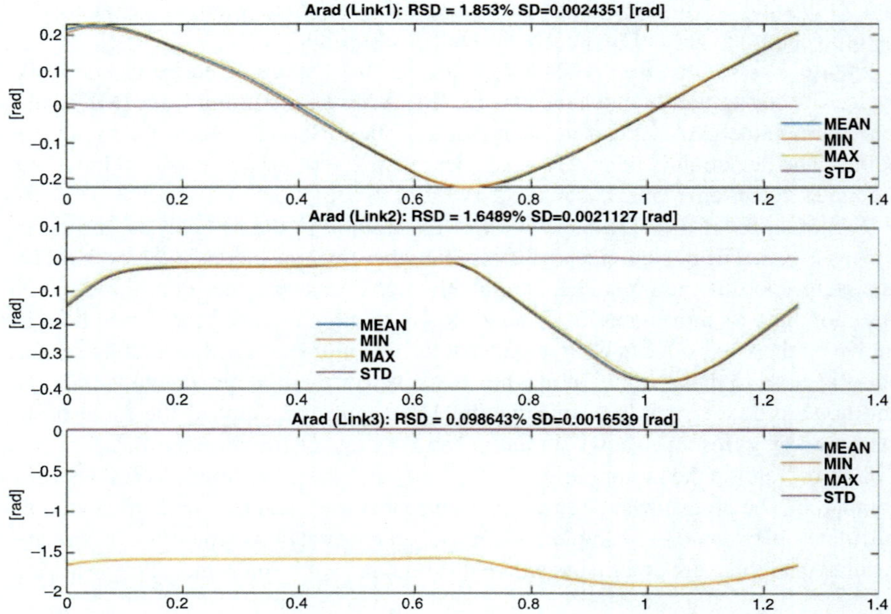

Fig. 3 Angle of links in a) Scenario Repeatability Analysis

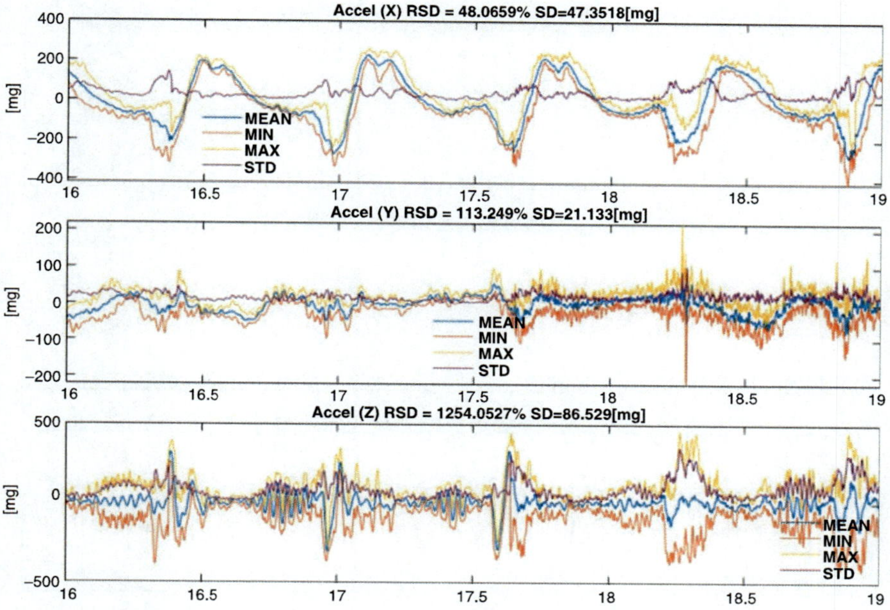

Fig. 4 3D Acceleration variable in b) Walk cycle Repeatability Analysis

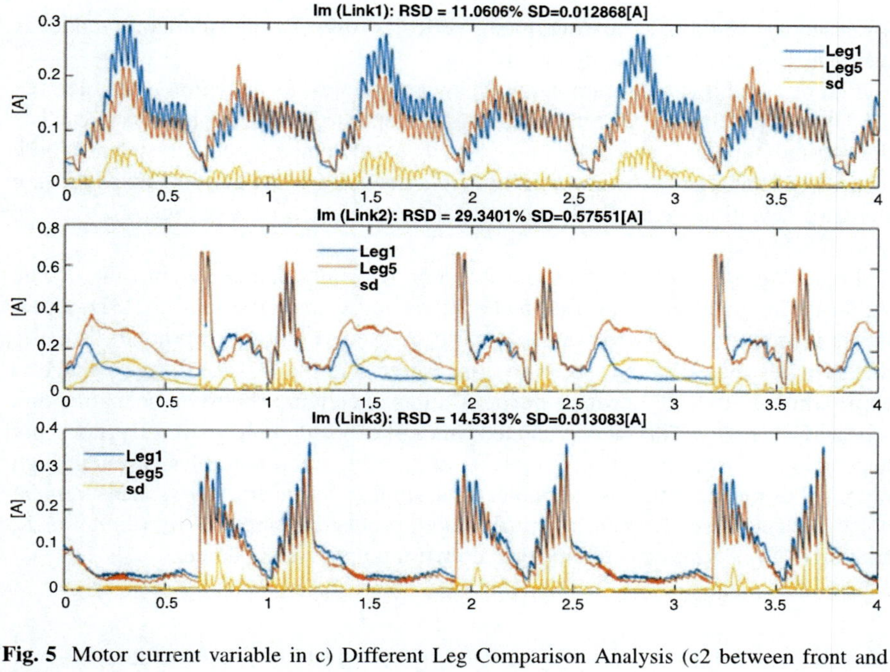

Fig. 5 Motor current variable in c) Different Leg Comparison Analysis (c2 between front and rear)

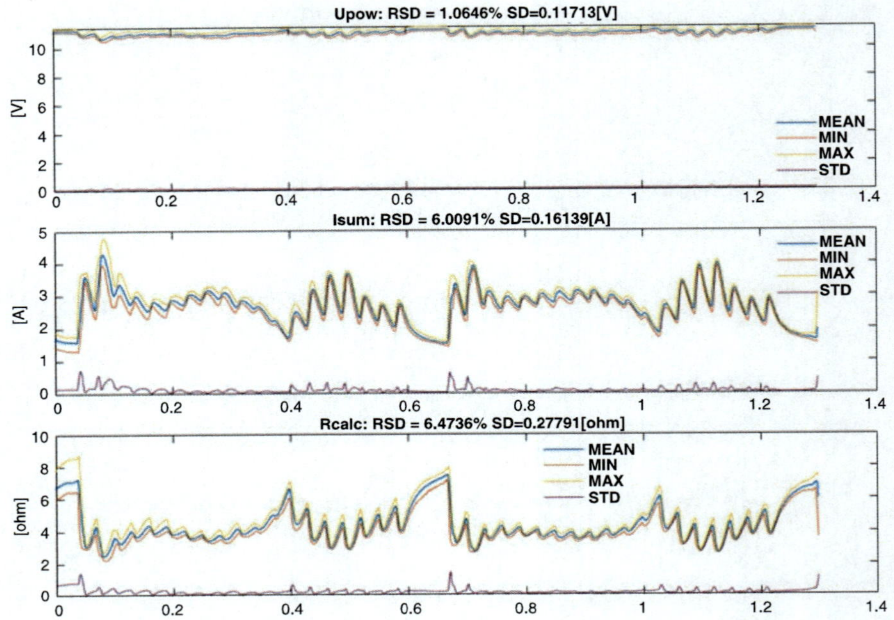

Fig. 6 All three power variables in e) Power Variable Analysis

is caused by the robot's "rearing horse"[1] effect, proved by the simulation model as well, and discussed previously in [4].

Figure 6 highlights the summarized power variables. Its deviation is smaller, but the highest uncertainty is in the moment of step, i.e., when the robot leg touches the ground (about 0 sec and 0.68 sec). The calculated system resistance both inherits and aggregates the uncertainties from the voltage and current and its relative standard deviation *Rcalc:RSD* = *6.5%* value becomes higher (*Upow:RSD* = *1%*, *Isum:RSD* = *6%*).

Figure 7 graphically shows all the uncertainty results documented in Table 1. The right-bottom graph shows the ration of a) repeatability and e) variability RSD values, which represent whether the same scenario (measured with "a)") produces smaller uncertainties than different scenarios (measured with "e)"). The power voltage is approximately to 100%, which means similar uncertainty between scenarios and repeated scenario. This was expected since the power voltage mostly does not depend on the scenario. The gyroscope and acceleration variables result in high ration. This means that those variables have significant uncertainty or measurement noise, except those channels where the robot produces some activity (direction X, and angle Y). Additional conclusions are written in the next chapter.

[1]Rearing occurs when a horse "stands up" on its hind legs with the forelegs off the ground

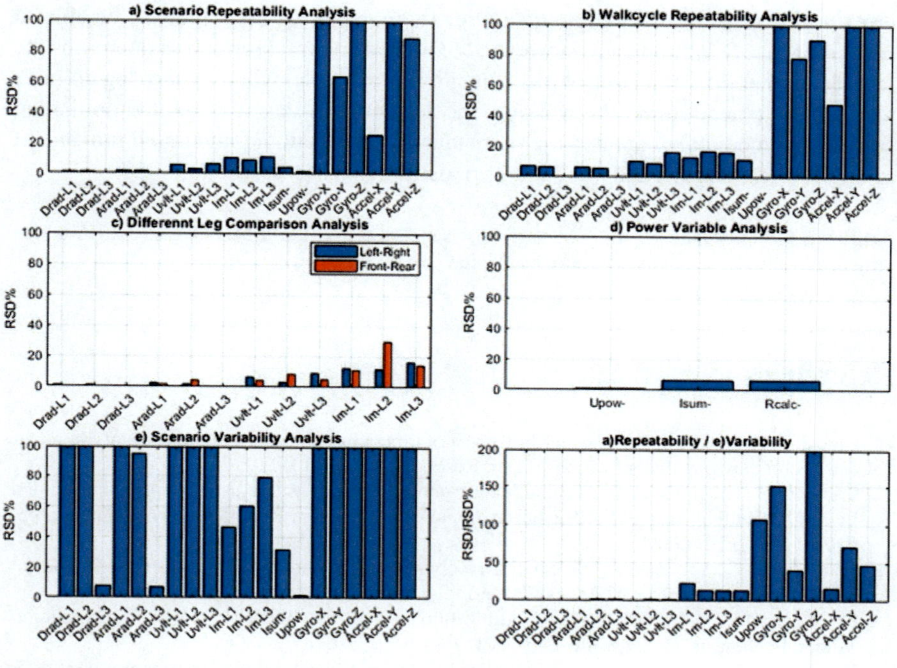

Fig. 7 Uncertainty analyses summary

4 Conclusion

The uncertainties of a hexapod walker robot are both quantified and evaluated through five different analyses in this research. The dynamic variables of robot walking were analyzed which were recorded previously for different walking scenarios. The evaluated variables can be grouped into three categories: (a) the angles of links, and control and power voltages having minor uncertainties (1–3%), (b) the motor currents having moderate uncertainties (10–20%) and (c) the 3D body movements – measured by accelerometer and gyroscope sensors – having major uncertainties (25–300%).

The derivation transformation on motion variables (both linear or angular) highlights the high frequency components, i.e., the jerk fluctuations and noises, thus velocity and acceleration variables show higher uncertainties compared to the base movements.

The robot uncertainty analysis provides important information for the robust control design research. The principal aim of the robust control is to minimize the variability of motion quality for both the robot motion and measurement variabilities.

The utilized navigation sensor cannot be used in a robot control loop without any preprocessing due to its high uncertainty. The acceleration and velocity should be transformed to movement/angle quantities (e.g., using Kalman filter) to be able to obtain robust 3D motion data. However, the integration transformation decreases the fast reaction ability of a control mechanism. For example, when a robot leg collides with a solid object during the walking, the acceleration sensor (mounted on the leg or body) immediately captures the particular moment, but the integrated movement or angle is less sensitive to this kind of registration processes.

Acknowledgments This work/publication is supported by the EFOP-3.6.1-16-2016-00003 project and is co-financed by the European Union.

References

1. Crespi, G.P., Kuroiwa, D., Rocca, M.: Robust optimization: Sensitivity to uncertainty in scalar and vector cases, with applications. Oper. Res. Perspect. **5**, 113–119 (2018)
2. Candido, S., Hutchinson, S.: Minimum uncertainty robot navigation using information-guided POMDP planning. In: 2011 IEEE International Conference on Robotics and Automation, pp. 6102–6108. IEEE (2011)
3. Latombe, J.-C., Lazanas, A., Shekhar, S.: Robot motion planning with uncertainty in control and sensing. Artif. Intell. **52**(1), 1–47 (1991)
4. Kecskes, I., Odry, P.: Drive Control Optimization of Walker Robots Using Dynamic Simulation Model. PhD thesis, Obuda University (2018)
5. Kecskés, I., Odry, P.: Multi-scenario multi-objective optimization of a fuzzy motor controller for the szabad (ka)-ii hexapod robot. Acta Polytechnica Hungarica. **15**, 7 (2018)
6. Kecskés, I., Burkus, E., Bazsó, F., Odry, P.: Model validation of a hexapod walker robot. Robotica. **35**(2), 419–462 (2017)
7. Kecskés, I., Odry, Á., Burkus, E., Odry, P.: Embedding optimized trajectory and motor controller into the szabad (ka)-ii hexapod robot. In: 2016 IEEE International Conference on Systems, Man, and Cybernetics (SMC), pp. 001417–001422. IEEE (2016)
8. Marijt, R.: Multi-objective robust optimization algorithms for improving energy consumption and thermal comfort of buildings. Master's thesis, Leiden Institute for Advanced Computer Science, Leiden University (2009)
9. Conrad, K.L., Shiakolas, P.S., Yih, T.: Robotic calibration issues: Accuracy, repeatability and calibration. In: Proceedings of the 8th Mediterranean Conference on Control and Automation (MED2000), vol. 1719. Rio, Patras (2000)
10. Şirinterlikçi, A., Tiryakioğlu, M., Bird, A., Harris, A., Kweder, K.: Repeatability and accuracy of an industrial robot: laboratory experience for a design of experiments course. Technol. Interface J. **9**(2), 1–10 (2009)
11. Goswami, A., Quaid, A., Peshkin, M.: Complete parameter identification of robot from partial pose information. In: [1993] Proceedings IEEE International Conference on Robotics and Automation, pp. 168–173. IEEE (1993)
12. Von Twickel, A., Hild, M., Siedel, T., Patel, V., Pasemann, F.: Neural control of a modular multi-legged walking machine: simulation and hardware. Robot. Auton. Syst. **60**(2), 227–241 (2012)

The Dynamics Analysis of a Spatial Linkage with Flexible Links and Imperfect Revolute Joints

Krzysztof Augustynek and Andrzej Urbaś

Abstract The algorithm for generating the dynamics equations of the two-dof spatial linkage is considered in the paper. The presented linkage is composed of the five rigid or flexible links which form a serial closed-loop kinematic chain. It is assumed that revolute joints can be imperfect. The joint coordinates together with homogeneous transformation matrices are applied to generate the equations of motion. The dynamics equations are derived using the Lagrange equations of the second kind. The presented algorithm gives the opportunity to generalize it for any linkages with a tree closed-loop kinematic structure. The flexible links are modelled by means of the Rigid Finite Element Method in the sense of the modified approach. The author's spatial model of the revolute joint with radial and axial clearance is applied to take into account clearance effects. In this model, a revolute joint is discretized by means of contact elements located on the cylindrical and frontal surfaces of the journal and bearing. Such an approach allows us to detect automatically collisions in many points of the contacting surfaces. The normal contact force is calculated using the Nikravesh-Lankarani formula which is an extension of the classic Hertz model because it additionally takes into account dissipation of the energy. The LuGre friction model is applied to model friction phenomenon in joints. In numerical simulations, an interaction between the links' flexibility and clearance in the joint during the motion of the linkage is analyzed.

Keywords Linkage · Dynamics analysis · Friction · Revolute joint · Clearance

---

K. Augustynek (✉) · A. Urbaś
Department of Mechanical Engineering Fundamentals, University of Bielsko-Biala, Bielsko-Biala, Poland
e-mail: kaugustynek@ath.bielsko.pl; aurbas@ath.bielsko.pl

J. Awrejcewicz (ed.), *Perspectives in Dynamical Systems I: Mechatronics and Life Sciences*, Springer Proceedings in Mathematics & Statistics 362,
https://doi.org/10.1007/978-3-030-77306-9_13

1 Introduction

The clearance in joints can be the result of structural assumptions or the effect of wearing parts. Simulation models of dynamics of linkages with clearance effects allow to better understand their behavior, as well as to estimate the values of impulse forces which increase the dynamic forces acting on the system. In addition, thanks to these models it is possible to determine the limit value of the clearance at which further operation can lead to damage of the system. Paper [1] shows that the flexibility of links has a significant impact on the behavior of linkages with the clearance in joints. The results presented there show that the flexibility can lead to a significant reduction of the impulse forces caused by the impact of the journal and bearing.

In the paper, it is assumed that the clearance can exist only in the revolute joints. There are many papers devoted to the clearance model of the revolute joints [1–7]. These models can be divided into two main groups: planar [2, 3] and spatial [1, 4–7], related to how the motion of the journal and bearing is described. Additionally, they can analyze only the radial clearance [2, 3] or they can take into account interaction between the radial and axial clearance [1, 4–7].

The dynamics model of the two-dof RPSUP linkage is presented in the paper. This model takes into account the flexibility of the link and the clearance in the revolute cut-joint. The kinematics of the linkage is described using the formalism of the joint coordinates and homogeneous transformation matrices. The Rigid Finite Element Method [8] is used to discretize the flexible coupler. The author's spatial model of the revolute joint with the radial and axial clearance is proposed. In these model contacting surfaces are discretized by means of the contact elements for which the normal and tangent contact forces are calculated. The impulse force in the clearance joint is modeled using the Nikravesh-Lankarani formula [9] and friction is modeled by means of the LuGre friction model [10]. In numerical simulations, the influence of the crank velocity and the clearance in the revolute cut-joint on the slider acceleration is analyzed.

2 Dynamics Model of the Two-dof RPSUP Spatial Linkage

The RPSUP linkage containing five links is shown in Fig. 1. The linkage is divided at cut-joint R and as a result, the two open-loop kinematic chains are obtained ($c \in \{1, 2\}$). The first chain contains three links $\left(n_l^{(1)} = 3\right)$ whilst the second chain forms two links $\left(n_l^{(2)} = 2\right)$. In the proposed model, it is assumed that the clearance is considered only in the cut-joint R.

It is also assumed that coupler (1, 3) can be flexible and the Rigid Finite Element Method is used to discretize its. As result, the coupler is replaced by the set of

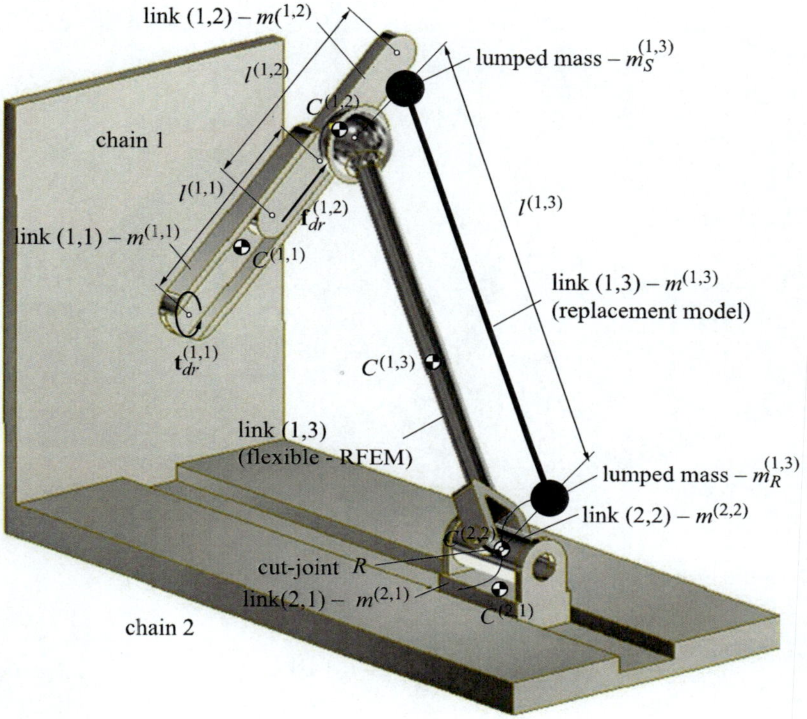

Fig. 1 Model of the RPSUP linkage

rigid finite elements (rfe) interconnected by means of the set of dimensionless and massless spring-damping elements (sde) (Fig. 2).

2.1 The Formalism of Generalised Coordinates and Homogeneous Transformation Matrices

The kinematics of the linkage considered is defined by means of the joint coordinates and homogeneous transformation matrices (Fig. 2).

The generalised coordinates vectors defined for each open-loop kinematic chain have the following form:

$$\mathbf{q}^{(1)} = \left(q_i^{(1)}\right)_{i=1,\ldots,n_{dof}^{(1)}} = \left[\psi^{(1,1)} \;\; z^{(1,2)} \;\; \psi^{(1,3,0)} \;\; \theta^{(1,3,0)} \;\; \varphi^{(1,3,0)} \;\; \tilde{\mathbf{q}}_f^{(1,3)}\right]^T, \tag{1.1}$$

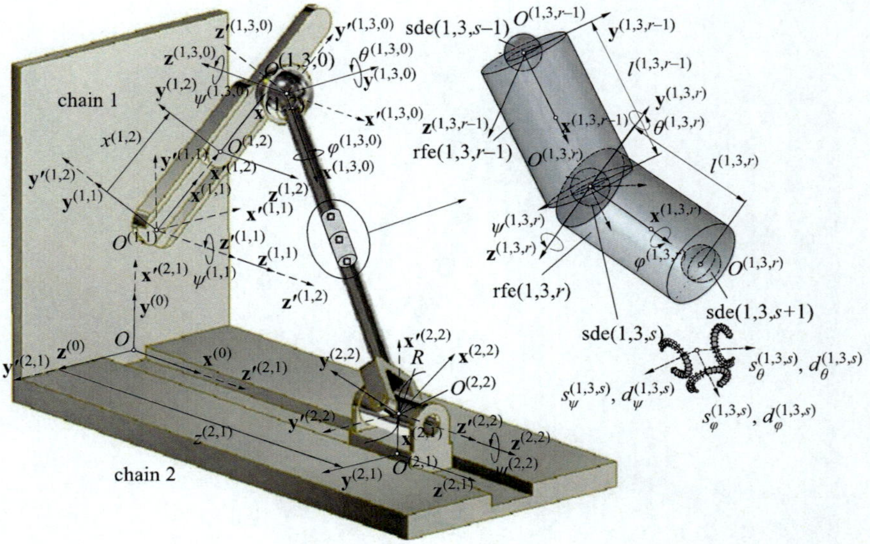

Fig. 2 Generalised coordinates

$$\mathbf{q}^{(2)} = \left(q_i^{(2)}\right)_{i=1,\ldots,n_{dof}^{(2)}} = \left[\, z^{(2,1)} \;\; \psi^{(2,2)} \,\right]^T, \tag{1.2}$$

$$\text{where: } \tilde{\mathbf{q}}_f^{(1,3)} = \begin{cases} \varnothing & \text{if coupler rigid,} \\ \left[\, \tilde{\mathbf{q}}^{(1,3,1)^T} \; \ldots \; \tilde{\mathbf{q}}^{(1,3,r)^T} \; \ldots \; \tilde{\mathbf{q}}^{\left(1,3,n_{rfe}^{(1,3)}-1\right)^T} \,\right]^T & \text{if coupler flexible,} \end{cases}$$

$$\tilde{\mathbf{q}}^{(1,3,r)} = \left[\, \psi^{(1,3,r)} \;\; \theta^{(1,3,r)} \;\; \varphi^{(1,3,r)} \,\right]^T.$$

The transformation matrices from the local reference frames defined for each link to the global reference frame are defined by:

$$\mathbf{T}^{(c,b)}\Big|_{\substack{c=1,2 \\ b=1,\ldots,n_l^{(c)}}} = \prod_{j=1}^{b} \tilde{\mathbf{A}}^{(c,j)} \tilde{\mathbf{T}}^{(c,j)}, \tag{2}$$

where $\tilde{\mathbf{A}}^{(c,j)} = \text{const}$ is the transformation matrix describing the initial position and orientation of link (c,j) with respect to the preceding link, $\tilde{\mathbf{T}}^{(c,j)}$ is the transformation matrix defining the actual position and orientation of link (c,j) with respect to the initial configuration.

2.2 Dynamics Equations of Motion

The dynamics equations for each open-loop kinematic chain are derived using the Lagrange equations of the second kind [8]:

$$\frac{\mathrm{d}}{\mathrm{d}t}\frac{\partial E_k^{(c)}}{\partial \dot{\mathbf{q}}^{(c)}} - \frac{\partial E_k^{(c)}}{\partial \mathbf{q}^{(c)}} + \frac{\partial E_p^{(c)}}{\partial \mathbf{q}^{(c)}} + \frac{\partial R^{(c)}}{\partial \dot{\mathbf{q}}^{(c)}} = \mathbf{Q}^{(c)}\Big|_{c=1,2}, \tag{3}$$

where $E_k^{(c)}$ is the kinetic energy of chain c, $E_p^{(c)} = E_{p,g}^{(c)} + E_{p,fl}^{(c)}$ is the sum of the potential energy of gravity forces and spring deformation energy of the flexible link, $R^{(c)}$ is the Rayleigh function defined for the flexible link and $\mathbf{Q}^{(c)}$ is the vector of non-potential generalised forces resulting from e.g. the contact forces acting in the clearance joint. The following sections present detailed formulas for determining components of the Eq. (3).

Kinetic Energy and Potential Energy of Gravity Forces

The kinetic energy of the particular subchains can be determined using the concept of the trace of the matrix:

$$E_k^{(1)} = \begin{cases} \frac{1}{2}\sum\limits_{l=1}^{n_l^{(1)}} \mathrm{tr}\left(\dot{\mathbf{T}}^{(1,l)}\mathbf{H}^{(1,l)}\dot{\mathbf{T}}^{(1,l)^T}\right) & \text{if coupler rigid,} \\ \frac{1}{2}\sum\limits_{l=1}^{n_l^{(1)}-1} \mathrm{tr}\left(\dot{\mathbf{T}}^{(1,l)}\mathbf{H}^{(1,l)}\dot{\mathbf{T}}^{(1,l)^T}\right) + \frac{1}{2}\sum\limits_{r=1}^{n_{rfe}^{(1,3)}-1} \mathrm{tr}\left(\dot{\mathbf{T}}^{(1,3,r)}\mathbf{H}^{(1,3,r)}\dot{\mathbf{T}}^{(1,3,r)^T}\right) & \text{if coupler flexible,} \end{cases} \tag{4.1}$$

$$E_k^{(2)} = \frac{1}{2}\sum_{l=1}^{n_l^{(2)}} \mathrm{tr}\left(\dot{\mathbf{T}}^{(2,l)}\mathbf{H}^{(2,l)}\dot{\mathbf{T}}^{(2,l)^T}\right), \tag{4.2}$$

where $\mathbf{H}^{(\bullet)}$ is the pseudo-inertia matrix of link or rfe.

After necessary transformations the Lagrange operator can be written in the matrix form as follows:

$$\frac{\mathrm{d}}{\mathrm{d}t}\frac{\partial E_k^{(c)}}{\partial \dot{\mathbf{q}}^{(c)}} - \frac{\partial E_k^{(c)}}{\partial \mathbf{q}^{(c)}} = \mathbf{M}^{(c)}\ddot{\mathbf{q}}^{(c)} + \mathbf{h}^{(c)}, \tag{5}$$

where: $\mathbf{M}^{(c)} = \left(\mathbf{M}_{i,j}^{(c)}\right)_{i,j=1,\ldots,n_l^{(c)}}$, $\mathbf{M}_{i,j}^{(c)} = \sum_{l=\max\{i,j\}}^{n_l^{(c)}} \mathbf{M}_{i,j}^{(c,l)}$, $\mathbf{M}_{i,j}^{(c,l)}\Big|_{i,j=1,\ldots,l} = \left(m^{(c,l)}_{n_{dof}^{(c,i-1)}+v,\,n_{dof}^{(c,j-1)}+w}\right)_{\substack{v=1,\ldots,\tilde{n}_{dof}^{(c,i)}\\ w=1,\ldots,\tilde{n}_{dof}^{(c,j)}}}$, $m_{i,j}^{(c,l)} = \operatorname{tr}\left\{\mathbf{T}_i^{(c,l)}\mathbf{H}^{(c,l)}\mathbf{T}_j^{(c,l)^T}\right\}\mathbf{h}^{(c)} = \left(\mathbf{h}_i^{(c)}\right)_{i=1,\ldots n_l^{(c)}}$, $\mathbf{h}_i^{(c)} = \sum_{l=i}^{n_l^{(c)}} \mathbf{h}_i^{(c,l)}$, $\mathbf{h}_i^{(c,l)}\Big|_{i=1,\ldots,l} = \left(h^{(c,l)}_{n_{dof}^{(c,i-1)}+v}\right)_{v=1,\ldots,\tilde{n}_{dof}^{(c,l)}}$, $h_i^{(c,l)} = \sum_{m=1}^{n_{dof}^{(c,l)}}\sum_{n=m}^{n_{dof}^{(c,l)}} \operatorname{tr}\left\{\mathbf{T}_i^{(c,l)}\mathbf{H}^{(c,l)}\mathbf{T}_{m,n}^{(c,l)^T}\right\}\dot{q}_m^{(c,l)}\dot{q}_n^{(c,l)}$.

The potential energy of the gravity forces for each subchain can be expressed in a similar way:

$$E_{p,g}^{(1)} = \begin{cases} \sum_{l=1}^{n_l^{(1)}} m^{(1,l)}\mathbf{j}_2\mathbf{T}^{(1,l)}\mathbf{r}_C^{(1,l)} & \text{if coupler rigid,} \\ \sum_{l=1}^{n_l^{(1)}-1} m^{(1,l)}\mathbf{j}_2\mathbf{T}^{(1,l)}\mathbf{r}_C^{(1,l)} + \sum_{l=1}^{n_{rfe}^{(1,3)}} m^{(1,l,r)}\mathbf{j}_2\mathbf{T}^{(1,l,r)}\mathbf{r}_C^{(1,l,r)} & \text{if coupler flexible,} \end{cases} \tag{6.1}$$

$$E_{p,g}^{(2)} = \sum_{l=1}^{n_l^{(2)}} m^{(2,l)}\mathbf{j}_2\mathbf{T}^{(2,l)}\mathbf{r}_C^{(2,l)}, \tag{6.2}$$

where $m^{(\bullet)}$ is the mass of link or rfe, $\mathbf{r}_C^{(\bullet)}$ is the vector of the centre of mass of link or rfe, $\mathbf{j}_2 = \begin{bmatrix} 0 & 1 & 0 & 0 \end{bmatrix}$.

The generalized forces due to the gravity forces can be calculated as follows:

$$\frac{\partial E_{p,g}^{(c)}}{\partial \mathbf{q}^{(c)}} = \mathbf{g}^{(c)}, \tag{7}$$

where: $\mathbf{g}^{(c)} = \left(\mathbf{g}_i^{(c)}\right)_{i=1,\ldots n_l^{(c)}}$, $\mathbf{g}_i^{(c)} = \sum_{l=i}^{n_l^{(c)}} \mathbf{g}_i^{(c,l)}$, $\mathbf{g}_i^{(c,l)}\Big|_{i=1,\ldots,l} = \left(g^{(c,l)}_{n_{dof}^{(c,i-1)}+v}\right)_{v=1,\ldots,\tilde{n}_{dof}^{(c,l)}}$,

$\mathbf{g}_i^{(c,l)}\Big|_{i=1,\ldots,l} = \left(g^{(c,l)}_{n_{dof}^{(c,i-1)}+v}\right)_{v=1,\ldots,\tilde{n}_{dof}^{(c,i)}}$, $g_i^{(p)} = m^{(c,l)} g \mathbf{j}_2\mathbf{T}_i^{(c,l)}\mathbf{r}_C^{(c,l)}$.

Modelling of the Flexibility of Coupler

The Rigid Finite Element Method [8] is used to discretize the coupler. The spring deformation energy and the Rayleigh function of the flexible link takes a form:

$$E_{p,fl}^{(1)} = E_{p,fl}^{(1,3)} = \frac{1}{2}\sum_{s=1}^{n_{sde}^{(1,3)}} \left(\mathbf{d}^{(1,3,s)}\right)^T \mathbf{S}^{(1,3,s)}\mathbf{d}^{(1,3,s)}, \tag{8.1}$$

$$R_{fl}^{(1)} = R_{fl}^{(1,3)} = \frac{1}{2}\sum_{s=1}^{n_{sde}^{(1,3)}} \left(\dot{\mathbf{d}}^{(1,3,s)}\right)^T \mathbf{D}^{(1,3,s)}\dot{\mathbf{d}}^{(1,3,s)}, \tag{8.2}$$

where $\mathbf{d}^{(1,3,s)} = \mathbf{q}^{(1,3,r)}$, $\mathbf{S}^{(1,3,s)}$, $\mathbf{D}^{(1,3,s)}$ are stiffness and damping matrices of sde(1,3,s).

These components are introduced to the dynamics equations as the generalized forces which can be calculated as follows:

$$\frac{\partial E_{p,fl}^{(1)}}{\partial \mathbf{q}} + \frac{\partial R_{fl}^{(1)}}{\partial \dot{\mathbf{q}}} = \mathbf{s}_{fl}^{(1)}, \tag{9}$$

$$\text{where: } s_{fl,i}^{(1)} = \begin{cases} \sum_{j=1}^{n_{sde}^{(1,3)}} \left(\frac{\partial \tilde{\mathbf{q}}^{(1,3,j)}}{\partial q_i}\right)^T \mathbf{S}^{(1,3,j)}\tilde{\mathbf{q}}^{(1,3,j)} \\ +\left(\frac{\partial \dot{\tilde{\mathbf{q}}}^{(1,3,j)}}{\partial \dot{q}_i}\right)^T \mathbf{D}^{(1,3,j)}\dot{\tilde{\mathbf{q}}}^{(1,3,j)} & \text{if } q_i \in \mathbf{q}_f^{(1,3)}, \\ 0 & \text{otherwise.} \end{cases}$$

Modelling of Contact Forces in Clearance Joint

In the presented approach, it is assumed that the clearance occurs only in the cut-joint R (Fig. 1). The spatial model of the revolute joint with clearance is proposed (Fig. 3). This model allows considering the radial and axial clearances. In order to take into account the radial clearance, the lateral surface of the journal is discretized into $n_{ce}^{(r)}$ radial contact elements (ce_r) located around the perimeter on n_r levels. In the case of the axial clearance, the frontal surface of the journal is discretized into $n_{ce}^{(a)}$ axial contact elements (ce_a) located around the perimeter on n_a levels. The contact force of $ce_\alpha(i,k)$ acting on the bearing and journal are determined as follows:

$$\mathbf{f}_{c,\alpha}^{(b,i,k)}\Big|_{\alpha\in\{r,a\}} = f_{n,\alpha}^{(i,k)}\mathbf{n}_\alpha^{(i,k)} + f_{t,\alpha}^{(i,k)}\mathbf{t}_\alpha^{(i,k)}, \tag{10.1}$$

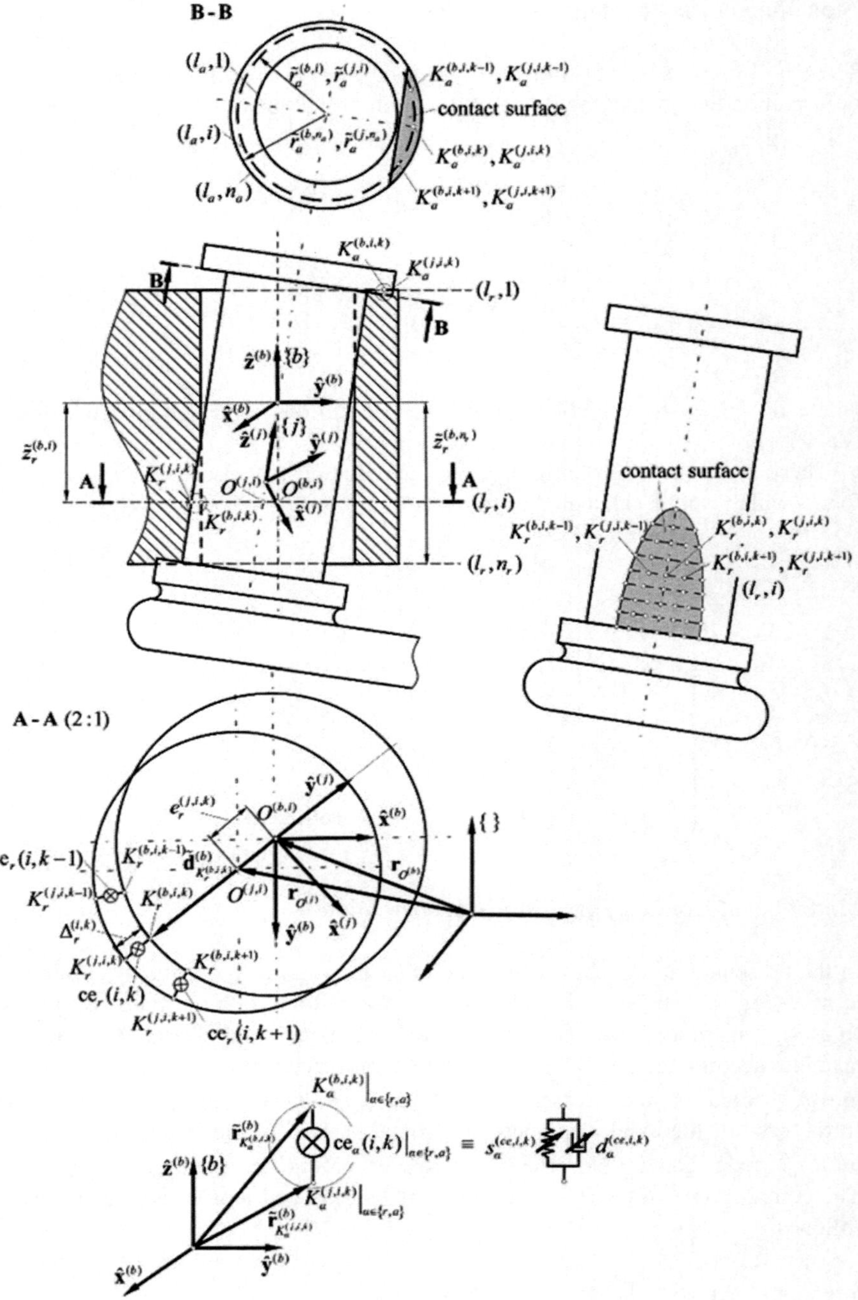

Fig. 3 Model of revolute joint with radial and axial clearance

$$\mathbf{f}_{c,\alpha}^{(j,i,k)}\Big|_{\alpha\in\{r,a\}} = -{}_{b}^{j}\mathbf{R}\mathbf{f}_{c,\alpha}^{(b,i,k)}, \tag{10.2}$$

where $f_{n,\alpha}^{(i,k)}$, $f_{n,\alpha}^{(i,k)}$ are the normal and tangent forces, $\mathbf{n}_{\alpha}^{(i,k)}$, $\mathbf{t}_{\alpha}^{(i,k)}$ are the normal and tangent unit vectors to the contact surfaces, ${}_{b}^{j}\mathbf{R}$ is the rotation matrix from bearing frame $\{b\}$ to journal frame$\{j\}$. Normal force $f_{n,\alpha}^{(i,k)}$ and tangent force $f_{t,\alpha}^{(i,k)}$ are calculated using the Lankarani-Nikravesh [9] and LuGre [10] formulas:

$$f_{n,\alpha}^{(i,k)}\Big|_{\alpha\in\{r,a\}} = s_{\alpha}^{(ce,i,k)}\Delta_{\alpha}^{(i,k)} + d_{\alpha}^{(ce,i,k)}\dot{\Delta}_{\alpha}^{(i,k)}, \tag{11.1}$$

$$f_{t,\alpha}^{(i,k)}\Big|_{\alpha\in\{r,a\}} = \left(\sigma_0 z_{\alpha}^{(i,k)} + \sigma_1 z_{\alpha}^{(i,k)} + \sigma_2 \upsilon_{t,\alpha}^{(i,k)}\right) f_{n,\alpha}^{(i,k)}, \tag{11.2}$$

where $s_{\alpha}^{(ce,i,k)}$, $d_{\alpha}^{(ce,i,k)}$ are stiffness and damping coefficients of the contact element [1, 9], $\Delta_{\alpha}^{(i,k)}$ is deformation of the contact element, $\sigma_0, \sigma_1, \sigma_2$ are stiffness, damping and viscous friction coefficients of the bristles, $z_{\alpha}^{(i,k)}$ is deformation of the bristle, $\upsilon_{t,\alpha}^{(i,k)}$ is a tangent velocity at the contact point.

The forces expressed by Eq. (10) are introduced to the dynamics equations in the form of the generalised forces as follows:

$$\begin{aligned}\mathbf{c}^{(1)} &= \sum_{i=1}^{n_r}\sum_{k=1}^{n_{ce}^{(r)}}\left(\mathbf{f}_{c,r}^{(b,i,k)}\frac{\partial \mathbf{r}_{K_r^{(b,i,k)}}}{\partial \mathbf{q}^{(1)}} + \mathbf{f}_{c,r}^{(j,i,k)}\frac{\partial \mathbf{r}_{K_r^{(j,i,k)}}}{\partial \mathbf{q}^{(1)}}\right)\\ &\quad + \sum_{i=1}^{n_a}\sum_{k=1}^{n_{ce}^{(a)}}\left(\mathbf{f}_{c,a}^{(b,i,k)}\frac{\partial \mathbf{r}_{K_a^{(b,i,k)}}}{\partial \mathbf{q}^{(1)}} + \mathbf{f}_{c,a}^{(j,i,k)}\frac{\partial \mathbf{r}_{K_a^{(j,i,k)}}}{\partial \mathbf{q}^{(1)}}\right),\end{aligned} \tag{12.1}$$

$$\mathbf{c}^{(2)} = \sum_{i=1}^{n_r}\sum_{k=1}^{n_{ce}^{(r)}}\left(\mathbf{f}_{c,r}^{(j,i,k)}\frac{\partial \mathbf{r}_{K_r^{(j,i,k)}}}{\partial \mathbf{q}^{(2)}}\right) + \sum_{i=1}^{n_a}\sum_{k=1}^{n_{ce}^{(a)}}\left(\mathbf{f}_{c,a}^{(j,i,k)}\frac{\partial \mathbf{r}_{K_a^{(j,i,k)}}}{\partial \mathbf{q}^{(2)}}\right). \tag{12.2}$$

2.3 Final Dynamics Equations

The dynamics equations of motion together with the state equations formulated for the LuGre friction model take the form:

$$\dot{\mathbf{z}} = \mathbf{LuGre}\left(t, \mathbf{v}, \mathbf{z}\right), \tag{13.1}$$

$$\begin{bmatrix}\mathbf{M}^{(1)} & \mathbf{0}\\ \mathbf{0} & \mathbf{M}^{(2)}\end{bmatrix}\begin{bmatrix}\ddot{\mathbf{q}}^{(1)}\\ \ddot{\mathbf{q}}^{(2)}\end{bmatrix} = \begin{bmatrix}-\mathbf{h}^{(1)} - \mathbf{g}^{(1)} - \mathbf{s}_{fl}^{(1)} + \mathbf{c}^{(1)}\\ -\mathbf{h}^{(2)} - \mathbf{g}^{(2)} + \mathbf{c}^{(2)}\end{bmatrix}, \tag{13.2}$$

where: $\mathbf{M}^{(c)}, \mathbf{h}^{(c)}|_{c \in \{1,2\}}$ are defined by Eq. (5), forces, $\mathbf{g}^{(c)}|_{c \in \{1,2\}}$ are described by Eq. (7), $\mathbf{s}_{fl}^{(1)}$ is defined Eq. (9), $\mathbf{c}^{(c)}|_{c \in \{1,2\}}$ are defined by Eq. (12).

3 Case Study

The presented model of the RPSUP spatial linkage is applied to analyze the influence of the crank's velocity on the dynamic response of the linkage. The geometrical and mass properties of the links are gathered in Table 1. It is assumed that the motion of crank and slider (1,2) have to change according to the assumed functions shown in Fig. 4.

It can be noted that after time 2 s the angular velocity of crank (1,1) is equal to $\dot{\varphi}_0^{(1,1)} \in \{5, 10, 12 \text{ rad s}^{-1}\}$. The motion of slider (1,2) is intermittent and after each complete cycle, there is a pause in movement, which duration time is equal to 0.5 s.

It is assumed:

Table 1 Parameters of RPSUP linkage

Parameters	link (1, 1)	link (1, 2)	link (1, 3)	link (2, 1)	link (2, 2)
$m^{(c,j)}$, kg	0.471	0.490	0.306	0.780	0.107
$l^{(c,j)}$, m	0.2	0.2	0.5	0.1	0.1
$I_x^{(c,j)}$, kgm^2	6.964×10^{-5}	2.451×10^{-5}	3.829×10^{-6}	8.126×10^{-4}	9.354×10^{-5}
$I_y^{(c,j)}$, kgm^2	6.309×10^{-3}	6.548×10^{-3}	2.553×10^{-2}	1.885×10^{-4}	8.377×10^{-6}
$I_z^{(c,j)}$, kgm^2	6.309×10^{-3}	6.548×10^{-3}	2.553×10^{-2}	1.885×10^{-4}	8.377×10^{-6}
$I_{yz}^{(c,j)}$, kgm^2	-1.080×10^{-5}	2.410×10^{-5}	0	0	0
$I_{xy}^{(c,j)}$, $I_{xz}^{(c,j)}$, kgm^2	0	0	0	0	0

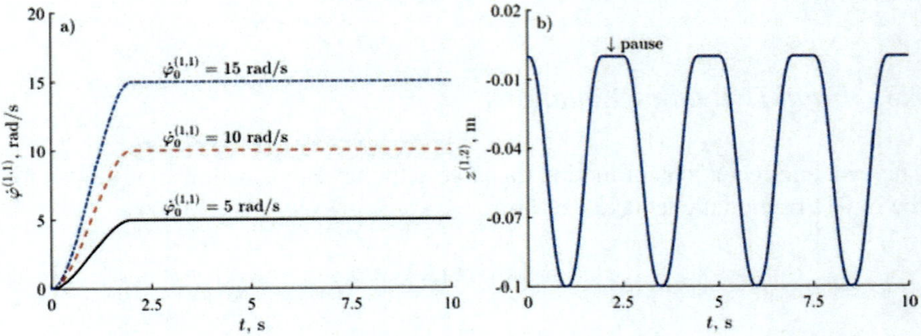

Fig. 4 The assumed time courses of the kinematic inputs: (**a**) crank (1,1), (**b**) slider (1,2)

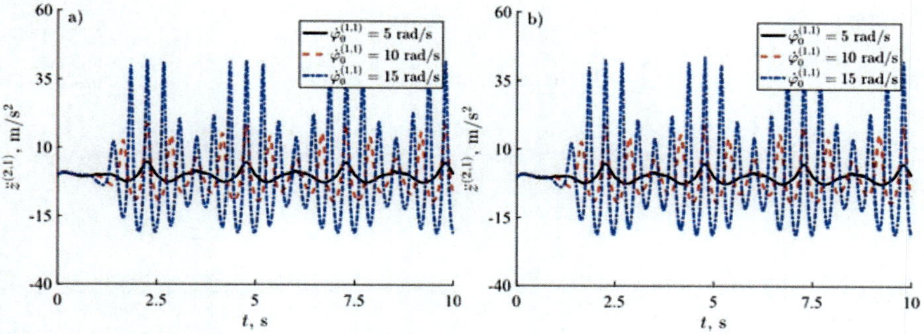

Fig. 5 Time courses of the acceleration of slider (2,1) calculated for $c_r = 0$ mm: (**a**) rigid coupler, (**b**) flexible coupler

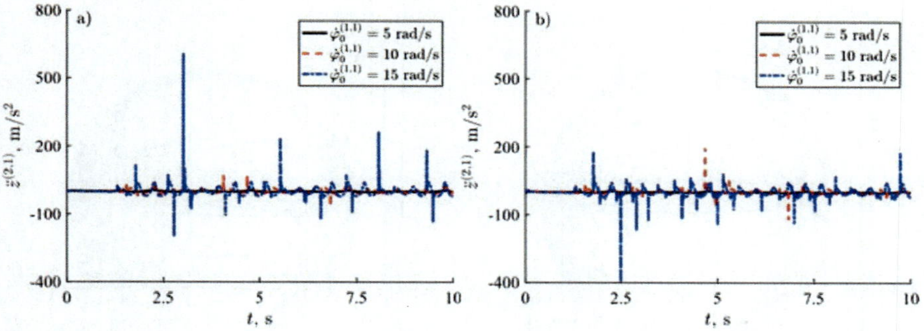

Fig. 6 Time courses of the acceleration of slider (2,1) calculated for: $c_r = 0.25$ mm (**a**) rigid coupler, (**b**) flexible coupler

- flexible link parameters: Young modulus$E = 2.1 \times 10^{11}$ Pa, Kirchhoff modulus$G = 0.8 \times 10^{11}$ Pa, number of rfes $n_{rfe}^{(1,3)} = 4$,
- clearance joint parameters: $\mu_s = 0.1$, $\mu_k = 0.2$, $\sigma_0 = 100\,\text{m}^{-1}$, $\sigma_1 = 5\ \text{s}\,\text{m}^{-1}$, $\sigma_2 = 0\ \text{s}\,\text{m}^{-1}$, $v_s = 1 \times 10^{-3}\text{m}\,\text{s}^{-1}$, restitution coefficient $k_r = 0.9$, radius of the bearing $r^{(b)} = 5 \times 10^{-3}$ m.

In simulations, it is assumed that axial clearance is equal to 0. The dynamics equations are integrated using the 4th order Runge-Kutte scheme. The Baumgarte stabilization method is applied to eliminate kinematic input constraints violations at position and velocity levels. Figure 5 shows acceleration courses of slider (2,1) obtained for the model without clearance effect. Analyzing the plots, it can be observed that the coupler's flexibility doesn't have a great impact on the motion of the linkage. As the crank velocity increases, the amplitude of the slider's acceleration becomes significantly larger. Further, the radial clearance of the revolute joint connecting coupler with link (2,1) is taken into account. Figures 6 and 7 show slider acceleration time courses obtained for the radial clearance equal

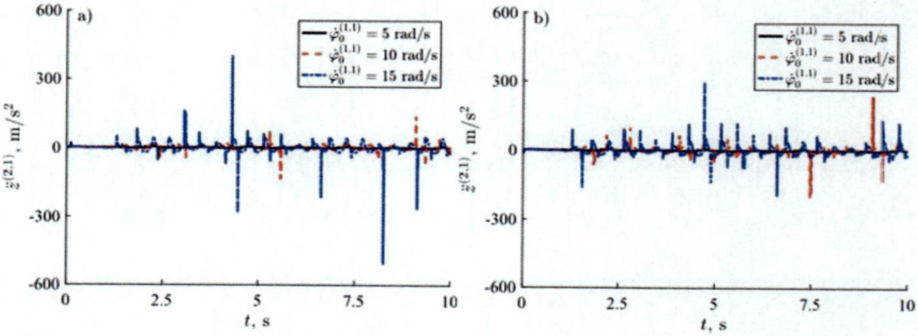

Fig. 7 Time courses of the acceleration of slider (2,1) calculated for $c_r = 0.50$ mm: **(a)** rigid coupler, **(b)** flexible coupler

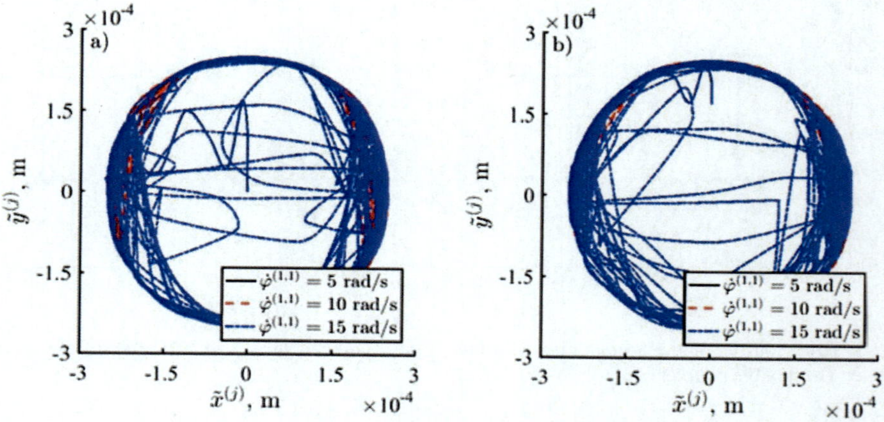

Fig. 8 The trajectory of the journal inside the bearing calculated for $c_r = 0.25$ mm: **(a)** rigid coupler, **(b)** flexible coupler

to $c_r = 0.25$ mm and $c_r = 0.5$ mm, respectively. The trajectories of the journal inside the bearing are presented in Figs. 8 and 9. Analyzing the results obtained for $c_r = 0.25$ mm it can be observed that as velocity $\dot{\varphi}_0^{(1,1)}$ increase from 10 to 15 rad s^{-1}, the maximum acceleration increases 8.5 times if the coupler is rigid and 4 times if the coupler is treated as a flexible.

In the case of $c_r = 0.5$ mm acceleration increases 3.5 times when the coupler is rigid and 1.25 times if the coupler is treated as a flexible. Additionally, values of acceleration peaks due to the collision between contacting bodies are smaller when the radial clearance equals to 0.5 mm. The presented simulation results confirm that the flexibility compensates for negative effects due to impulse forces existing in the revolute joint.

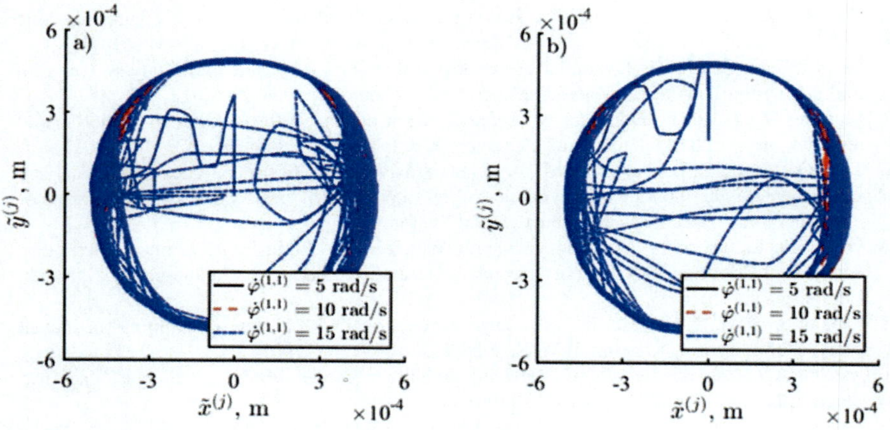

Fig. 9 The trajectory of the journal inside the bearing calculated for $c_r = 0.50$ mm: (**a**) rigid coupler, (**b**) flexible coupler

4 Concluding Remarks

The paper presents the mathematical model of the two-dof RPSUP linkage with the flexible coupler and clearance in the revolute joint. An essential feature of the presented approach is that it can be easily generalized to model dynamics of linkages with a serial open-loop kinematic structure composed of flexible links and revolute frictional joints with clearance, not only at cut-joints. The main advantage of the proposed model of the revolute joint with the clearance is that it allows us to analyze different combinations of contact between the journal and bearing. The simulation results show that the coupler's flexibility has a great impact on the motion of the linkage with the clearance joint. When the flexibility of the coupler is taken into account the acceleration of the slider is significantly smaller because some part of the energy resulting from the impulse force is transformed into the link's spring deformation.

References

1. Augustynek, K., Urbaś, A.: Analysis of the influence of the links' flexibility and clearance effects on the dynamics of the RUSP linkage. In: Kecskeméthy, A., Geu Flores, F. (eds.) Multibody Dynamics 2019. ECCOMAS 2019 Computational Methods in Applied Sciences, vol. 53, pp. 104–111. Springer, Cham (2020)
2. Flores, P., Ambrósio, J.: Revolute joints with clearance in multibody systems. Comput. Struct. **82**(17–19), 1359–1369 (2004)
3. Xiang, W., Yan, S., Wu, J.: Dynamic analysis of planar mechanical systems considering stick-slip and Stribeck effect in revolute clearance joints. Nonlinear Dyn. **95**(1), 321–341 (2019)

4. Dubowsky, S., Deck, J.F., Costello, H.: The dynamic modeling of flexible spatial machine systems with clearance connections. J. Mech. Transm. Autom. Des. **109**, 87–94 (1987)
5. Liu, C., Tian, Q., Hu, H.: Dynamics and control a spatial rigid-flexible multibody system with multiple cylindrical clearance joints. Mech. Mach. Theor. **52**, 106–129 (2012)
6. Marques, F., Isaac, F., Dourado, N., Flores, P.: An enhanced formulation to model spatial revolute joints with radial and axial clearance. Mech. Mach. Theory. **116**, 123–144 (2017)
7. Tian, Q., Flores, P., Lankarani, H.M.: A comprehensive survey of the analytical, numerical and experimental methodologies for dynamics of multibody mechanical systems with clearance or imperfect joints. Mech. Mach. Theory. **116**, 123–144 (2017)
8. Wittbrodt, E., Szczotka, M., Maczyński, A., Wojciech, S.: Rigid Finite Element Method in Analysis of Dynamics of Offshore Structures. Ocean Engineering & Oceanography. Springer, Heidelberg (2013)
9. Lankarani, H.M., Nikravesh, P.E.: A contact force model with hysteresis damping for impact analysis of multibody systems. J. Mech. Des. **112**, 369–376 (1990)
10. Åström, K.J., Canudas-de-Witt, C.: Revisiting the LuGre model. IEEE Control Syst. Mag. Inst. Electr. Electron. Mag. **28**(6), 101–114 (2008)

Application of Homogenous Transformations in the Dynamic Analysis of Truck Trailers

Andrzej Harlecki, Adam Przemyk, and Szymon Tengler

Abstract Results of an analysis of dynamics of a truck with a trailer are presented in the paper. A mathematical model of this combination of vehicles, constituting a multi-body system, was developed by using formalism of Lagrange's equations, based on the joint coordinates and homogeneous transformations taken from robotics. Within the computer simulations performed behaviour of the modelled system was studied in the road traffic conditions while performing typical road manoeuvres, and changing design parameters and load of the trailer. The developed mathematical model can be treated as a virtual prototype of the system in question. According to the authors, the proposed method can have practical significance and it can be used in designing the truck trailers. Many design proposals can be formulated on the basis of the results of the performed simulations, and they can become the basis for making real trailer prototypes.

Keywords Truck trailer · Dynamics · Lagrange's equations · Joint coordinates

1 Introduction

The subject matter of the considerations presented in this paper is a truck trailer with the central axle. This axle is placed in half the length of the trailer (see Fig. 1), what provides that at uniform distribution of loads on its surface only a small part of this load is transferred through its drawbar to the towing truck. For these reasons the vehicles with trailers having the central axle are characterized by a great ease of manoeuvring. The trailers with the central axle are used, first of all, for transporting light goods. In practice, this type of the trailer is often equipped with not one, but two or more axles.

A. Harlecki (✉) · A. Przemyk · S. Tengler
Faculty of Mechanical Engineering and Computer Science, University of Bielsko-Biala, Bielsko-Biała, Poland
e-mail: aharlecki@ath.bielsko.pl

J. Awrejcewicz (ed.), *Perspectives in Dynamical Systems I: Mechatronics and Life Sciences*, Springer Proceedings in Mathematics & Statistics 362,
https://doi.org/10.1007/978-3-030-77306-9_14

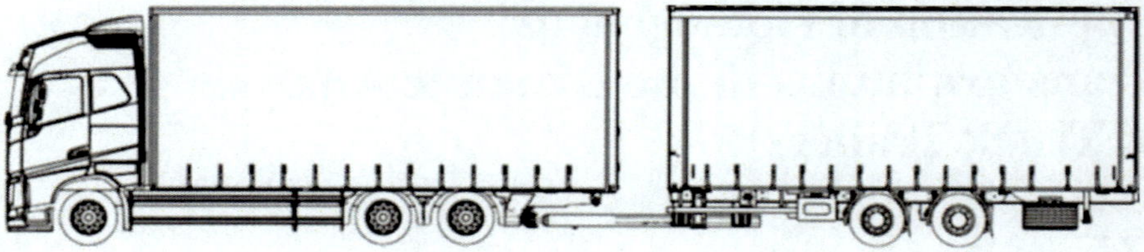

Fig. 1 The analysed combination of vehicles

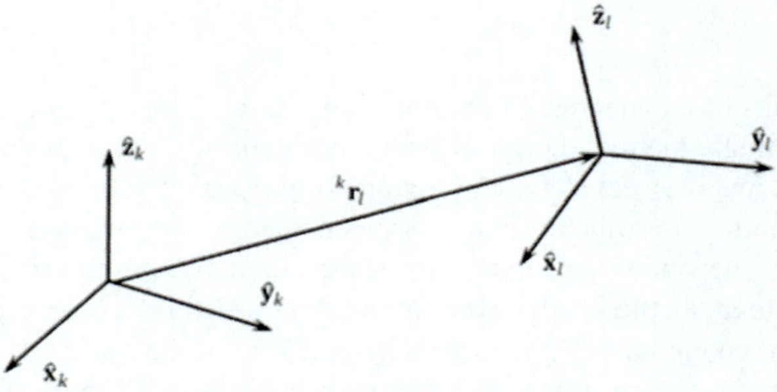

Fig. 2 A position of the coordinate systems in question

2 Mathematical Model of Combination of Vehicles

Mathematical formalism used in the analysis of dynamic systems in a form of open kinematic chains (mainly for modelling robot manipulators) is applied in the adopted procedure. It is based on use of joint (relative) coordinates – as generalised coordinates and homogenous transformations [1, 5].

In the case of two adjacent Cartesian coordinate systems $\hat{\mathbf{x}}_k\hat{\mathbf{y}}_k\hat{\mathbf{z}}_k$ and $\hat{\mathbf{x}}_l\hat{\mathbf{y}}_l\hat{\mathbf{z}}_l$ (see Fig. 2), made by versors, the matrix of the transformation from system $\hat{\mathbf{x}}_l\hat{\mathbf{y}}_l\hat{\mathbf{z}}_l$ to system $\hat{\mathbf{x}}_k\hat{\mathbf{y}}_k\hat{\mathbf{z}}_k$ has a form:

$$ {}_l^k\mathbf{T} = \begin{bmatrix} {}_l^k\mathbf{R} & {}^k\mathbf{r}_l \\ 0\ \ 0\ \ 0 & 1 \end{bmatrix}. \tag{1} $$

The position vector ${}^k\mathbf{r}_l$ determines the position of origin of coordinate system $\hat{\mathbf{x}}_l\hat{\mathbf{y}}_l\hat{\mathbf{z}}_l$ in system $\hat{\mathbf{x}}_k\hat{\mathbf{y}}_k\hat{\mathbf{z}}_k$.

Elements of the rotation matrix from coordinate system $\hat{\mathbf{x}}_l\hat{\mathbf{y}}_l\hat{\mathbf{z}}_l$ to system $\hat{\mathbf{x}}_k\hat{\mathbf{y}}_k\hat{\mathbf{z}}_k$ of a general form:

$$ {}_l^k\mathbf{R} = \begin{bmatrix} \hat{\mathbf{x}}_l\hat{\mathbf{x}}_k & \hat{\mathbf{y}}_l\hat{\mathbf{x}}_k & \hat{\mathbf{z}}_l\hat{\mathbf{x}}_k \\ \hat{\mathbf{x}}_l\hat{\mathbf{y}}_k & \hat{\mathbf{y}}_l\hat{\mathbf{y}}_k & \hat{\mathbf{z}}_l\hat{\mathbf{y}}_k \\ \hat{\mathbf{x}}_l\hat{\mathbf{z}}_k & \hat{\mathbf{y}}_l\hat{\mathbf{z}}_k & \hat{\mathbf{z}}_l\hat{\mathbf{z}}_k \end{bmatrix} \tag{2} $$

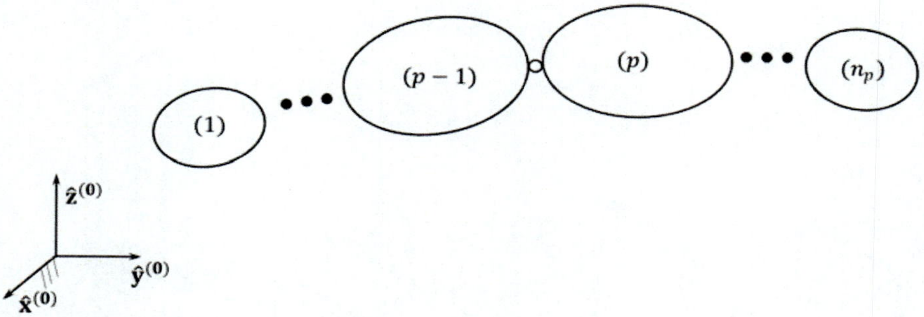

Fig. 3 The multi-body system as the open kinematic chain

constitute direction cosines expressed in a form of scalar products of appropriate versors.

Orientation of the versors of system $\hat{\mathbf{x}}_l\hat{\mathbf{y}}_l\hat{\mathbf{z}}_l$ in relation to the versors of system $\hat{\mathbf{x}}_k\hat{\mathbf{y}}_k\hat{\mathbf{z}}_k$ can be determined by Euler's angles ψ, θ, φ of type $\hat{\mathbf{z}}\ \hat{\mathbf{y}}\ \hat{\mathbf{x}}$ [1].

In the case of the multi-body systems considered in the form of open kinematic chains (see Fig. 3) the vector of generalised coordinates of body (p) can be presented in the following form:

$$\mathbf{q}^{(p)} = \begin{bmatrix} \mathbf{q}^{(p-1)} \\ \overset{\sim(p)}{\mathbf{q}} \end{bmatrix}, \tag{3}$$

where:

$\mathbf{q}^{(p-1)}$ – the vector of global generalised coordinates describing the motion of body (p-1) preceding body (p),

$\overset{\sim(p)}{\mathbf{q}} = \left[q_1^{(p)}, \ldots, q_{\tilde{n}_p}^{(p)}\right]^{\mathrm{T}}$ – the vector of $\tilde{n}_p$ local generalised coordinates defining the relative motion of body (p) in relation to body (p-1).

The multi-body system modelling the combination of vehicles consists main sub-assemblies of the truck (see Fig. 4) and the trailer (see Fig. 5), which were considered as rigid bodies. The rear wheels of the truck modelled in a form of bodies (12L) and (12P) are driven by torques $\boldsymbol{\tau}^{(12L)}$ and $\boldsymbol{\tau}^{(12P)}$. The system includes selected spring and damping elements (not shown in the figures) occurring in the vehicles modelled.

The multi-body system in question is an open kinematic chain of a branched structure (a tree-structure) – see Fig. 6. The model of the truck is assumed as the root of this chain. The end parts of the chain in a form of models of vehicle wheels are loaded by road reactions (not shown in the figures).

As stated in the paper [6], in the case of the multi-body system considered in a form of the tree-structure open kinematic chain (see Fig. 7) motion of body (p) depends on the global generalised coordinates of its predecessor in the chain, called base body (u_p), and its local coordinates in relation to this body.

Vector of $N^{(p)}$ generalised coordinates of body (p) can be written as follows:

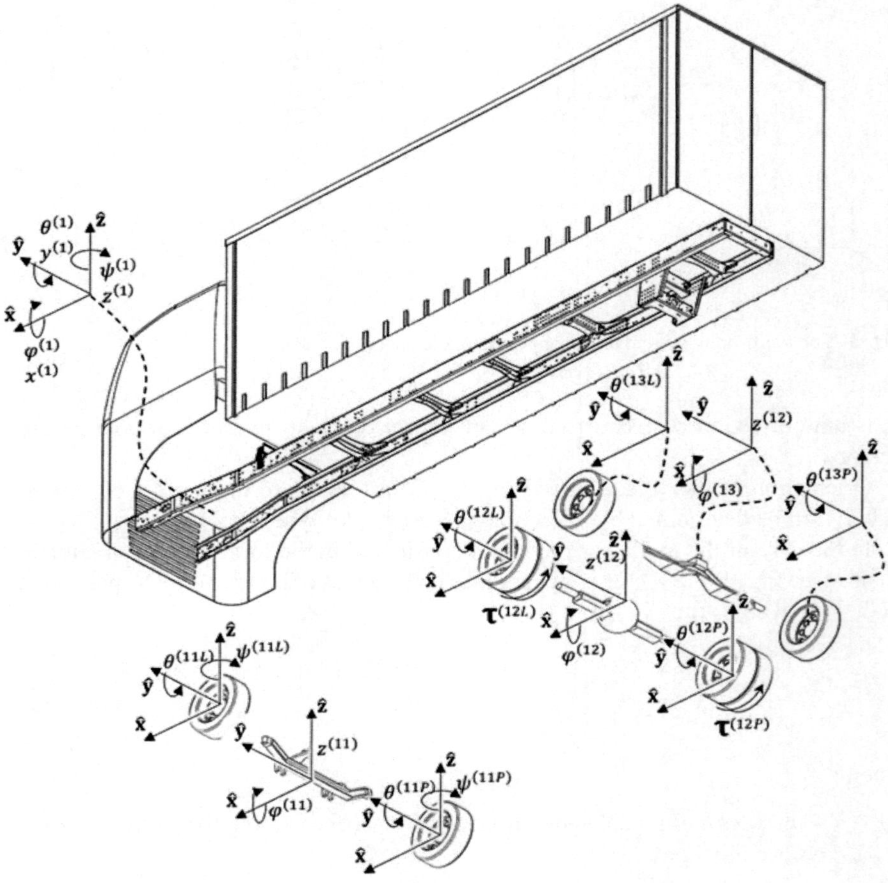

Fig. 4 The truck model

$$\mathbf{q}^{(p)} = \left[\mathbf{q}^{(u_p)\mathrm{T}}, \tilde{\mathbf{q}}^{(p,u_p)\mathrm{T}}\right]^{\mathrm{T}} = \left(q_i^{(p)}\right), \text{ for } i = 1, \ldots, N^{(p)}, \tag{4}$$

where:

$\mathbf{q}^{(u_p)} = \left(q_i^{(u_p)}\right)$, for $i = 1, \ldots, N^{(u_p)}$ – vector of $N^{(u_p)}$ global generalised coordinates of base body (u_p),

$\tilde{\mathbf{q}}^{(p,u_p)} = \left(\tilde{q}_i^{(p,u_p)}\right)$, for $i = 1, \ldots, \tilde{N}^{(p,u_p)}$ – vector of $\tilde{N}^{(p,u_p)}$ local generalised coordiantes of body (p) determining its motion in relation to base body (u_p).

The following relationship is true:

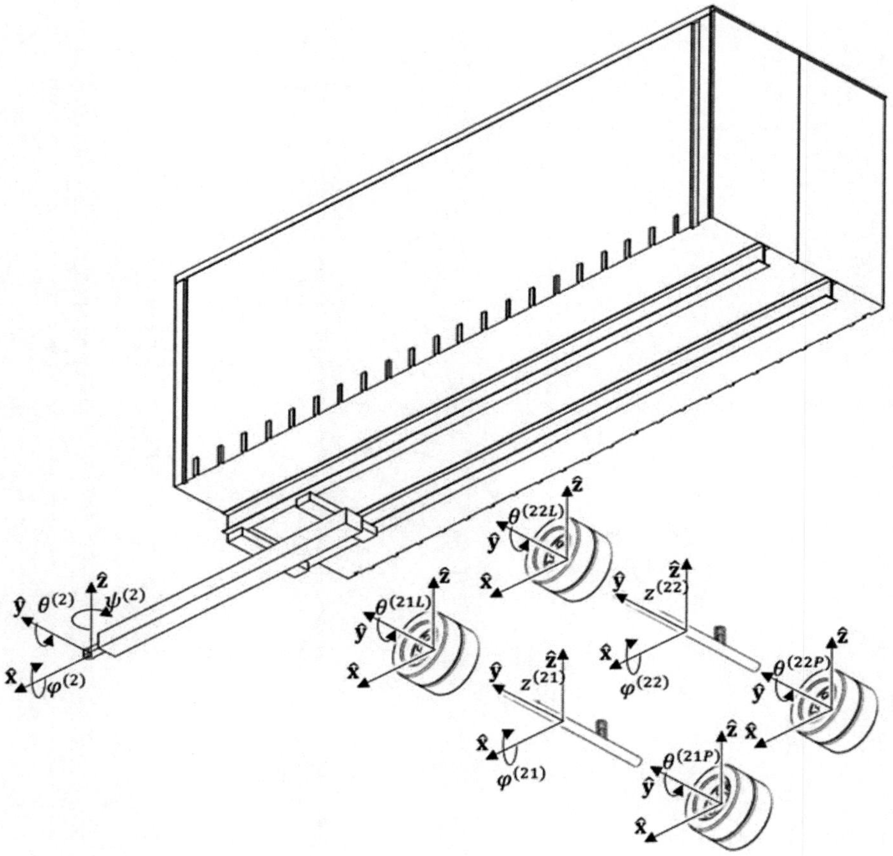

Fig. 5 The trailer model

$$N^{(p)} = N^{(u_p)} + \tilde{N}^{(p,u_p)}. \tag{5}$$

Following further the guidelines provided in the work quoted, for each body of the tree-structure chain in question an ordered set containing numbers of the bodies preceding it including its own number (p) is defined:

$$M^{(p)} = \left\{ m_1^{(p)}, \ldots, m_i^{(p)}, \ldots, m_{s^{(p)}}^{(p)} \right\}, \tag{6}$$

where:

$s^{(p)}$ – a number of bodies preceding body (p) including it.
Vector $\mathbf{q}^{(p)}$ can be presented now in the following form:

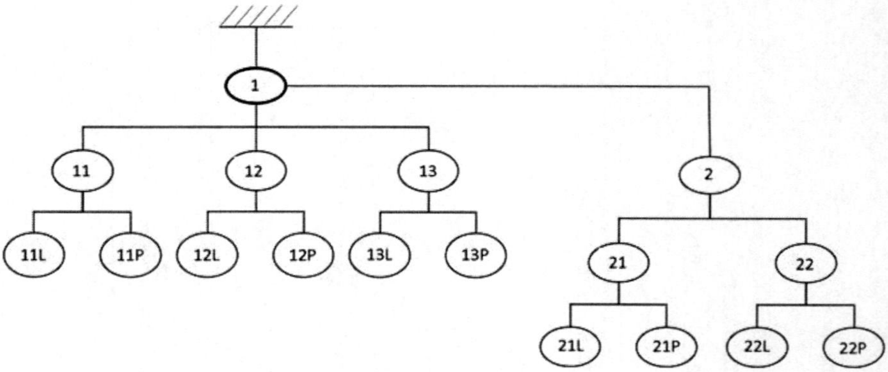

Fig. 6 The tree-structure of the multi-body system considered

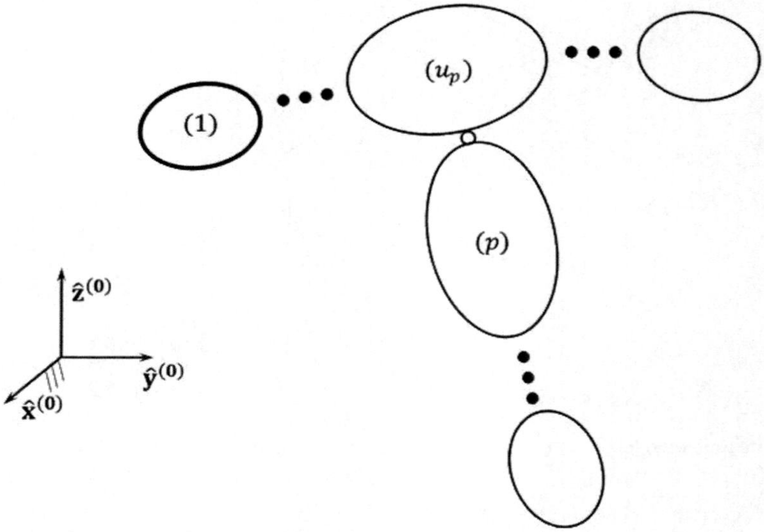

Fig. 7 The multi-body system as the tree-structure open kinematic chain

$$\mathbf{q}^{(p)} = \left[\tilde{\mathbf{q}}^{\left(m_1^{(p)}\right)^{\mathrm{T}}}, \ldots, \tilde{\mathbf{q}}^{\left(m_i^{(p)}\right)^{\mathrm{T}}}, \ldots, \tilde{\mathbf{q}}^{\left(m_{s(p)}^{(p)}\right)^{\mathrm{T}}} \right]^{\mathrm{T}}. \tag{7}$$

The model of the truck, i.e. body (1), has six degrees of freedom (three displacements and three rotations), which can be presented by a vector of generalised coordinates: $\tilde{\mathbf{q}}^{(1)} = \left[\tilde{x}^{(1)}, \tilde{y}^{(1)}, \tilde{z}^{(1)}, \tilde{\psi}^{(1)}, \tilde{\theta}^{(1)}, \tilde{\varphi}^{(1)} \right]^{\mathrm{T}}$. The next sub-assemblies connected with the model of the truck are three axles modeled as bodies (11), (12), (13) of two degrees of freedom, thus their vectors of generalised coordinates have a form of: $\overset{\sim(k)}{\mathbf{q}} = \left[\overset{\sim(k)}{z}, \overset{\sim(k)}{\varphi}\right]^{\mathrm{T}}$, where $k = 11, 12, 13$. Each axle is equipped with two wheels as bodies (11 L), (11P), (12 L), (12P), (13 L), (13P) of which the vectors of generalised coordinates have a form of: $\overset{\sim(k)}{\mathbf{q}} = \left[\overset{\sim(k)}{\psi}, \overset{\sim(k)}{\theta}\right]^{\mathrm{T}}$, where $k =$ 11L, 11P (for the front steering wheels) and $\overset{\sim(k)}{\mathbf{q}} = \left[\overset{\sim(k)}{\theta}\right]^{\mathrm{T}}$, where $k =$ 12L, 12P, 13L, 14P (for the wheels of the other axles). The drawbar of the model of the trailer, being its integral part, is joined with the model of the truck by a spherical joint. Therefore, the model of the trailer, i.e. body (2), has three degrees of freedom (rotations) in relation to the model of the truck, thus its vector of generalised coordinates has a form of: $\overset{\sim(2)}{\mathbf{q}} = \left[\overset{\sim(2)}{\psi}, \overset{\sim(2)}{\theta}, \overset{\sim(2)}{\varphi}\right]^{\mathrm{T}}$. Two axles modelled as bodies (21), (22) are joined with the model of the trailer and their vectors of generalised coordinates have a form of: $\overset{\sim(k)}{\mathbf{q}} = \left[\overset{\sim(k)}{z}, \overset{\sim(k)}{\varphi}\right]^{\mathrm{T}}$, where k = 21, 22. Two wheels as bodies (21 L), (21P), (22 L), (22P) of which the vectors of generalised coordinates have form of: $\overset{\sim(k)}{\mathbf{q}} = \left[\overset{\sim(k)}{\theta}\right]^{\mathrm{T}}$, where $k =$ 21L, 21P, 22L, 22P, are attached to each axle. The vector of generalised coordinates of the entire multi-body system considered has a form of: $\mathbf{q} = \left[\overset{\sim(1)\mathrm{T}}{\mathbf{q}}, \overset{\sim(11)\mathrm{T}}{\mathbf{q}}, \overset{\sim(12)\mathrm{T}}{\mathbf{q}}, \overset{\sim(13)\mathrm{T}}{\mathbf{q}}, \overset{\sim(11L)\mathrm{T}}{\mathbf{q}}, \overset{\sim(11P)\mathrm{T}}{\mathbf{q}}, \overset{\sim(12L)\mathrm{T}}{\mathbf{q}}, \overset{\sim(12P)\mathrm{T}}{\mathbf{q}}, \overset{\sim(13L)\mathrm{T}}{\mathbf{q}}, \overset{\sim(13P)\mathrm{T}}{\mathbf{q}}, \overset{\sim(2)\mathrm{T}}{\mathbf{q}}, \overset{\sim(21)\mathrm{T}}{\mathbf{q}}, \overset{\sim(22)\mathrm{T}}{\mathbf{q}}, \overset{\sim(21L)\mathrm{T}}{\mathbf{q}}, \overset{\sim(21P)\mathrm{T}}{\mathbf{q}}, \overset{\sim(22L)\mathrm{T}}{\mathbf{q}}, \overset{\sim(22P)\mathrm{T}}{\mathbf{q}}\right]$. This vector has 31 elements what is the number of degrees of freedom of the system in question.

Based on Lagrange's equations and using the algorithms presented in the monograph [4], equations of motion of the multi-body system in question can be formulated as:

$$\mathbf{A}\ddot{\mathbf{q}} = \mathbf{f}, \tag{8}$$

where:

A – mass matrix,
f – vector of external (generalised), Coriolis and centrifugal forces.

In the adopted method a Dugoff-Uffelmann tire model [2] was used to present an interaction of the road on wheels of the vehicles. Detailed algorithms of this model are also presented in the monograph [3]. Values of forces acting on the wheels and lying in the road plane and value of the aligning torque were determined by using these algorithms. These values were dependent on the values of deformations of the tires in the normal direction to the road. The reaction values determined in such a way were included in the equations of motion of the vehicle combination model in a form of generalised forces, using the algorithms given in the monograph [3].

It is assumed that generalised coordinates referring to steering angles $\tilde{\psi}^{(11L)}$ and $\tilde{\psi}^{(11P)}$ of the truck front wheels, modelled in a form of bodies (11 L) and (11P), are the given time function. Therefore, the equations corresponding to them should be removed from the system of the equations of motion. This system should be completed with two equations of constraints which are twice differentiated with respect to time. Consequently, the dynamic equations of motion of the multi-body system considered can be written in the matrix form as:

$$\begin{cases} \mathbf{A}\ddot{\mathbf{q}} - \mathbf{\Phi}\mathbf{r} = \mathbf{f} \\ \mathbf{\Phi}^{\mathbf{T}}\ddot{\mathbf{q}} = \mathbf{w} \end{cases}, \tag{9}$$

where:

$\mathbf{\Phi}$ – constraint matrix,
r – vector of unknown constraint reactions corresponding to torques acting on front wheels of truck,
w – vector of right sides of constraint equations.

3 Computer Simulations and Results

The selected results of the numerical calculations were verified experimentally. For this purpose the real vehicle combination was equipped with motion sensors which allowed determining time courses of its selected kinematic parameters while performing typical road manoeuvres. The experimental courses were compared with the computing courses, obtaining their good qualitative and quantitative compliance. While performing numerical simulations the values of the main geometrical and physical parameters of the trailer were changed, and their influence on stability of motion of the vehicle combination was analysed. The position of its centre of gravity was changed as a result of including a variable load distribution in the trailer. Figure 8 presents one of the cases investigated, namely driving of the towed trailer over a speed bump. The motion sensors were attached to the trailer in points A and B (see Fig. 9).

In the following figures some examples of numerical calculations compared with the experimental results are presented. They relate to the time courses of vertical accelerations of the trailer frame points in which the sensors were mounted – A (see Fig. 10a) and B (see Fig. 10b). The first observed increase of the acceleration value (point A) occurred when the rear wheels of the truck were driving over the speed bump, and the second one (point B) when the wheels of the trailer were driving over the obstacle. Sufficient compliance of the computing results and measurements can be observed.

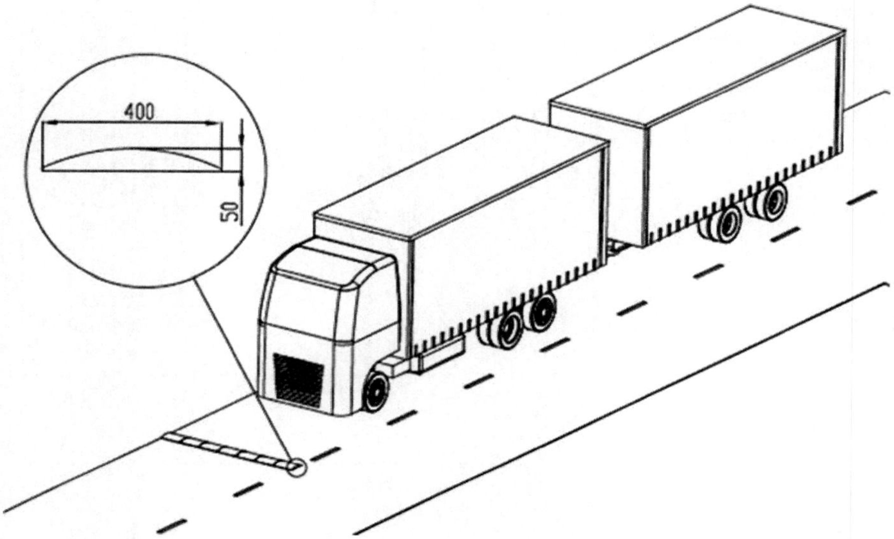

Fig. 8 Driving over a speed bump

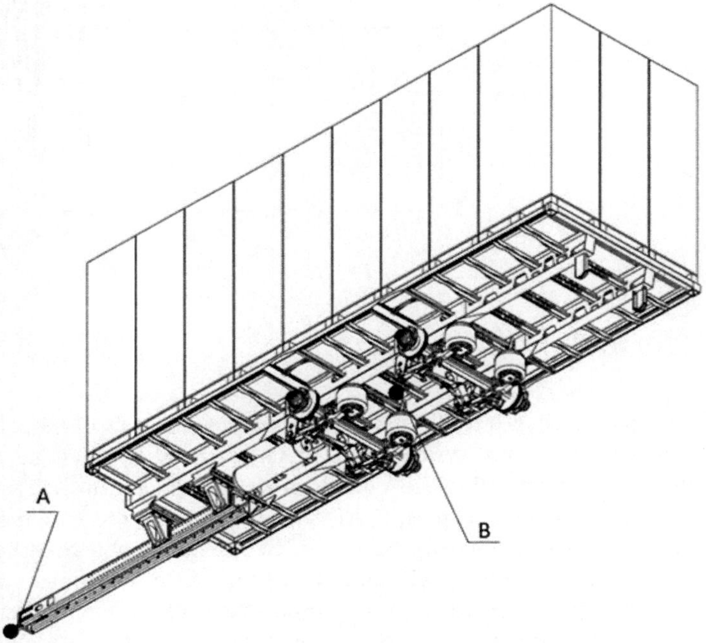

Fig. 9 Placement of the sensors

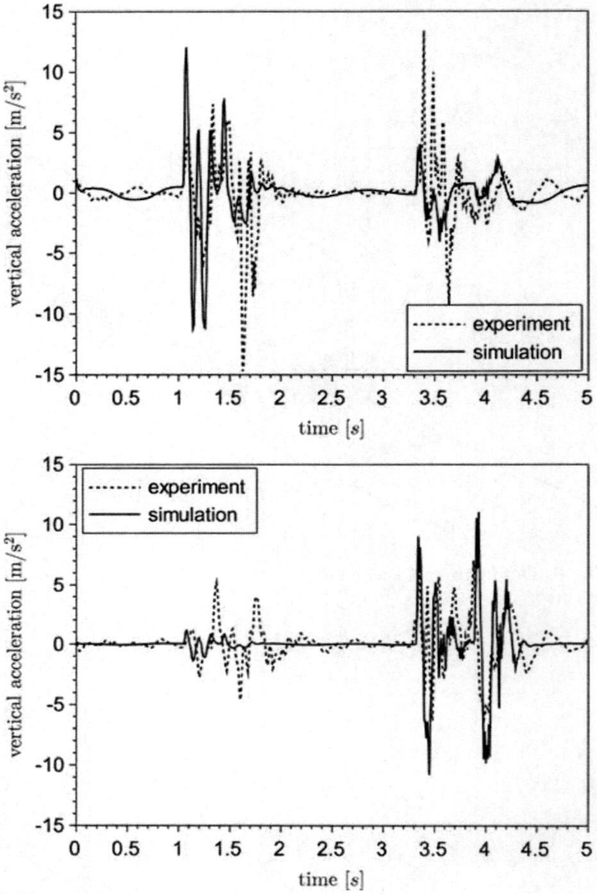

Fig. 10 Vertical accelerations of points A and B – the comparison of simulation and experimental results: (**a**) the course for point A, (**b**) the course for point B

4 Conclusions

The conclusions obtained as a result of a quick numerical analysis of the relatively simple considered multi-body systems can be used as initial guidelines in the analysis of the advanced model of this combinations of vehicles, made by authors of this paper by use of commercial MSC ADAMS program. In this case the frame of the trailer of complicated geometry was modelled as a deformable body by using the finite element method. The developed advanced model of the trailer could be treated as its virtual prototype which can be used effectively in the process of its design. It became basis to build a real prototype of the trailer of the structure near to its final version to be implemented into the production. The procedure presented enabled to shorten significantly design time.

References

1. Craig, J.J.: Introduction to Robotics. Mechanics and Control. Pearson Education, Inc (2005)
2. Dugoff, H., Fancher, P.S., Segel, L.: An analysis of tire traction properties and their influence on vehicle dynamics performance. SAE Technical Paper 700377
3. Grzegożek, W., Adamiec-Wójcik, I., Wojciech, S.: Computer Modeling of Dynamics of Motor Vehicles, (in Polish). Publisher of Cracow University of Technology, Cracow (2003)
4. Jurevič, E.I. (ed.): Dynamics of robot control, (in Russian). "Nauka", Moscow (1984)
5. Paul, R.P.: Robot Manipulators: Mathematics, Programming, and Control. MIT Press, Cambridge (1981)
6. Warwas, K.: Analysis and control of motion of multibody vehicles taking into account flexibility of links, (in Polish). Ph.D. thesis, Faculty of Mechanical Engineering and Computer Science, University of Bielsko-Biała, Poland (2007)

Improving Capabilities of Constitutive Modeling of Shape Memory Alloys for Solving Dynamic Problems Via Application of Neural Networks

Adam Martowicz, Mikołaj Żabiński, Jakub Bryła, and Jakub Roemer

Abstract The paper addresses an issue of improving capabilities of the constitutive models elaborated for shape memory alloys (SMA) to solve dynamic problems. Artificial neural networks (ANN) are utilized to simulate the experimentally identified complex behavior of the mentioned type of smart materials. Although SMA are known and widely used in various engineering applications for many decades, both understanding and, therefore, modeling of their physical behavior suffer continuous limitations regarding accuracy and performance. The present work reports the results of the properties assessment carried out for the proposed ANN based constitutive model for SMA. As presented, the application of ANN allows to reliably model the hysteretic character of the stress-strain relationship observed by the authors for the experimentally tested SMA material — a wire made of Nitinol. The work is complemented with the results of a study on the influence of an ANN structure and training method on the quality of numerical results. The combined ANN-finite element method code is used to provide solutions for the given dynamic problems. Finally, improvement perspectives regarding SMA constitutive modeling are discussed making a reference to the identified capabilities of the ANN based material model.

Keywords Constitutive model · Shape memory alloy · Smart material · Superelasticity · Artificial neural network · Numerical simulation · Model validation

A. Martowicz (✉) · M. Żabiński · J. Bryła · J. Roemer
Department of Robotics and Mechatronics, AGH University of Science and Technology, Krakow, Poland
e-mail: adam.martowicz@agh.edu.pl

J. Awrejcewicz (ed.), *Perspectives in Dynamical Systems I: Mechatronics and Life Sciences*, Springer Proceedings in Mathematics & Statistics 362,
https://doi.org/10.1007/978-3-030-77306-9_15

1 Introduction

Modern engineering requires applications of the materials which provide advantageous capabilities. Applied materials should exhibit demanded constructional features, by assuring necessary load strength, and, possibly, functional properties, which stand for additional extraordinary characteristics as in the case of smart materials [1–4]. Simultaneously, a continuous demand for new, more efficient and accurate computational tools that, first, may aid the designing process for the above mentioned types of materials, and second, help to successfully bring new products into the market, is observed as well [5–8]. New numerical approaches are made available to better the description of the material behavior. Consequently, even the already known types of materials may still extend their application areas and, based on the innovation provided, new technical solutions are continuously presented. The newly launched products can be more reliable, lighter and compacted due to effective use of the properties of the structural parts.

Smart materials are of a particular interest due to their capabilities of acting as both actuators and sensors in various engineering applications [2]. They allow for the measurements of many physical quantities and can efficiently modify the structural properties of the designed constructions. Hence, the simulation methods as well as the respective computational tools dedicated for smart materials are especially required. New applications of smart materials considerably depend on the quality of the nowadays provided design methodologies and computational approaches.

In the present work the authors focus on the development of a new numerical tool used for modeling the behavior of shape memory alloys (SMA) [1]. These unique smart materials exhibit interesting properties of memorizing the geometric shapes and withstand high strains when comparing with other metallic materials. Extraordinary characteristics of SMA are allowed due to the changes of their crystal structure occurred after mechanical and thermal loads are provided [9]. Specifically, the reversible martensitic transition enables large macroscale deformations. Solid phase transitions, i.e., the two-way transitions carried out between austenite and martensite phases, respectively result in the one-way and two-way memory effects and the phenomenon of superelasticity.

Advantageous properties of SMA motivate many practical applications of this type of smart materials, even though the manufacturing processes are costly and teaching of geometric shapes is quite complicated. SMA are used as constructional parts, to significantly modify the structural properties, and sensors and actuator in the field of automation and robotics. The main application areas of SMA are: medical staples, stents, clamps and other chirurgical tools, orthodontic wires, actuators and various control components, e.g., used in aircrafts to change the geometry of wings for better air flow. SMA are also used in valves and gears as the components that control oil lubrication process. Temperature activation of SMA leads to their considerable change of the geometric shape, therefore providing the source of mechanical force. SMA can considerably reduce mechanical vibrations due to the hysteretic character of their behavior. It should be mentioned that this capability is of the authors' interest considering the application of SMA to control the operational parameters of gas foil bearings [10]. Moreover, the hysteresis observed for the constitutive relationship allows for construction of the SMA actuators or springs, which are capable of generation almost constant mechanical load, irrespectively form the strain level. Finally, a lack of movable parts in the SMA based actuators means that they can be exploited in very demanding conditions and in risky environments, e.g., in the objects localized in the space to control the motion of their structural parts.

Considering the wide application area of SMA, the authors of the present work, being motivated by the existing deficiency regarding the properties of the available modeling tools, decided to present preliminary numerical results for the application of artificial neural networks (ANN) to model constitutive relation of SMA and discuss the quality of the investigated approach. The phenomenon of superelasticity is of the authors' particular concern.

The work consists of the following parts. After present introductory Sect. 1, where the authors' overall motivation is formulated, Section 2 provides a more detailed description of the studied phenomenon of superelasticity, followed by the results of the experimental tests gathered in Sect. 3. The investigated numerical approach for modeling SMA via ANN is described in Sect. 4. The results of numerical simulations and experimental validation are reported in Sect. 5. The work is summarized in Sect. 6, where the final conclusions and the directions for the future works are also presented.

2 Superelasticity in SMA

Superelasticity is one of the advantageous effects observed in SMA. It stands for large reversible elastic deformations, characterized by the strains up to 8%. As visualized in Fig. 1, the macroscale deformation reflects the nanoscale changes of solid phases (austenite and martensite) occurred when a mechanical load is applied. The respective hysteretic constitutive characteristics is presented in Fig. 2.

As shown in Fig. 1, the stresses at which the phase transistors are observed change with the temperature. Complementarily, it may be stated that the characteristic temperatures at which the phases swap, i.e., A_s, A_f, M_s, and M_f depend on

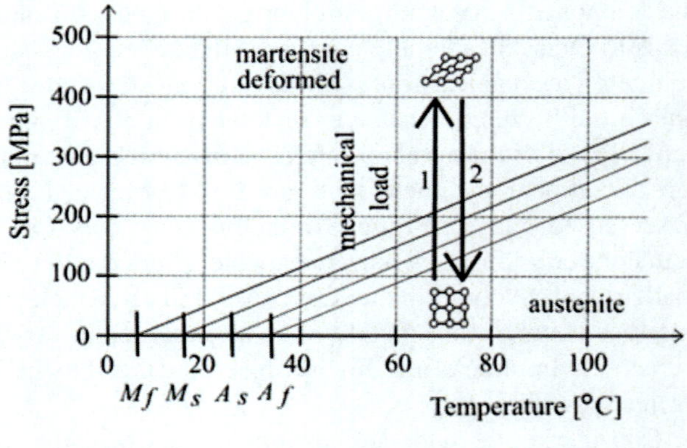

Fig. 1 Effect of superelasticity in SMA – characteristic temperatures for phase transitions [11]

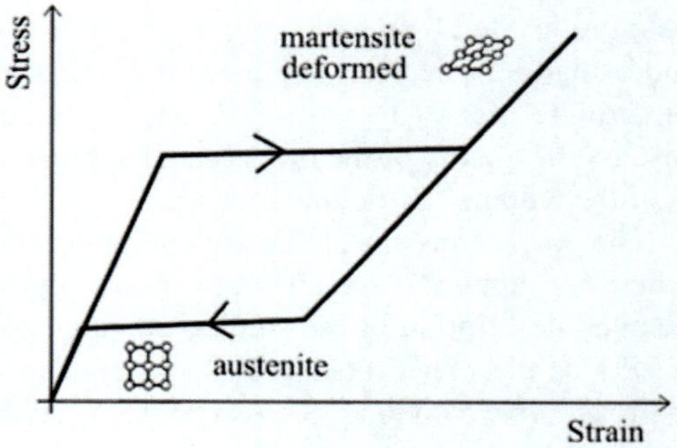

Fig. 2 Effect of superelasticity in SMA – hysteretic constitutive relationship [11]

the stress (the subscripts 's' and 't' respectively denote start and finish of the phase change). An exemplary simulated force-displacement relationship obtained by the authors using the software MSC.Software/Marc is presented in Fig. 3.

As already mentioned, the effect of superelasticity is entirely reversible unless the maximal allowed stresses are exceeded. The area covered be the hysteresis loop corresponds to the amount of energy which is dissipated in an SMA component when being mechanically loaded and unloaded.

Modeling the superelasticity is a challenging task due to complicated character of the involved physical phenomena [12–15]. Basically, the observations carried out at macroscale deal with the resultant SMA behavior being a projection of the changes that occur in the crystal structure. A very popular and efficient approach to handle the behavior of SMA is the use of phenomenological models, which stand for an acceptable compromise between the complexity of the formulas used and the accuracy of the results [1, 8]. The approach, which is presented in Sect. 4, also provides a convenient method of modeling the behavior of SMA.

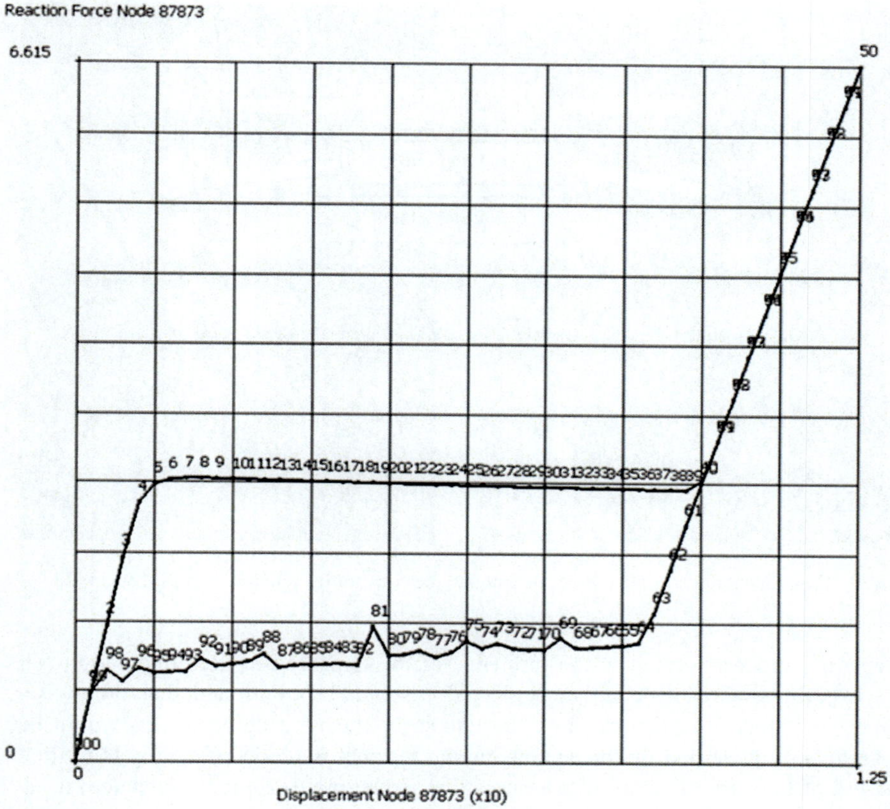

Fig. 3 Effect of superelasticity observed in an SMA finite element (FE) model. Consecutive points of the characteristics are numbered

3 Experimental Tests

Experimental tests have been carried out by the authors to provide exemplary results required for validation of the developed ANN-FE model. Fatigue testing machine Instron 8872 has been used to extract the elastic properties for an SMA specimen. Figure 4 shows the experimental test stand.

A dedicated clamps have been used during experiments to prevent from additional stress concentrations in SMA, which would lead to the undesired effects in the tested wire, i.e., initiation of the phase transitions untimely. In the used test stand, the SMA wire compasses the purposely mounted discs instead of being caught by typical jaws. The stretching speed for the wire has been assumed according to the guides presented in the work [1] and based on the authors' previous research [11]. Specifically, slow stretching and relaxation at the elongation rate of 0.005 mm/s have been carried out to assure near-isothermal character of the phase

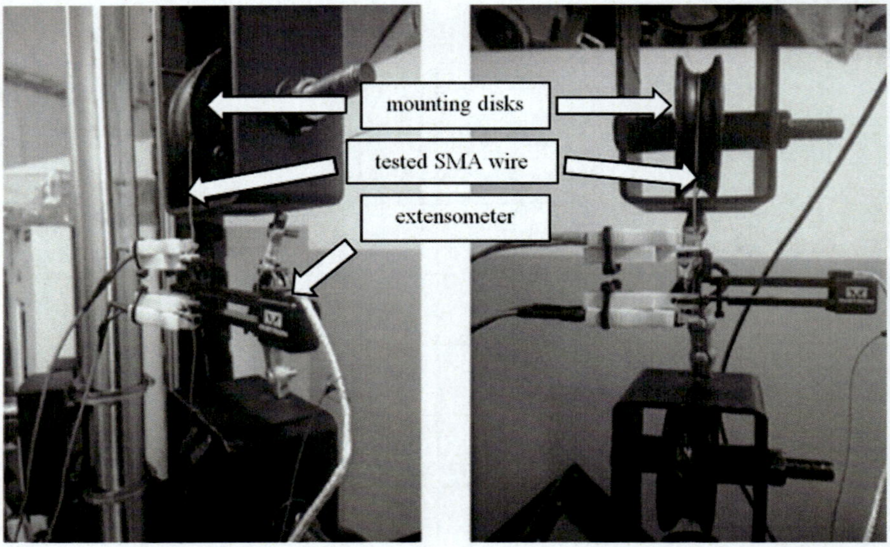

Fig. 4 Experimental test stand used to investigate the constitutive relation of an SMA sample

transition processes. Hence, the influence of the temperature field fluctuations on the experimental results could be neglected. It should be mentioned, that, otherwise, kinetics of phase transitions (i.e., the course of the wire deformation) would be significantly governed by the accompanying thermal issues. Sufficiently long time period of the experimental tests assures that isothermal (or, precisely stated, near-isothermal) conditions are satisfied. The experimental results have been obtained for various variants of the stress-strain hysteresis, i.e., for the total elongations: 2.5 mm, 5 mm, 26 mm, 36 mm and 43 mm. These results are visualized in Fig. 5.

The performed tests allow to deliver the data for experimental model validation, as reported in Sect. 5, including the learning process of ANN.

4 Application of ANN to Constitutive Modeling of SMA

Below, the concept of an application of ANN to model the constitutive relation of SMA is discussed. Principally, the ANN is used to provide the properties of the modeled material for the FE code while its execution for static and dynamic analyses for the modeled mechanical structure.

Before its usage, ANN is subject to the learning process based on the data collected in experiments [16]. The ANN learns the specificity of the constitutive relations based on the list of the input parameters. While learning, ANN cyclically makes use of the previous data generated by the FE code to determine the elastic moduli of the SMA material. Iteratively, within the consecutive simulations steps,

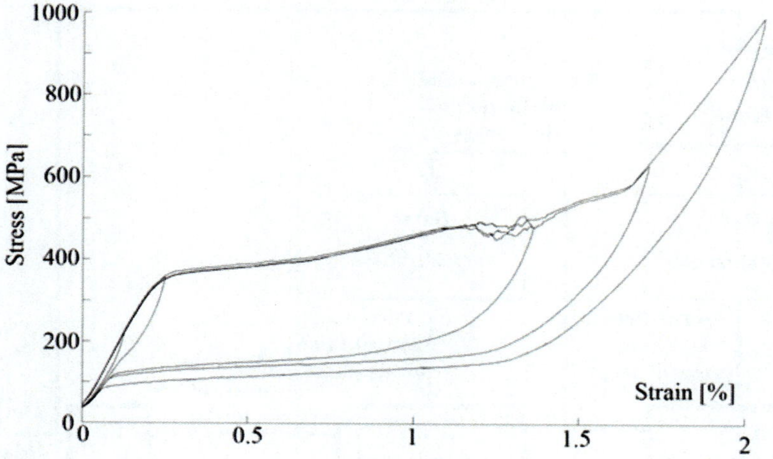

Fig. 5 Experimentally identified stress-strain relationships for the tested SMA wire. Various variants of the hysteretic constitutive relation have been obtained when changing the maximum value of the total elongation

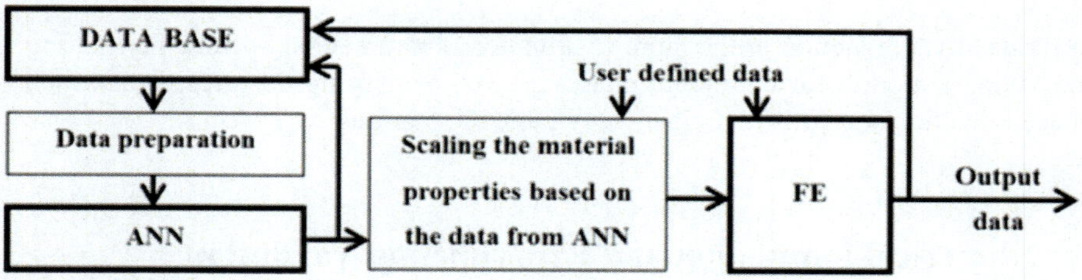

Fig. 6 Computational workflow for experimental validation of the ANN-FE model

the ANN updates its parameters. When the acceptable results' convergence is obtained, the complete ANN-FE model undergoes final experimental validation, during which the simulated and experimentally found stress-strain curves are compared.

Figure 6 presents the scheme of the computational framework used for model validation. During simulations, which are performed after validation, the ANN-FE model provides the output data (material response) mapping the previously loaded experimental results. Consequently, the constitutive relation may be recreated via numerical simulations, for given arbitrarily set mechanical excitations [17, 18]. The respective ANN-FE computational framework, which is applied after learning and validation processes are completed, is shown in Fig. 7, in turn.

ANN based mathematical model of an SMA material contains all the data required for FE simulations, and may be considered as an alternative way of the description of material properties. Additionally, it should be noted that in case of quasi-static studies farther simplifications regarding the behavior of SMA may be

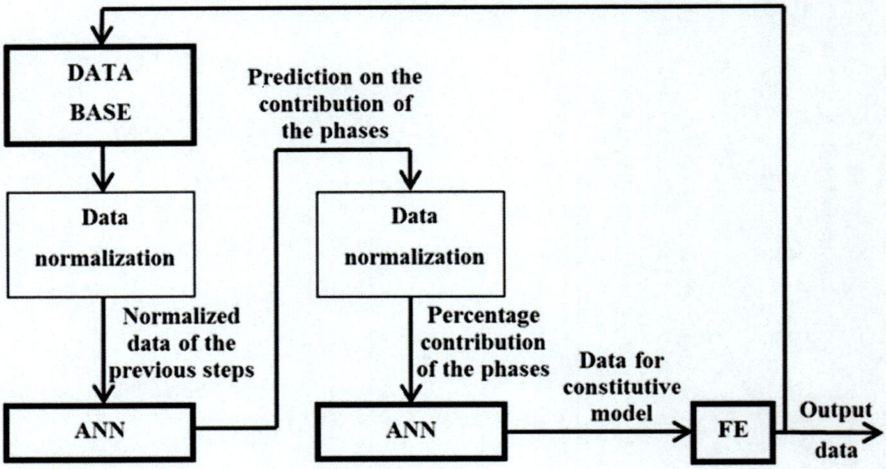

Fig. 7 Computational workflow for the ANN-FE model after validation. ANN provides data regarding elastic moduli for the FE model of an SMA component

introduced via reduction of the number of material parameters. Specifically, setting the Young's moduli for both solid phases as well as defying the stresses at which phase transitions occur is sufficient for reliable simulations.

5 Numerical Simulation and Experimental Validation

FE model of an SMA wire, which has been used during numerical simulations consists of approximately 100,000 FEs (type Quad4 in MSC.Software/Marc). Figure 8 shows a cross-sectional view for the SMA wire and an exemplary simulated constitutive relationship. The identified values of the Young's moduli are 76PGa and 35GPa, respectively for austenite and martensite. The Poisson's ratio equals 0.3. ANN considers 21 independent input variables, including the series of 20 subsequent values of the stress and a single output value of the ANN determined during the previous time step.

ANN uses sigmoidal functions in all layers except the last one, which features a linear one. ANN is parameterized using the Broyden–Fletcher–Goldfarb–Shanno (BFGS) method, which is a type of the quasi-Newton methods. Figure 9 visualizes exemplary results of the ANN-FE model gained for the first part of the hysteretic stress-strain relationship.

The provided numerical results should be considered as the preliminary ones, successfully obtained by the authors for the elaborated model of SMA. As confirmed during testing the properties of the investigated model, it should be highlighted that creation of a reliable SMA model is not a trivial task. It requires many trials

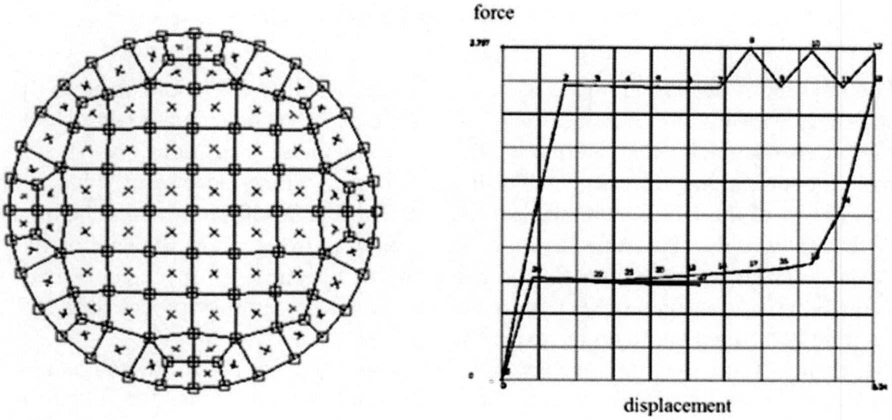

Fig. 8 FE model and simulations: cross-sectional view for the modelled SMA wire (on left), exemplary course of the force-displacement (stress-strain) relationship (on right)

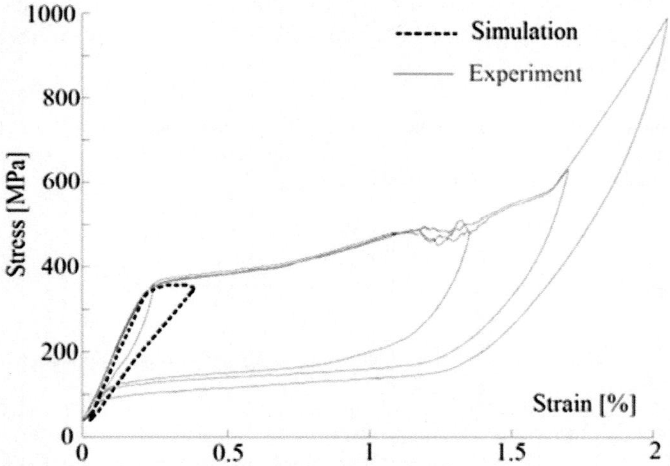

Fig. 9 Comparison between the experimental and simulation results – preliminary results obtained using the ANN-FE model for the first part of the hysteretic stress-strain relationship

before its acceptable behavior is observed. The presented results confirm that the combination of ANN and a FE code allows for mapping nonlinear elastic properties of the modeled materials, including recreation of the parts of the hysteretic constitutive relationships. The above stated capability is maintained even though an approximate description of the material is introduced (as mentioned in Sect. 4). The model correctly memorizes the stress at which the martensitic transition should start. Similarly, a part of the plateau region (i.e., the region of the stress-strain relationship where excessive strain changes occur) is also represented in the model response.

6 Summary, Final Conclusions and Future Directions

The reported ANN-FE approach has confirmed its usability for modeling strongly nonlinear constitutive relationships, as for the case of metallic materials, which exhibit mechanically and thermally induced solid phase transitions. As shown with the preliminary results, the elaborated model, being developed with the application of the BFGS method, correctly recreates parts of the hysteretic stress-strain relationship. The time series of the previous stresses is sufficient for the model to generate correct material response for an arbitrarily set mechanical load. Reduction of the domain of the required model parameters aids the process of ANN learning performed based on the experimental data.

The presented study is considered as the initial step for farther investigations regarding hybrid ANN-FE models. It is scheduled by the authors to add new functionalities for the elaborated model, taking into account more comprehensive analysis of the experimental data used to learn ANN and widening the scope of the handled types of material properties, including variable temperatures of the phase transitions. Finally, the quality of the experimental test stand will be increased by introduction of the newly designed 3D printed specialized clamps to assure better fixation of the SMA wires.

Acknowledgments This study was funded by National Science Center, Poland (Grant No. OPUS 2017/27/B/ST8/01822 Mechanisms of stability loss in high-speed foil bearings — modeling and experimental validation of thermomechanical couplings).

References

1. Lagoudas, D.: Shape Memory Alloys: Modeling and Engineering. Applications, Springer, Berlin (2008)
2. Araujo, A.L., Mota Soares, C.A.: Smart Structures and Materials. Computational Methods in Applied Sciences, ECCOMAS. Springer (2017)
3. Rusinek, R., Warminski, J., Weremczuk, A., Szymanski, M.: Analytical solutions of a nonlinear two degrees of freedom model of a human middle ear with SMA prosthesis. Int. J. Non Lin. Mech. **98**, 163–172 (2018)
4. Bryła, J., Martowicz, A.: Shape memory materials as control elements used in a dot braille actuator. Mech. Contr. **33**(4), 83–89 (2014)
5. Kaltenbacher, M.: Numerical Simulation of Mechatronic Sensors and Actuators Finite Elements for Computational Multiphysics. Springer, New York (2015)
6. Martowicz, A., Ciszewski, M., Buratowski, T., Gallina, A., Rosiek, M., Seweryn, K., Teper, W., Zwierzyński, A. J., Uhl, T.: Mechatronic approach in application to solution of research and design problems. Mechatronics. **36**, 1–17 (2016)
7. Besson, J., Cailletaud, G., Chaboche, J.-L., Forrest, S., Bletry, M.: Non-Linear Mechanics of Materials Solid Mechanics and Its Applications 167. Springer, Heidelberg (2010)
8. Lagoudas, D.C., Bo, Z.C., Qidwai, M.A.: A unified thermodynamic constitutive model for SMA and finite element analysis of active metal matrix composites. Mech. Compos. Mater. Struct. **3**(2), 153–179 (1996)

9. Meyers, M., Chawla, K.: Mechanical Behavior of Materials, 2nd edn. Cambridge University Press (2009)
10. Lubieniecki, M., Roemer, J., Martowicz, A., Wojciechowski, K., Uhl, T.: A muli-point measurement method for thermal characterization of foil bearings using customized thermocouples. J. Electron. Mater. **45**(3), 1473–1477 (2016)
11. Martowicz, A., Bryła, J., Staszewski, W.J., Ruzzene, M., Uhl, T.: Nonlocal elasticity in shape memory alloys modeled using peridynamics for solving dynamic problems. Nonlinear Dynamics. **97**(3), 1911–1935 (2019)
12. Martowicz, A., Staszewski, W.J., Ruzzene, M., Uhl, T.: Nonlocal elasticity theory for solving dynamic problems via peridynamics. In: Awrejcewicz, J., et al. (eds.) Mathematical and Numerical Aspects of Dynamical System Analysis. 14th Conference Dynamical Systems—Theory and Applications—DSTA 2017, pp. 345–356. Łódź, Poland, 11–14 December 2017
13. Auricchio, F., Bonetti, E., Scalet, G., Ubertini, F.: Theoretical and numerical modeling of shape memory alloys accounting for multiple phase transformations and martensite reorientation. Int. J. Plast. **59**, 30–54 (2014)
14. Cisse, C., Zaki, W., Zineb, T.B.: A review of modeling techniques for advanced effects in shape memory alloy behavior. Smart Mater. Struct. **25**(10), 103001 (2016)
15. Auricchio, F., Taylor, R.L.: Shape-memory alloy: modeling and numerical simulations of the finite-strain superelastic behavior. Comput. Methods Appl. Mech. Eng. **143**, 175–194 (1997)
16. Geron, A.: Hands-on Machine Learning with Scikit-Learn and TensorFlow. O'Reilly Media, Inc (2017)
17. Neamt, L., Matei, O., Chiver, O.: Finite element method combined with neural networks for power system grounding investigation. Int. J. Adv. Comput. Sci. Appl. **8**(2), 187–192 (2017)
18. Christiansen, N.H., Høgsberg, J., Winther, O.: Artificial Neural Networks for nonlinear dynamic response simulation in mechanical systems. In: Proceedings of the 24th Nordic Seminar on Computational Mechanics, Helsinki, Finland, 3–4 November (2011)

Modeling of Electro-Hydraulic Servo-Drive for Advanced Control System Design

Jakub Możaryn, Arkadiusz Winnicki, and Damian Suski

Abstract The paper describes the synthesis of a mathematical model of the electro-hydraulic servo-drive. Because of the complexity of the electro-hydraulic servo-drive system and the difficulty in determining all system's coefficients, the simplification of the mathematical model is proposed. The model includes different non-linearities such as the friction or the pressure-dependent oil bulk module. The simulation results are presented, and the comparison with the data collected from the real servo drive is discussed. With the proposed method it was possible to choose the values of physical parameters such that the real electro-hydraulic servo-drive is modeled with the accuracy suitable for the fast prototyping and design of the advanced control system.

Keywords Hydraulics · Servo-drive · Nonlinear dynamics

1 Introduction

Hydraulic drives are fluid drives in which the oil medium acts as the energy carrier between the generator and the receiver. These drives are devices used to generate forces, torques, and perform displacements. Depending on the actuation element used, the hydraulic actuators may provide linear or rotary movement. The electro-hydraulic system is a combination of the hydraulic system for energy transfer and electric power equipment for transmitting control signals. In this way, a synergy effect is obtained, which results in better static and dynamic properties of the electro-hydraulic system compared to pure hydraulic systems.

Nowadays electro-hydraulic servo-drive systems play an important role in industrial plants because of their high power to weight ratio, stiffness, and pay-

---

J. Możaryn (✉) · A. Winnicki · D. Suski
Institute of Automatic Control and Robotics, Warsaw University of Technology, Warsaw, Poland
e-mail: jakub.mozaryn@pw.edu.pl
https://iair.mchtr.pw.edu.pl

J. Awrejcewicz (ed.), *Perspectives in Dynamical Systems I: Mechatronics and Life Sciences*, Springer Proceedings in Mathematics & Statistics 362,
https://doi.org/10.1007/978-3-030-77306-9_16

load capability. However, the control of these systems is a difficult task since their dynamics are highly nonlinear. Therefore, producers of electro-hydraulic components provide proprietary, specialized software for preliminary simulations e.g. HYVOS (Bosh-Rexroth), or FluidSIM (Festo). The efficient modeling and simulation of electro-hydraulic servo-drive is still being studied [1–3].

The investigation of the accurate, and computationally efficient models of the electro-hydraulic actuators is an interesting task from both academic and industrial perspectives, because preliminary analysis and fast prototyping techniques based on simulations can improve the design of the advanced state-space control system. In [4] authors present the Embedded Model Control (EMC) when a hierarchical control architecture, that takes into account multi-domain description of the actuator, is implemented on an electro-hydraulic proportional valve. In [5] there is described the experimental evaluation of the Linear-Quadratic-Gaussian (LQG) controller of electro-hydraulic servo-drive with the model-based analysis of the problem.

The article is organized as follows. In Sect. 2, the laboratory test stand used to gather experimental data is described. In Sect. 3, mathematical models of electro-hydraulic servo-system components are given, taking into account their nonlinearities and proposed simplifications. In Sect. 4, model parameters are given, and results of the experimental evaluation of the model are described. Finally, concluding remarks are given.

2 Laboratory Test Stand

The laboratory test stand (Fig. 1) comprises the following elements: hydraulic pump, pressure relief valve, servo-valve, cylinder, linear position encoder and PC computer with MATLAB/Simulink software and the control-card dSpace DS1104.

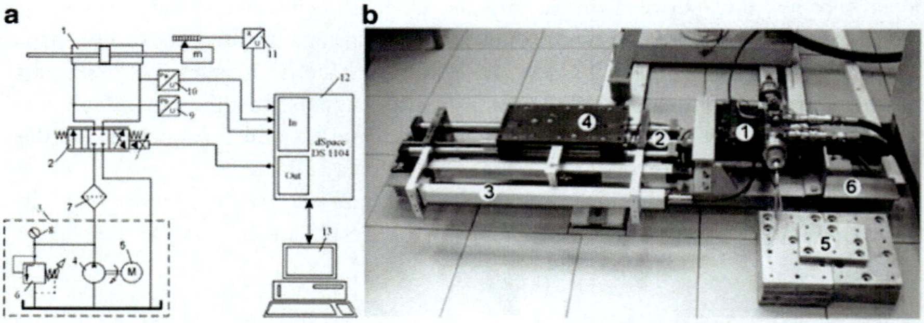

Fig. 1 (**a**) Electro-hydraulic scheme of the laboratory test stand. a-1—double-acting hydraulic actuator, a-2—electro-hydraulic servo-valve, a-3—power station, a-4—hydraulic pump, a-5—electric motor, a-6—overflow valve, a-7—high pressure filter, a-8—manometer, a-9, a-10—pressure transmitters, a-11—magnetostrictive position transmitter, a-12—measurement and control card DS1104, 13—PC computer; (**b**) picture of the laboratory test stand: b-1—servo-valve, b-2—cylinder, b-3—position encoder, b-4—load platform, b-5—mass, b-6—support

The system contains the double-acting actuator). In order to stabilize the movement of the piston rod, the platform is positioned on slideways. The position of the piston rod is changed by the servo-valve, controlled by the voltage signal in the range [-10; 10] [V]. The position of the actuator piston rod is measured with the magnetostrictive transducer. The data transfer between the position transmitter, the controller, and the servo-valve is performed via the 1104 dSPACE controller-card. The controller algorithm and the data acquisition are realized with a PC computer and Matlab/Simulink software.

3 Mathematical Modeling of the Electro-Hydraulic Servo-Drive

The electro-hydraulic servo-drive can be presented in a form of a block diagram with a set of interconnected components between which there is a flow of electrical, mechanical and hydraulic signals. Such a representation of the system allows for a detailed analysis of its structure and parameters influence on the entire system.

The set of differential and algebraic equations stating the model of the electro-hydraulic servo-drive has been implemented as a block diagram in the Simulink environment with partial models of system components.

3.1 Modeling of Double-Acting Hydraulic Actuator

The mathematical description of the piston motion in the double-acting hydraulic actuators is based on the analysis of the Newton's second law of motion and the equations of the oil flow balance. The following assumptions were made: servo valve is installed close to the actuator, such that the losses of pressure between valve and actuator can be omitted, pipes and all system elements are non-deformable against the pressure, there are no internal leakages in the system, and density of the working fluid is constant.

The differential equation of the piston displacement x [m], resulting from the Newton's second law, is

$$m\frac{d^2x}{dt^2} + F_t(v) + F_o = A_A p_A - A_B p_B \tag{1}$$

where: $m = m_t + m_o$ [kg]—a total mass (including: m_t—the mass of the piston and the piston rod and m_o—the mass of the platform and the load); $F_t(v)$ [N]—the Stribeck's friction force acting on the piston, the piston rod and the platform, expressed as a function of the piston velocity v [m/s]; $F_o(v)$]—the external force acting on the platform (in our case we take F_o=0 [N]); A_A,A_B[m^2]—the piston active areas in the actuator A and B chambers, for double-sided piston rod from

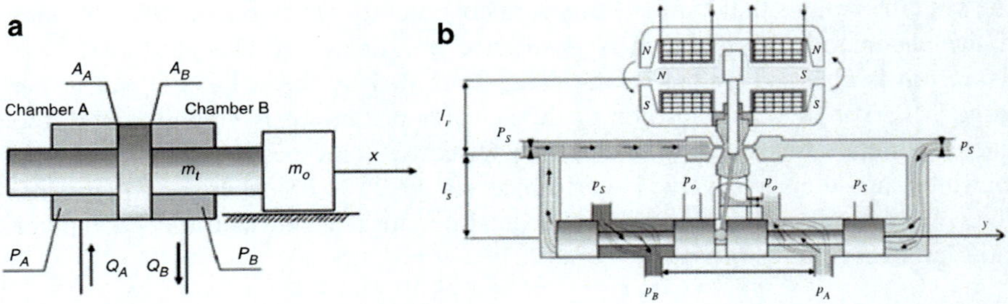

Fig. 2 (**a**) Double-acting hydraulic actuator, (**b**) Servo-valve arrangement

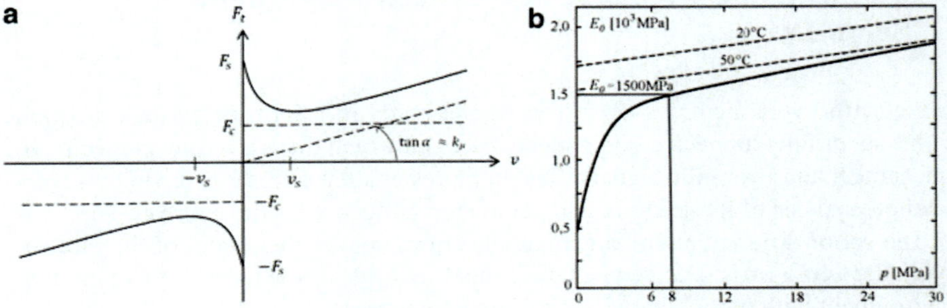

Fig. 3 (**a**) Stribeck's friction force, (**b**) oil bulk modulus in pressure function [6]

stand $A_A = A_B = A$; p_A,p_B [Pa]—the absolute pressures in the actuator A and B chambers (see Fig. 2).

While modeling the friction force, in the simulation the Stribeck's formula was used (Fig. 3a) described as

$$F_t(v) = \begin{cases} \pm F_s, \text{ for, } v = 0 \\ F_{t2} = \left(F_c + (F_s - F_c) \cdot e^{-\left(\frac{v}{v_{st}}\right)^2} + k_\mu |v| \right) \mathrm{sgn}(v), \text{for, } v \neq 0 \end{cases} \tag{2}$$

where: F_t [N]—the modeled friction force; F_s [N]—the static friction force; F_c [N]—the Coulomb friction force; v_{st} [m/s]—the Stribeck's velocity; k_μ [Ns/m]—the viscous friction coefficient.

The friction between the following elements must be considered: piston and cylinder, piston rod and cylinder heads, platform and slide-ways.

To model the hydraulic actuator, the oil flow balance equation can be used, that describes the volume flow into or out of the cylinder chamber as the sum of the flow caused by the piston motion and the flow covering the oil compressibility. The flow balance equations for both chambers (A and B) are

$$\begin{aligned} Q_A &= A\frac{dx}{dt} + \frac{V_{Ap} + Ax(t)}{E_0}\frac{dp_A}{dt} \\ Q_B &= -A\frac{dx}{dt} + \frac{V_{Bp} + A(x_{\max} - x(t))}{E_0}\frac{dp_B}{dt} \end{aligned} \tag{3}$$

where: V_{Ap},V_{Bp} [m^3]—the inactive volumes in chambers A and B; x_{max} [m]—the maximum piston displacement; E_0 [MPa]—the oil bulk modulus.[1]

3.2 A Servo-Valve Model

In the partial model of the operation of the servo valve, by changing the position of the spool y [m], the oil flow direction and intensity can be changed. The one-way flow of an oil in a valve can be described as [7, 8]

$$Q = yK_Q\sqrt{\Delta p} \tag{4}$$

where: Q[m^3/s]—the oil flow; K_Q[(m^5/kg)$^{1/2}$]—the flow gain coefficient; Δp [Pa]—the difference of pressures at the valve way's endings.

Depending on the spool position, the chambers A and B of the cylinder are connected through the servo valve to supply with pressure p_s [Pa] and outlet with pressure p_o [Pa]. The equations describing the fluid flow through the servo valve to cylinder chambers are

$$\text{for } y \geq 0 \quad \begin{cases} Q_A = yK_Q\sqrt{p_s - p_A} \\ Q_B = -yK_Q\sqrt{p_B - p_o} \end{cases} \tag{5}$$

$$\text{for } y < 0 \quad \begin{cases} Q_A = yK_Q\sqrt{p_A - p_o} \\ Q_B = -yK_Q\sqrt{p_s - p_B} \end{cases} \tag{6}$$

The position of the valve spool is controlled by the torque motor connected with flapper-nozzle arrangement (see Fig. 2b).

To derive a mathematical model, it is necessary to consider the equations describing the magnetic circuit and the dynamics equations of the diaphragm and the slider of the hydraulic amplifier.

The equation of the spool movement can be written as

$$(k_{ms} \cdot i - k_{mh} \cdot y)\,\mathrm{k}_{mp} \cdot A_S = m_s\frac{d^2y}{dt^2} + F_{ts}(v_s) + F_h \tag{7}$$

where: k_{ms} [Nm/rad]—torque coefficient of the drive (represented as hysteresis of magnetic circuit); k_{mh} [N/rad]—coefficient of the mechanical feedback, k_{mp} coefficient of the flapper-nozzle arrangement; A_s[m^2]—the spool face area; m_s [kg]—the mass of the spool; $F_{ts}(v_s)$[N]—the friction force acting on the spool; F_h[N]—the hydrodynamic force.

[1]The oil bulk modulus (Fig. 3b) depends nonlinearly on the oil pressure and the oil aeration level.

Transfer function describing the current of the coils i [A] as a function of the control voltage u [V] is given as follows

$$i(s) = \frac{\frac{1}{R}}{\frac{L}{R}s + 1} \cdot u(s) \tag{8}$$

where: R [Ω]—resistance of coils; L [H]—inductance of coils.

The hydrodynamic force works in the direction opposite to the direction of the spool movement, and its value can be determined from the equation

$$F_h = 0,36Q\sqrt{2\rho\Delta p} = k_h Q\sqrt{\Delta p} \tag{9}$$

where: $k_h = 0,36\sqrt{2\rho}[(\text{kg/m}^3)^{1/2}]$—coefficient of the hydrodynamic forces; Δp [Pa]—pressure drop in the gap.

3.3 A Constant Displacement Pump Model

Using the balance of flow rates and assuming that the lines connecting the pump with the drive system do not deform under the influence of pressure, the relation binding the pump's flow Q_p with the supply pressure p_s delivered to the system can be written as [9]

$$Q_p = \frac{V_{ps}}{E_0}\frac{dp_s}{dt} + K_v p_s \tag{10}$$

where: $V_{ps}[\text{m}^3]$—volume of the oil between the pump and the servo-valve; $K_v = \frac{\pi dh^3}{12\mu l}$ [m^5/Ns]—volumetric loss factor.

4 Experimental Verification of the Electro-Hydraulic Servo-Drive Model

The complete model of the electro-hydraulic servo drive, containing the elements and blocks described in the previous sections, is presented in Fig. 4. The model contains 34 parameters gathered in Table 1, estimated using catalog data, literature studies or laboratory tests, and approximate estimates.

During verification there were compared the piston positions calculated with the mathematical model and the positions measured at the test stand as a result of the same extortion. The verification experiments made it possible to check the correctness of the initially estimated values of the model parameters and confirmed the correctness of the analytical solutions. For the simplified servo valve model, not

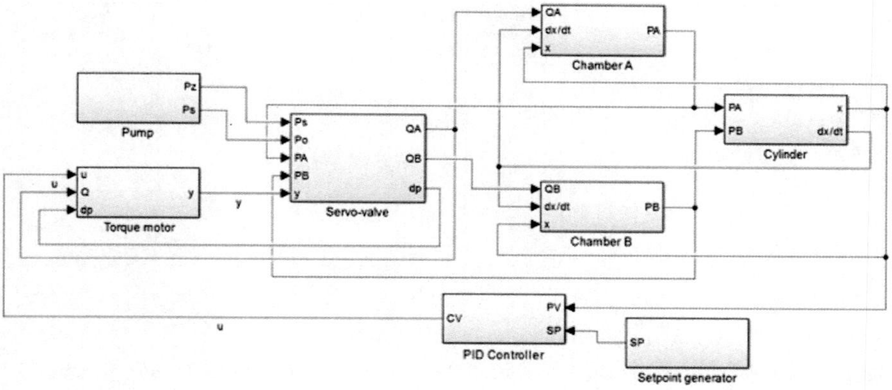

Fig. 4 The schematic block diagram of the electro-hydraulic servo-drive implemented in Simulink software

Table 1 The parameters of the model

Parameter	Value	Parameter	Value	Parameter	Value
Oil					
ρ	865 kg/m^3				
Hydraulic cylinder					
m_t	4 kg	μ_{s2}	0.16	v_{st}	0.3 m/s
m_o	6 kg	μ_{c1}	0.009	V_{mA}	10^{-6} m^3
A_A,A_B	0.001 m^2	μ_{c2}	0.1	V_{mB}	10^{-6} m^3
μ_{s1}	0.1	k_μ	0.5 Ns/m	x_{max}	1.378 m
Servo-valve					
α	0.7	y_{max}	0.001 m	$k_{\mu s}$	0.5
d	0.005 m	y_d	$0.015 \cdot 10^{-3}$ m	k_h	14.974 (kg/m^3)$^{1/2}$
K_Q	$5.29 \cdot 10^{-4}$ (m^5/kg)$^{1/2}$	A_s	$1.963 \cdot 10^{-5}$	k_{ms}	0.0009 Nm/rad
L_c	1 H	μ_{cs}	0.009	k_{mh}	22 N/rad
R_c	330 Ω	μ_{ss}	0.1	k_{mp}	$2.7 \cdot 10^9$ Pa/m
m_s	0.10 kg				
Constant displacement pump					
p_0	$81 \cdot 10^5$ Pa	V_{ps}	0.001 m^3	K_v	$4.6 \cdot 10^{-12}$ m^5/Ns
p_s	10^{-5} Pa	Q_p	$1.33 \cdot 10^{-12}$ m^3/s		

taking into account all non-linear phenomena occurring in the actual system, newly selected parameter values guarantee satisfactory results only in a small range of amplitude values of step excitations.

The test stand with a position control algorithm (PID controller tuned using Ziegler-Nichols method) implemented has been tested in position tracking tasks (see Fig. 5). The comparison of the actual system and the model gives the maximum relative error less than 6.8% and the mean relative error less than 1.5%. These results show that the nonlinear model is valid and the coefficients of the model has been identified with a satisfying accuracy.

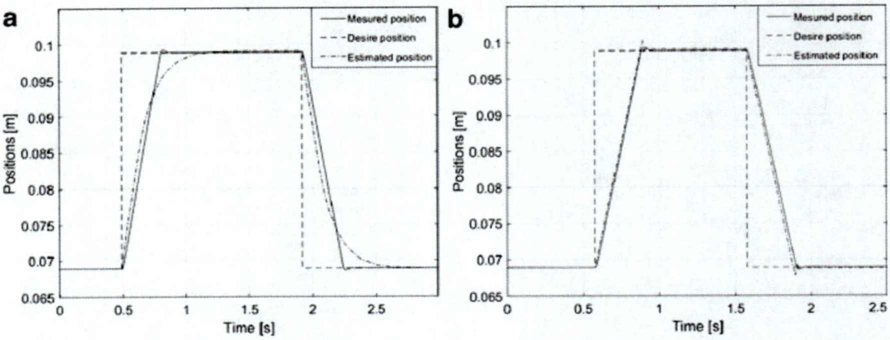

Fig. 5 Results for position tracking experiments: (**a**) before parameters fitting; (**b**) after parameters fitting

5 Conclusions

The verification of the developed mathematical model with the real electro-hydraulic servo-drive show that the model with high accuracy reflects the operation of the object near the operating point. The model includes such nonlinearities as friction model, characteristics of the modulus of the elasticity, dependence of flow intensity on pressure drop at control edges of the valve slide, dependence of hydrodynamic force, and characteristics of the volumetric loss factor in the pump.

The system parameters have been initially estimated on the basis of professional literature and components data-sheets. The work presents the set of the model parameters used for experimental evaluation, where the step excitations of large amplitudes were performed, at which the operation of the object satisfactorily maps the operation of the actual object. The selected parameter values allowed to get a maximum relative error of less than 6.8%, for different amplitudes of excitation.

Proposed simplified models can be used in fast prototyping of the nonlinear state-space control systems for the electro-hydraulic servo-drive e.g. [7]. The presented modelling approach was successfully used in [5, 10]. Currently authors of the paper work on the systematic methods of the parameters identification, to improve the accuracy of the model.

References

1. Aboelela, M.A.S., Essa, M.E.M., Hassan, M.A.M.: Modeling and identification of hydraulic servo systems. Int. J. Model. Simul. **3**(38),139–149 (2018). https://doi.org/10.1080/02286203.2017.1405713
2. Bimal, R., Mate, N.S., Sanjawadmath, V.G., Rao, A.N.V.: Computationally efficient non-linear electro-hydraulic actuation system model for real-time simulation. IFAC-PapersOnLine **1**(51), 243–248 (2018). https://doi.org/10.1016/j.ifacol.2018.05.059

3. Grosschmidt, G., Harf, M.: Multi-pole modeling and simulation of an electro-hydraulic servo-system in an intelligent programming environment. Int. J. Fluid Power **1**(17), 1–13 (2016). https://doi.org/10.1080/14399776.2015.1110093
4. Acuña-Bravo, W., Canuto, E., Agostani, M., Bonadei, M.: Proportional electro-hydraulic valves: an embedded model control solution. Control. Eng. Pract. **62**, 22–35 (2017). https://doi.org/10.1016/j.conengprac.2017.01.013
5. Winnicki, A., Mozaryn, J., Micewicz, M.: Experimental verification of discrete linear-quadratic-gaussian control system of electro-hydraulic servo-drive. In: Proceedings of the23rd International Conference on Methods and Models in Automation and Robotics, Miedzyzdroje, Poland, pp. 241–246 (2018). https://doi.org/10.1109/MMAR.2018.8486081
6. Findeisen, D., Helduser, S.: Ölhydraulik- Handbuch der Hydraulischen Antriebe und Steuerungen. Springer, Zurich (2015). https://doi.org/10.1007/978-3-642-54909-0
7. Chapple, P.: Principles of Hydraulic System Design. Coxmoor Publishing, Oxford (2003)
8. Jelali, M., Kroll, A.: Hydraulic Servo-System Modeling, Identification and Control. Springer, London (2004). https://doi.org/10.1007/978-1-4471-0099-7
9. Totten, G., De Negri, V.: Handbook of Hydraulic Fluid Technology. CRC Press, New York (2012). https://doi.org/10.1201/b11225
10. Winnicki, A., Mozaryn, J., Suski, D.: Impulse identification and discrete P/PD control of electro-hydraulic servodrive. J. Autom, Mobile Robotics Intell. Syst. **4**(12), 51–57 (2018). https://doi.org/10.14313/JAMRIS_4-2018/25

Assessment of Implementation of Neural Networks in On-Board Dynamic Payload Weighing Systems

Andrzej Kosiara, Aleksander Skurjat, and Jakub Chołodowski

Abstract While loading loose materials onto dump trucks, freight wagons or any other vehicles, a crucial issue is not to exceed their maximum permissible load. The on-board payload weighing systems, installed in modern earthmoving machinery such as single bucket excavators or loaders, are very efficient tools for monitoring the overall weight of the material loaded onto the vehicles mentioned above. Conventional systems of this type are typically based on the mathematical models deriving from the equations describing dynamic equilibrium of the machines and their manipulators. Unfortunately, estimating the parameters of those models sometimes brings some difficulties. This paper presents a discussion on replacing the conventional models implemented in the most up to date payload weighing systems with the ones based on the neural networks. A number of payload weighing systems involving neural networks varying in terms of structure and neuron types were designed by the authors and tested using a sample excavator. The article presents a comparison of the performance of those systems, including a discussion on the influence of training dataset size on the accuracy of the system.

Keywords Payload weighing system · Neural network · Earthmoving machine

1 Introduction

Single bucket excavators are often utilized to load different types of loose materials onto dump trucks. If the weight of the material loaded onto a truck is monitored, the overall weight of the laden truck can be set very close to the maximum permissible load with no risk of overload. Consequently, the amount of the material transported by the truck is maximized with no violation of the regulations pertaining to the

A. Kosiara (✉) · A. Skurjat · J. Chołodowski
The Department of Off-Road Machine and Vehicle Engineering, Wrocław University of Science and Technology, Wrocław, Poland
e-mail: andrzej.kosiara@pwr.edu.pl

J. Awrejcewicz (ed.), *Perspectives in Dynamical Systems I: Mechatronics and Life Sciences*, Springer Proceedings in Mathematics & Statistics 362,
https://doi.org/10.1007/978-3-030-77306-9_17

maximum permissible axle loads. For this reason, the on-line weighing of the material loaded onto dump trucks is an issue of a great importance nowadays.

In practice, the weight of a dump truck payload might be controlled by two methods. Firstly, an on-board payload weighing system might be integrated with the machine loading the material onto the truck. Secondly, the weighbridges, i.e., the stationary, stand-alone weighing devices, might be installed on construction sites. The weighbridges are less efficient than the on-board weighing systems because they are usually situated remotely from the places where the excavators operate. For this reason, the on-board payload weighing systems have recently become very desirable.

The on-board payload weighing systems estimate the mass of every portion of the goods loaded with the excavator's bucket onto the truck's bed. The overall load of the truck is calculated as a sum of the masses estimated during every work cycle of the machine. The modern on-board weighing systems can be classified into two basic groups, i.e., the static and the dynamic systems. The main difference between them is as follows. While using the static system, the excavator needs to be brought to a standstill for a short period of time during every work cycle in order to determine the bucket payload with acceptable precision. On the other hand, the dynamic systems, do not require any machine holdups, which improves the productivity and makes the operator's work easier.

1.1 Dynamic Payload Weighing Systems for Single Bucket Excavators – The Overview

Numerous dynamic on-board payload weighing systems for single bucket excavators are available nowadays. Some of them have been discussed in [1]. The X2350 by Trimble Loadrite Auckland Ltd. and The Loadex 100 by RDS Technology Ltd. are the great examples of the systems of this type.

The dynamic payload weighing systems typically include sensors that measure the angular position of the excavator's boom and arm, and the oil pressure supplied to the hydraulic cylinders supporting the boom (see Fig. 1.). Inclination sensors are also attached to the excavator's house to eliminate the payload estimation error arising from the ground inclination. In order to improve the accuracy of the systems, the excavators are sometimes fitted with some optional sensors for determining the angular position of the work tool.

The majority of the dynamic weighing system manufacturers estimate the accuracy of their products at approx. +/− 3%, however, this might be achieved if the excavator moves smoothly, the bucket is located in the so-called weighing zone (see Fig. 2) and the system calibration procedure is periodically repeated during machine operation.

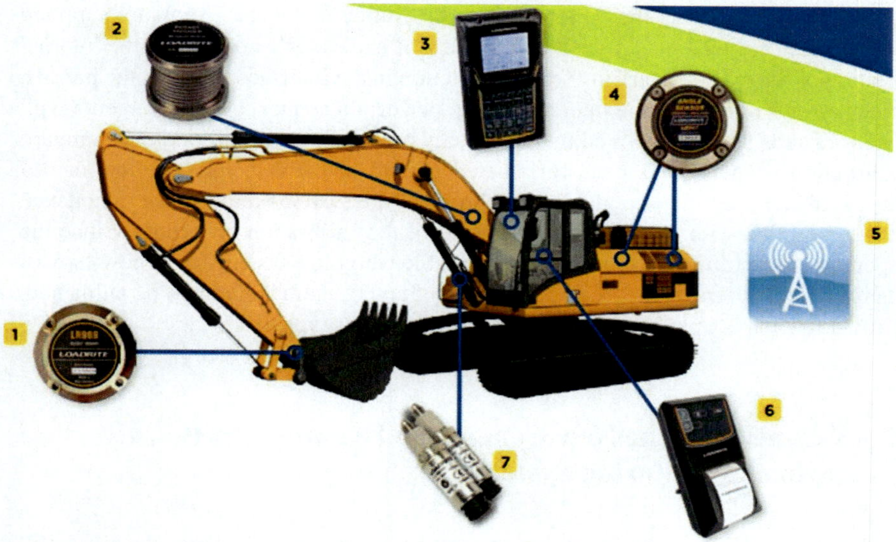

Fig. 1 Components of the Load Rite X2350 payload monitoring system by Trimble [3]: 1, 2 – inclination sensors for determining the angular position of the bucket and boom, 3 – control panel, 4 – ground inclination sensors, 5 – wireless data transmission device (wi-fi, GPRS, radio communication), 6 – printer, 7 – oil pressure sensors

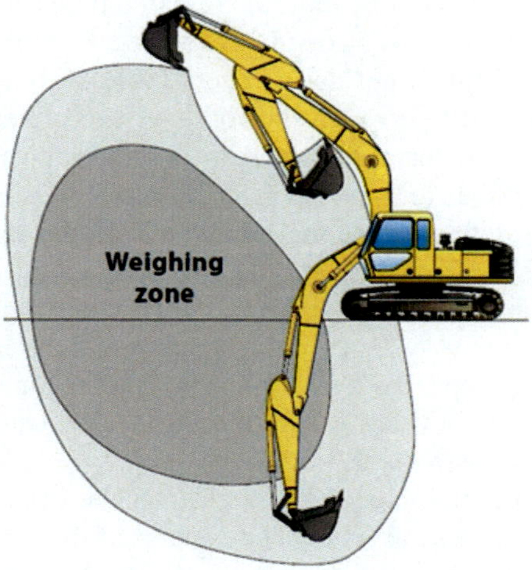

Fig. 2 Weighing zone of the Trimble Load Rite X2350 payload monitoring system [3]

In order to calibrate the system, i.e., to determine how the oil pressure supplied to the boom cylinders is affected by the mass of excavator's bodies and the internal friction of the cylinders arising from the friction in the interface between the barrels, pistons, rods and seals, the operator has to perform a sequence of movements with the excavator's bodies defined by the system's manufacturer. Although no standard calibration sequence has been formulated yet, all of them include the motions of the excavator's boom, arm and bucket. Some of the systems require that the bucked has to be filled with payload on some stage of the calibration procedure. Since the operation algorithms of commercially available weighing systems are a now-how of their manufacturers, no detailed description of any of them is likely to be published. Only a few papers pertaining to the algorithms of this type exist [2].

2 A Concept of the New Generation Dynamic On-Board Payload Monitoring System

The commercially available on-board payload monitoring systems do not fulfill the requirements of their users. There is a strong need to increase their accuracy, whereas the prices should preferably stay level. The most severe drawback of the present payload monitoring systems is that the calibration procedure needs to be repeated several times during every shift of machine operation in order to maintain satisfactory accuracy. Consequently, research and development on the payload monitoring systems is carried out by multiple research units all around the world.

The authors of this article have already developed a system where the bucket payload is estimated multiple times during every cycle of excavator operation and the ultimate bucket payload for a given work cycle is computed by averaging the values determined over the cycle time. In order to enhance the accuracy of the system, the ultimate bucket payload is estimated involving only the samples determined in strictly defined dynamic conditions of the excavator [4]. It has been revealed that the performance of this system is good enough to compete with its commercially available counterparts. Unfortunately, the payload estimation algorithm implemented in the system is quite complex and the calibration procedure needs to be periodically repeated over system operation time to compensate the fluctuations in the bucket payload estimation error. Consequently, a decision to redevelop the system has been made.

In the very beginning, the causes of the drift in the accuracy of the payload monitoring systems have been surveyed. The systems of this type typically estimate the bucket payload on the basis of the forces delivered by the excavator's boom cylinders. Those forces are determined by measuring the oil pressure supplied to the boom cylinders. Unfortunately, the oil pressure is affected by the internal friction of the cylinders, which, in practice, might be estimated at approx. 10% of the overall force developed by the cylinder. Furthermore, the exact amount of the cylinder internal friction depends on the oil temperature. The oil temperature gradually increases over the excavator operation time and is influenced by the ambient temperature. Apparently, in order to avoid quick deterioration in the accuracy of the system developed by the authors in [4], the computational algorithm implemented in the system should have been upgraded with a model for estimating the internal friction of the boom cylinders.

According to the literature, the internal friction of hydraulic cylinders might be basically described with the LuGre model. Unfortunately, upgrading the payload monitoring systems with the classical LuGre model does not significantly improve their accuracy. The internal friction of hydraulic cylinders is strongly affected by the thickness of the lubrication film in the interface between the cylinder tube and the piston seals, whereas the thickness of the film changes over the operation time of the cylinder. According to [5], in order to describe this phenomenon, the classical LuGre model needs to be replaced with the extended one that includes at least 12 parameters. In practice, it would be extremely difficult to estimate those parameters because they might vary depending on whether the cylinder is being extended or retracted. Furthermore, neither the classical nor the extended LuGre model describes the internal friction of hydraulic cylinders as a function of the force delivered by the cylinders and the oil temperature [6].

In the face of those difficulties, a conclusion was made that the internal friction of boom cylinders of excavators cannot be described by a reasonably simple, conventional model. Consequently, a completely new approach was adopted. Namely, a neural network based computational model was implemented in the new generation payload monitoring system presented herein, so that the parameters of the model could have been easily determined by means of standard algorithms for neural networks training implemented, for example, in the MATLAB environment. In order to estimate the model parameters, the kinematic structure of the sample excavator investigated in the article does not need to be known, which is another advantage of this approach. The only problem was to find out what type of neural network would provide low payload estimation error for a relatively long period of time.

It should be noted, that some neural network based algorithms for the payload monitoring systems have been already created [7], however, the vast majority of the systems identified by the authors during the literature review were developed for single bucket loaders.

3 Acquisition of the Neural Network Training Dataset

The data for training the neural networks implemented in the operation algorithm of the bucket payload monitoring system developed herein were obtained by the experiments involving the test stand presented in Fig. 3. In general, the stand was developed to carry out the research on various types of operator assistance systems for single bucket excavators, including bucket payload monitoring systems.

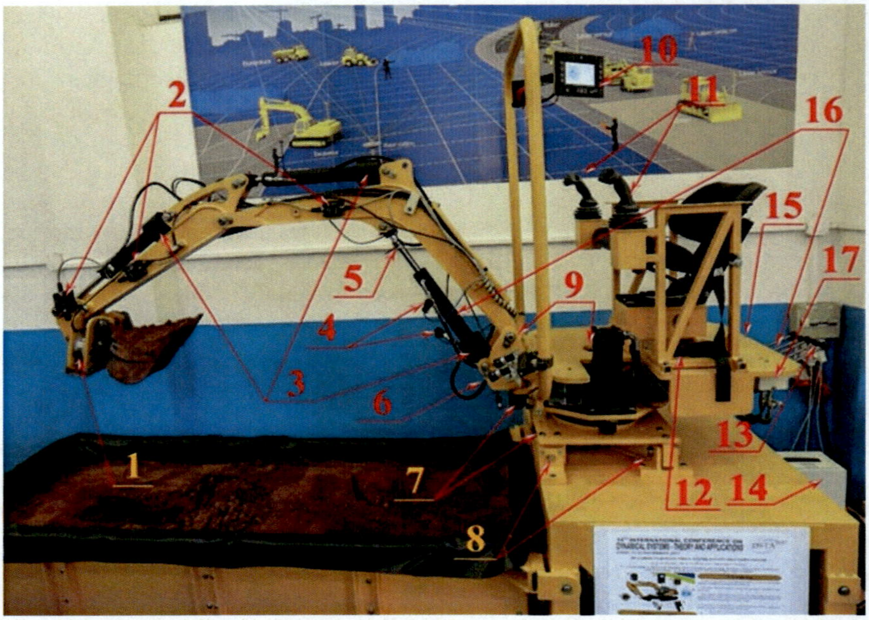

Fig. 3 The test stand involved in obtaining the data for training the neural networks investigated in the following article; 1 – six component force-torque sensor, 2 – single-axis inclinometers, 3 – magnetostrictive transducers of hydraulic cylinders displacement, 4 – temperature compensated pressure sensors, 5 – single-axis force sensor, 6 – stereometric camera, 7 – optical angular position sensors, 8 – single-axis load cells embedded in bolts, 9 – single-axis gyroscopes, 10 – controller of the tip over stability monitoring system integrated with HMI, 11 – joysticks, 12 – two-axis inclinometer, 13 – valve manual control, 14 – real-time CPU with data acquisition cards, 15 – PLUS+1 SC050–020 controller, 16 – temperature transducers, 17 – proportional hydraulic valves

During the tests, the bucket was moved along the trajectories representing typical work cycles of excavators. Namely, it was lifted from the lower area of the manipulator working zone, where it is typically located right after the digging process is accomplished, to the upper area, where the operator is usually about to empty the bucket. 105 experiments were carried out over 2 days. 49 lifting cycles were performed on the first day and 56 tests were carried out on the second one. Since the manipulator was manually controlled, every trajectory performed with the bucket was unique.

In the opposite to the situation presented in Fig. 3., the bucket was not filled with loose material dug from the ground while conducting the experiments. Weights of a priori known mass were collected in the bucket instead of the material. The overall weight of the weights placed inside the bucket in the consecutive trials was changed according to the following pattern: 0 kg – 1 kg – 2 kg – 3 kg – 4 kg – 5 kg – 10 kg –

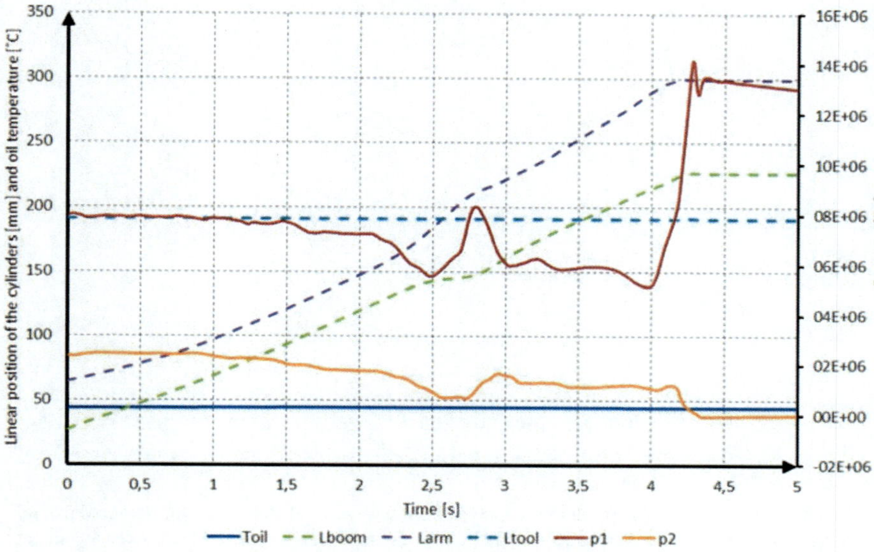

Fig. 4 The results of the measurements carried out during a work cycle where the bucket with 10 kg payload was lifted

11 kg – 12 kg – 13 kg – 15 kg – 20 kg – 0 kg – 1 kg – .... Furthermore, the temperature of the oil supplied to the boom cylinders gradually increased during the experiments, as depicted in Fig. 5.

The article presents only a preliminary research that was carried out in order to assess in general terms the applicability of the algorithms based on the neural networks to the payload monitoring systems for single bucket excavators. For this reason, the influence of the excavator upper structure swing motion on the performance of the system presented herein was out of the scope of this study. The tests were conducted at fixed angular position of the excavator's upper structure with respect to the undercarriage. The following quantities were measured during the experiments: the length of the hydraulic cylinders driving the boom, the arm, and the attachment (L_{boom}, L_{arm} and L_{tool}, respectively, see Fig. 4.), the oil pressure inside the cap-end and the rod-end chamber of the boom cylinder (p_1, p_2) and the temperature of the oil supplied to the boom cylinder (T_{oil}). Moreover, numerical differentiation of the signals representing the length of the cylinders was carried out in order to estimate the linear velocity and acceleration of the cylinders. Figure 4 presents the results recorded during a sample lifting cycle performed during the tests.

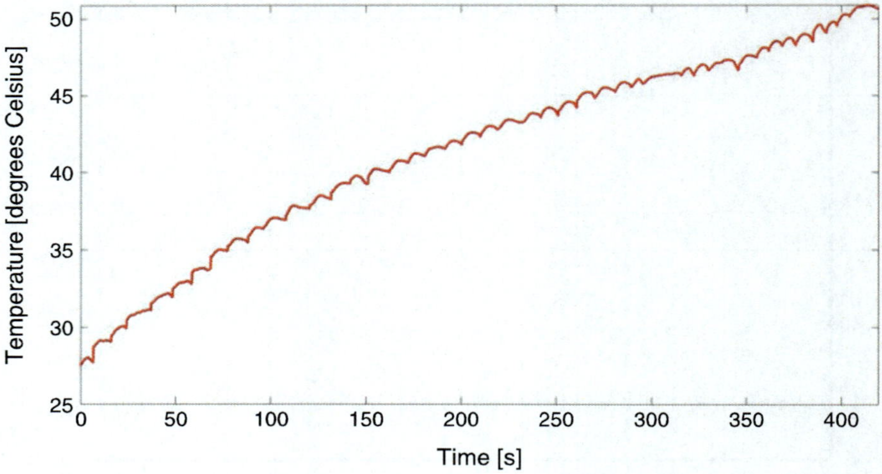

Fig. 5 The temperature of the oil delivered to the boom cylinder of the excavator depicted in Fig. 3. as a function of time – the data collected on the first day of the experimental tests, over 49 bucket lifting cycles

4 Test Results and Conclusions

In the article the applicability of the unidirectional neural networks to the bucket payload monitoring systems for single bucket excavators was assessed. The networks consisting of 2, 3, 4 and 5 hidden layers were investigated. The influence of the number of the neurons in the consecutive layers on the learning and the generalization capabilities of the networks was investigated.

The script for training the neural networks investigated in the article was implemented using the MATLAB environment. 99 experimental datasets were uploaded to the MATLAB's workspace and automatically divided into the three subsets. 70% of the data were used to carry out the neural network learning procedure, another 15% of the data were involved in the validation and the remaining 15% were used to carry out the ultimate testing. 6 of the datasets collected in the experiments were not uploaded to the program at all so that the network could have been independently evaluated in terms of generalization capabilities (Fig. 5).

The neural networks considered herein were learned using three different algorithms: the Levenberg – Marquardt, the Scaled Conjugate Gradient and the Bayesian Regularization. The best results were obtained with the Bayesian Regularization, whereas the worst with the Levenberg – Marquardt algorithm. In the majority of cases, the networks trained with the Bayesian Regularization predicted bucket payload with the best accuracy, regardless of whether the input data were involved in the network learning process or not (Fig. 6).

The networks consisting of 4 hidden layers and at least 25 neurons in the most sophisticated layer basically suffered from noticeable overfitting, i.e. the predictions

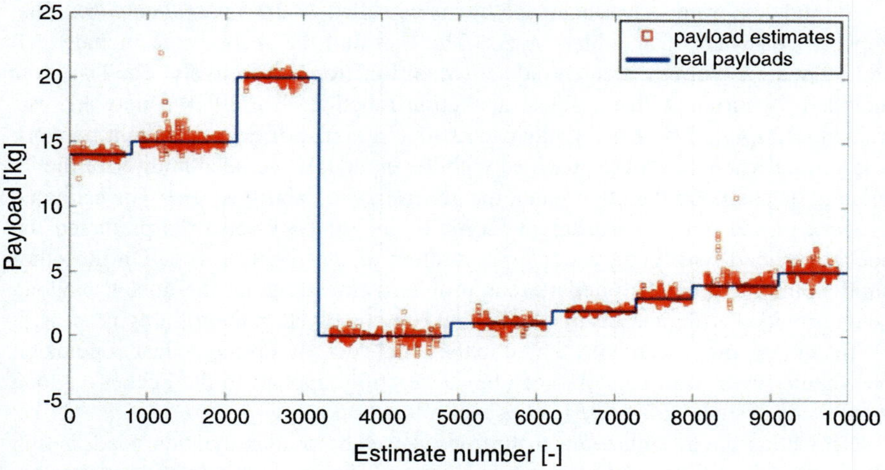

Fig. 6 Bucket payload estimated by the 20-10-5 neural network on the basis of the data involved in training and validation of the network

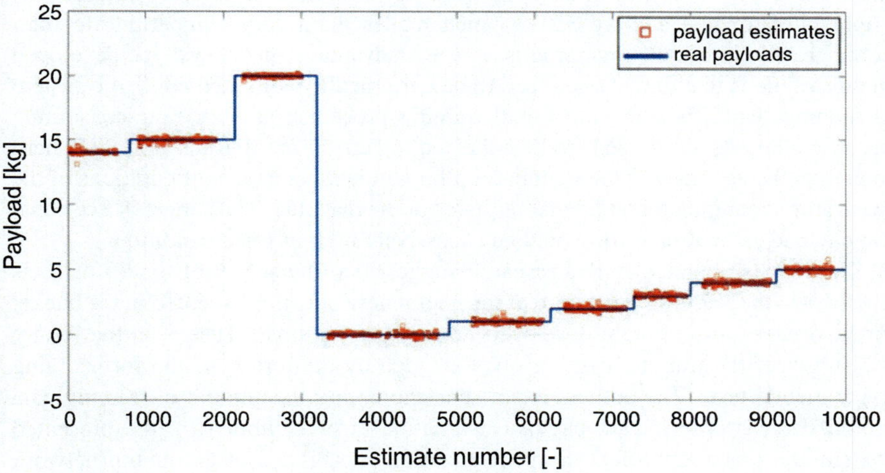

Fig. 7 Bucket payload estimated by the 35-25-11-7 neural network on the basis of the data involved in training and validation of the network

made by those networks were exceptionally good only if the input data were previously included in the training dataset (see Fig. 7.). Unfortunately, if the input data were not involved in the training process, estimation error was high. The opposite behavior was observed for the more simple networks investigated in this research (see Fig. 6.).

A good compromise was achieved by the so-called 20-10-5 neural network. This network consisted of 3 hidden layers. The first and the second one included 20 and 10 neurons with the tangentoid activation function, respectively. The last layer included 5 neurons with the linear activation function. The 20-10-5 network was trained multiple times. Bucket payload estimation error of every resultant network was similar when they were provided with the input data included in the 1 of the 99 training datasets. On the other hand, the generalization ability of every investigated network turned out to be different. Some of the 20-10-5 networks predicted the bucket payload with good accuracy regardless of the input dataset. On the other hand, some of them exhibited serious problems in estimating the bucket payload when provided with datasets that had never been involved in the training process.

To sum up, the research presented in the article does not bring a clear conclusion on whether the neural networks can be successfully applied to the bucket payload monitoring systems for single bucket excavators or not.

Neglecting the instantaneous distortions caused by sudden dynamic loads acting on the investigated excavator, which could have been easily eliminated by filtering, bucket payload estimation error exhibited by some of the neural networks developed in this research did not exceed +/− 1 kg. Since the lifting capacity of the excavator involved in this research is 30 kg, the relative estimation error of those networks might be estimated at only 3.3%, which makes them very competitive to their commercially available counterparts. A great advantage of the systems developed in this article is that they do not need to be periodically calibrated while the system is running. Unfortunately, the initial training procedure is very time-consuming because a large dataset needs to be collected to carry it out. Furthermore, it is hard to predict the accuracy of the system for all possible kinematic configurations of the excavator's manipulator and dynamic states of the machine. Thus, there is a risk that the payload estimation error would be excessively high in some conditions.

The results obtained in this research suggest that the subset of the 99 datasets involved in the learning procedure of the neural networks implemented in the bucket payload monitoring system developed herein was too small. Hence, in the future, the subset of the training data will be extended by the results of the forthcoming experimental tests. The authors expect that increasing the amount of training data would decrease the bucket payload estimation error exhibited by the presented system and positively affect the generalization capabilities of the neural network implemented in the system. The neural network based algorithm presented herein will be also modified so that the bucket payload could be successfully predicted by the system in dynamic conditions, namely, while performing swing motions with the upper structure of the excavator.

References

1. Leski, M.: Design of system for dynamic weighing of material in the excavator bucket. Master thesis supervised by Andrzej Kosiara, Wrocław University of Science and Technology, Wrocław (in Polish) (2017)
2. Bennett, N., Walawalkar, A., Schindler, Ch.: Payload estimation in excavators: model-based evaluation and comparison of current payload. In: 3rd Commercial Vehicle Technology Symposium (CVT 2014), Kaiserslautern, Germany, pp. 333–346. Shaker Verlag, Aachen, Germany (2014)
3. Trimble Loadrite Auckland Ltd., technical note, https://www.halomec.com/uploads/4/5/2/1/45211295/trimble_x2350_brochure.pdf, last accessed 2020/11/20
4. Kosiara, A., Chołodowski, J.Z., Skurjat, A.R.: Development of a prototype dynamic weighing system for single bucket excavator. In: Awrejcewicz, J. (eds.) Dynamical Systems in Applications, Łódź, Poland, December 11–14, 2017, Springer Proceedings in Mathematics & Statistics, vol. 249, pp. 217–228. Springer, Heidelberg (2018)
5. Yanada, H., Sekikawa, Y., Takahashi, K., Matsui, A.: Unsteady friction characteristic of hydraulic actuator and its mathematical model. In: Proceedings of the 7th JFPS International Symposium on Fluid Power, TOYAMA 2008, Vol. 3, pp 635–640. Toyama, Japan (2008)
6. Tran, X.B., Matsui, A., Yanada, H.: Effects of viscosity and type of oil on dynamic behaviors of friction of hydraulic cylinder. JFPS Int. J. Fluid Power Syst. **3**(2), 16–23 (2010)
7. Hindman, J.J.: Dynamic Payload Estimation in Four Wheel Drive Loaders. Doctoral thesis, University of Saskatchewan, Saskatoon, Canada (2008)

Lower Limb Rehabilitation Exoskeleton with a Back Support – Mechanical Design

Bartosz Stańczyk, Olga Jarzyna, Wojciech Kunikowski, Dariusz Grzelczyk, Jerzy Mrozowski, and Jan Awrejcewicz

Abstract Mobility impairment is a serious medical and social problem with an increasing prevalence. Therefore, it is important to develop devices that would help both the patients and the therapists. In this paper, a literature overview of available constructions of lower limb exoskeletons was presented. Degrees of freedom (DOFs), kinematics and drive systems of regarded devices were studied. The outcomes of the investigation of the literature regarding active and passive DOFs and torque requirements were used as guidelines for the development of the conceptual design and a prototype of the exoskeleton. The proposed device is intended to be used for gait rehabilitation for patients suffering from different types of gait abnormality. It has 11 DOFs that can be actuated, or not, depending on the needs of the wearer. These DOFs are accompanied by two complex movements, which are flexion/extension of the patient's upper body as well as flexion/extension of toes.

Keywords Human gait · Exoskeleton · Lower limb · Gait · Rehabilitation · Powered orthosis · Paraplegia

1 Introduction

According to the World Report on Disability, published by the World Health Organization (WHO), more than 1 billion people (which is about 15% of the world population at the time of publication of the Report) live with some form of disability, with about 200 million having considerable difficulties in functioning (see [1]). The same report says that not only health outcomes of the disabled but also their educational achievements and financial status are lower than of people without

B. Stańczyk (✉) · O. Jarzyna · W. Kunikowski · D. Grzelczyk · J. Mrozowski · J. Awrejcewicz
Department of Automation, Biomechanics and Mechatronics, Lodz University of Technology, Lodz, Poland
e-mail: bartosz.stanczyk@p.lodz.pl

J. Awrejcewicz (ed.), *Perspectives in Dynamical Systems I: Mechatronics and Life Sciences*, Springer Proceedings in Mathematics & Statistics 362,
https://doi.org/10.1007/978-3-030-77306-9_18

disabilities. According to the National Center for Health Statistics (see [2]), 7.1% of the U.S. adults cannot of find it very difficult to walk a quarter of a mile. In Poland, 15% of population declare some sort of disability, 59% of which is motor disability [3]. Also 13% of Polish population declare that walking a distance of 500 m without assistance (walkers, canes) is very difficult or impossible.

The number of people suffering from different forms of disability, including mobility impairments resulting from past injuries and various diseases, is constantly growing mainly due to the increase in the average life expectancy followed by age-related diseases, which are accompanied by the increase in the number of traffic accidents or sports injuries [4, 5]. Other reasons of motor disability are various diseases such as stroke, cerebral palsy, multiple sclerosis, partial or complete spinal cord injury, to mention a few [6–9].

In spite of the causes, dysfunctions of the locomotor and neural system yield negative consequences, both medical and social, by limiting the mobility and social activity of the disabled. Mobility impairment is a serious health, social, psychological, and economic problem, which can be improved by removing the barriers in accessing various aspects of life, for instance by gait rehabilitation.

Nowadays, the most frequently used and effective form of rehabilitation of patients with the mobility impairment is physiotherapy [10]. Handling of patients, however, is labour-intensive and often requires sustained demanding postures or lifting and transferring patients, which results in occupational conditions such as lower back problems [11–13]. Therefore, it is of great significance to use scientific and technological advances to develop devices that can be used to restore motor and cognitive functions of the disabled patients, increase capabilities of physiotherapists and other specialists, and improve the available therapeutic methods and techniques [14].

According to numerous studies, motor functions of patients can be successfully facilitated by external stimulation with the use of lower limb exoskeletons (LLEs) [15–20]. Hence, these robotic devices have become the subject of numerous studies in the last decades.

The term "exoskeleton" or "active orthosis" is usually used to describe a device intended to increase the mobility of a person suffering from lower limb dysfunction. Namely, an exoskeleton is a machine that increases power in one or more joints, which results in an increase in the operator's efficiency. In contrast to passive orthoses, active exoskeletons can not only provide simple mechanical reinforcement but also actively control movement in the joints of the device. Despite many active orthoses, including LLEs, have been developed, this area still has great potential for research [21–25].

In this paper, the state-of-the-art of the strategies used in rehabilitation LLEs is presented, after which the conceptual design of the developed exoskeleton (further referred to as a "K11 exoskeleton") is given. The name "K11" originates from authors' affiliation – it is the symbol of the Department of Automation, Biomechanics and Mechatronics at the Lodz University of Technology. The main focus has been put on the mechanical design of the exoskeleton, including feet elements and a novel element for back support, which is aimed at improving patient's posture.

2 Available Solutions (State-of-the-Art)

We focus on a review of available exoskeleton kinematic solutions. The information was sought on the number of DOFs, the number of driven DOFs, and the types of actuators used. Although many commercial devices have become available recently, this overview focuses mainly on academic studies. Detailed information about commercial products are usually regarded as trade secrets and are not available to the public. A summary of the information collected is provided in Table 1 at the end of the section.

The first of the considered constructions was the Vanderbilt exoskeleton [9, 26]. It was created in order to facilitate the movement of people suffering from paraplegia and also as a support for the rehabilitation therapy. The Vanderbilt exoskeleton allows for the movement in an upright position with the support of a pair of crutches and performing sit-down and sit-up manoeuvres. This LLE provides support in the sagittal plane in both the hip and knee joints. Lithium-polymer battery is used for

Table 1 Summary of the reviewed exoskeletons

Exoskeleton	Total number of DOFs	Number of actuated DOFs	Actuators
Vanderbilt [9, 26]	4	4	BLDC motors with reduction gears
BLEEX [27, 28]	14	6	Hydraulic servomechanisms and actuators
HAL3 [29, 30]	6	4	DC servomechanisms with wave gears
LOPES [31]	10	8	AC servomotors with reduction gears; linear servomechanisms
HEXAR [32]	15	4	BLDC servomotors with wave gears
MINDWALKER [33, 34]	12	6	BLDC motors with screw nut gear
IHMC [35]	10	6	BLDC servomotors with wave gears

powering DC motors that drive the joints through reduction gears. Each actuator is able to produce a continuous torque of 20 N·m and a maximum torque of 80 N·m. Knee joints are additionally equipped with normally clamped brakes in order to avoid knee hyperextension in the event of a power failure.

The Berkeley Lower Extremity Exoskeleton (commonly known as BLEEX) is a much more advanced construction. It is a system created to provide the user with the ability to carry considerable loads on the back with minimal effort relative to any type of terrain [27, 28]. BLEEX has seven independent DOFs per leg: 3 DOFs in the hip, 1 DOF in the knee (pure rotation in the sagittal plane), and 3 DOFs in the ankle. The construction of the exoskeleton allows the user to freely modify the length of individual modules. For safety reasons, mobility in all joints has been reduced to the normal human range. The actuated degrees of freedom are: ankle, knee, and hip, all in the sagittal plane. Hydraulic servomechanisms were used to drive the device. On the basis of the BLEEX exoskeleton, many similar constructions were created in the Berkeley Robotics & Human Engineering Laboratory: The Human Universal Load Carrier (HULC), ExoHiker, ExoClimber, eLEGS.

The best known medical exoskeleton is the HAL-3 system [29, 30]. The first prototype was completed in 1997, and since 2008, the manufacturer, Cyberdyne, started to rent it on the territory of Japan. The exoskeleton supports walking in an upright position, climbing stairs, and the movement of getting up and sitting down. HAL-3 has three degrees of freedom for each leg, suitable for hips, knees, and ankles. All these joints work in the sagittal plane. To protect the user in the event of power failure, the mobility in each joint was mechanically limited. To drive hip and knee joints, DC servomechanisms with wave gears were used.

A different approach is presented in LOPES [31]. LOPES consists of two main parts: the exoskeleton itself and a static frame connected to the pelvic segment of the exoskeleton. The system uses a treadmill for gait recreation. This solution allows to apply supporting forces to the pelvic segment, thus relieving the leg segment. The exoskeleton itself has two actuated DOFs in the pelvis (frontal and sagittal rotation) and one actuated DOF in the knee (knee rotation in the sagittal plane). The ankle element consists of a frame that allows one to mount an external ankle foot orthosis. The place of connection of the pelvis to the frame has two driven DOFs (linear actuators) in the transverse plane and one non-actuated DOF in the frontal plane. The engines are mounted outside the exoskeleton, and the drive is transmitted through a set of flexible bowden cables.

Another example of the human strength augmenting construction is the Hanyang Exoskeleton Assistive Robot (HEXAR) [32]. In principle, the exoskeleton was supposed to be able to carry, in addition to its (21 kg) and user's mass, an additional mass of 35 kg. HEXAR has 15 DOFs: three DOFs in the hip segment, rotation in the knee, and three DOFs in the ankle. An additional DOF is the rotation in the frontal plane at the level of the spine in the lower back section. Turning back is possible because the hip segment frame is not a rigid structure but a serial connection of four torsion elements. Such a chain combines a joint that allows for rotation in the lower back with a joint that realizes the hip abduction and adduction. In the foot segment, the structure allows one to perform movements in three DOFs: flexion/extension, internal/external rotation as well as inversion and eversion. Actuators that drive flexion in the hip and knee are brushless DC motors with 100:1 reduction gearboxes.

One of the most kinematically advanced rehabilitation constructions is the MINDWALKER exoskeleton [33, 34]. The equipment was created for patients after severe spinal injuries in order to allow moving in an upright position and the sitting down and up manoeuvres. The exoskeleton has ten DOFs, six of which are driven. DOFs with own actuators are: knee flexion, hip flexion and hip abduction and adduction. The remaining non-driven DOFs (hip rotation, ankle flexion) are equipped with spring elements whose deformation is zero in the neutral position. The exception is the inversion and eversion of the foot, the mobility of which results from the elastic construction of the exoskeleton. The actuators that drive each of the aforementioned DOFs consist of: BLDC (brushless direct-current motor), screw-nut system, torsion spring with high rigidity, and encoder. The motor drives the linear movement of the screw which deflects the torsion spring. Thanks to the measurements from the encoder, it is possible to calculate the torque generated in a given joint. Thus, the actuator becomes a controllable source of torque.

A similar construction to the MINDWALKER is the IHMC exoskeleton [35]. The total number of DOFs and the number of those that are driven is almost the same as in the MINDWALKER (the IHMC design does not allow for the inversion and eversion of the foot). Like the MINDWALKER, IHMC uses actuators equipped with spring elements and encoders to measure torques in the joints. The difference is in the construction of elastic elements. The IHMC drives the joint directly via the BLDC motor with the wave gear and a steel rope wrapped around the driven axis, the ends of which are attached symmetrically to two linear springs. Based on the data from the encoder and the spring stiffness constants, the torque is determined.

In recent years, commercial exoskeletons have become increasingly popular rehabilitation equipment. Their great advantage is the possibility of extending the time of the patient's rehabilitation session and relieving the physiotherapist carrying out the procedure.

The LLE HANK, produced by the Spanish GOGOA Mobility Robots, is used to rehabilitate people with gait disorders and in cases of complete paralysis. It has six driven DOFs, responsible for the rotation in the hip, knee, and ankle, all in the sagittal plane [36].

Another popular example is the ReWalk Rehabilitation System produced by ReWalk Robotics [37]. It has four driven DOF – rotation of the hip and the knee in the sagittal plane. The gait is initiated by tilting the trunk forward. Then, ReWalk generates a series of movements that reproduce the movement during a normal walk.

One of the most advanced rehabilitation exoskeletons is HAL MEDICAL, produced by Cyberdyne [38]. It is an exoskeleton with actuated DOFs in the sagittal plane of the knees and hips. An advanced control system detects the intentions of the user and allows for free gait as well as the sit-to-stand and stand-to-sit movements. The device is the next generation of exoskeletons HAL-3 and HAL-5.

Another commercialized university construction is the INDEGO exoskeleton, previously developed under the name Vanderbilt [39]. This version does not significantly differ from the previously mentioned exoskeleton in terms of construction.

EksoGT is a rehabilitation equipment supporting the movement of the lower limb in the sagittal plane [40]. Similarly to the previous example, the only actuated DOFs are flexion/extension in the hip and knee. The exoskeleton supports the walking motion, but the patient is forced to use a pair of crutches or a walking frame. Some parameters of motion generated by the machine can be adjusted in real time during a rehabilitation session.

Based on the literature review above, one can observe that the essential joints for enabling walking in the upright position, which require driving, are the hip and the knee. Driving the rotation in the ankle can be omitted in order to reduce the mass and demand for power (as can be seen in commercial devices). It was observed that the BLDC motors equipped with wave or reduction gears are the most common types of actuators due to their power to weight ratio.

The outcomes of the above overview were used as guidelines for construction of the K11 exoskeleton, described in the next section of the paper.

3 Conceptual Design of the Developed LLE

In what follows, we describe the design and construction of the developed exoskeleton. Wherever it is possible, we include a justification for the choice of particular solutions. We give additional arguments concerning price, ability to block certain DOFs (depending on the therapy), or modularity. The exoskeleton is intended to be used for gait rehabilitation of patients suffering from paraplegia resulting, for instance, from a spinal cord injury. To fulfil its purpose, it is planned to be mounted to a mobile frame so as to assure the safety of the patient.

One of the most important features of an exoskeleton is the number of DOFs and the number of those DOFs that can be powered. It determines to what extent the natural pattern of the human gait can be recreated. The prototype of the K11 LLE has a total number of 11 DOFs accompanied by the capability of performing the motion in other segments, i.e. the back and feet. The exoskeleton allows for fundamental movements of the lower limb such as foot dorsi- and plantarflexion, foot inversion/eversion, knee flexion/extension, hip extension/flexion, abduction and adduction (which make a total of 6 DOFs per side). These movements have been supplemented by the capability of rotating the upper body and performing flexion/extension of the upper body as well as performing flexion/extension of toes. Hence, gait rehabilitation is possible. Actuation of particular movements (DOFs) can be modified, i.e. abled/disabled (allowed/blocked), based on the decision of the physiotherapist, depending on the specific treatment strategy for the patient.

By design, mobility in particular joints has been mechanically limited to the range presented in Table 2 so as to allow for normal changes in angles expected during gait as well as the sit-to-stand (and stand-to-sit) movement. The main purpose of limitation of the range of motion is to avoid knee hyperextension in the event of a power failure. The maximum angles have been retrieved from the source literature

Table 2 Range of motion of the most important movements of the developed exoskeleton

Segment/Motion	Flexion – extension	Abduction – adduction
Hip	90° – 20°	10° – 10°
Knee	90° – 0°	–
Ankle	15° (DF) – 25° (PF)	5° (INV) – 5° (EV)
Back (upper body)	45° – 0°	–

DF dorsiflexion, *PF* plantarflexion, *INV* inversion, *EV* eversion

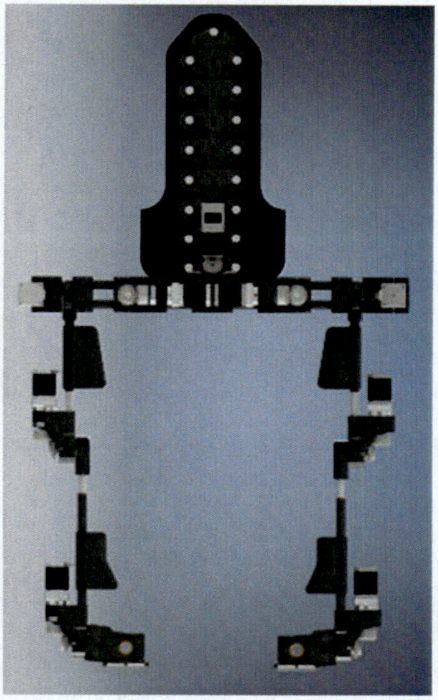

Fig. 1 3D model of the developed exoskeleton – rear view

[41–43]. However, as many patients may have the range of motion reduced, the mobility of the K11 exoskeleton can be also modified by the control system.

Rapid prototyping was used for the development of the model. For this purpose, Autodesk Inventor software was used for 3D computer-aided design, and then the parts were fabricated by 3D printing by the Zortrax M300 device. The CAD model of the exoskeleton in presented in Figs. 1 and 2. The lengths of the modules (hip width, thighs, shanks, feet) can be easily adjusted by virtue of the employment of telescopic elements.

The main purpose of facilitation of extension of the upper body is to increase the stability of the patient during the rehabilitation process by means of correcting the inclination of the pelvis. Facilitation of upright posture is also beneficial for the hip extension range and loading [43]. In the literature, one can find a few studies aimed at assisting spine flexion/extension [44, 45]. However, the motivation of the authors of the present study was to implement a simple and inexpensive solution. Thus, the back support element (see Fig. 3) was inspired by a back protector for snowboarders.

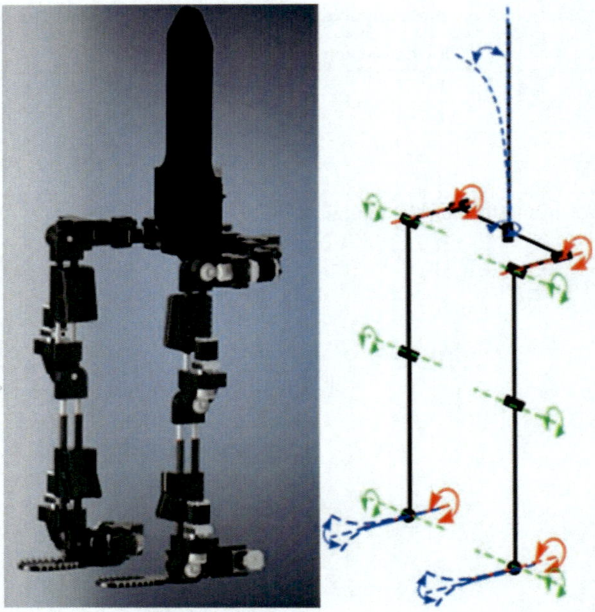

Fig. 2 3D model of the developed exoskeleton (left) and DOFs (right)

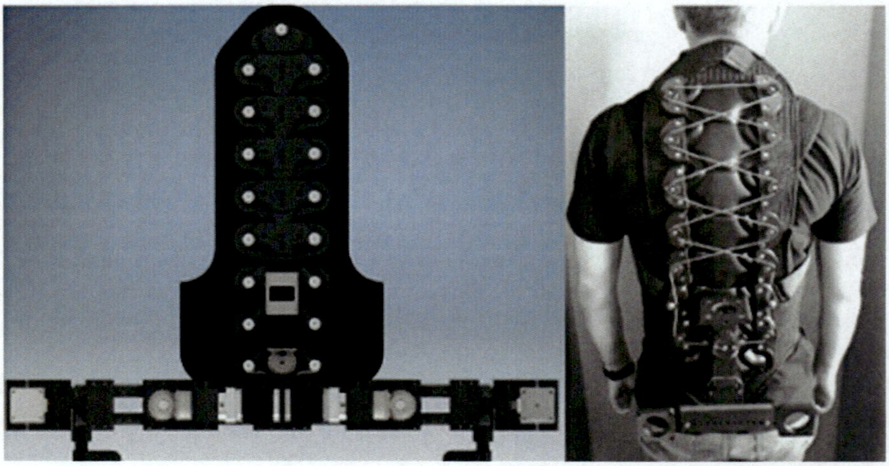

Fig. 3 3D model (left) and a real-photo (right) of the back support segment

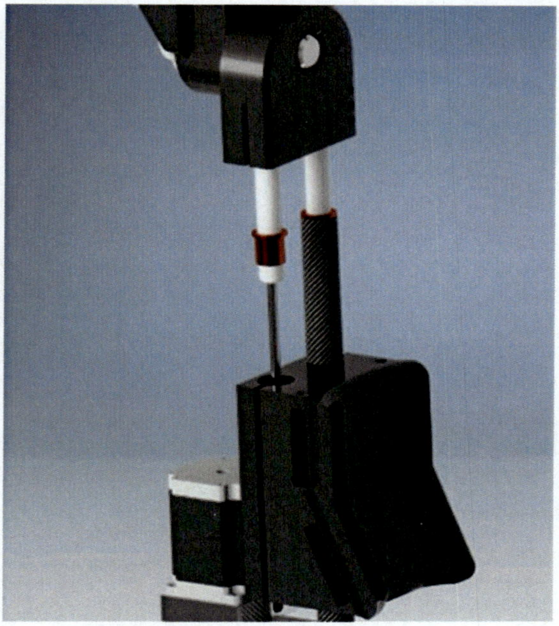

Fig. 4 3D model of the knee segment

It is supplemented by rollers attached to each segment of the element and connected via a cable to an additional electric motor, the role of which is to manage tension so as to keep the patient's upper body properly extended. The extension of the upper body changes the position of the centre of mass (centre of gravity) of the human body and shifts it towards the spine, which results in the improvement in balance. As the developed exoskeleton is intended to be used only for rehabilitation, and not for force augmentation purposes, such a simple construction is believed to be satisfactory.

To assist motion in the hip joints, four electric motors (two per side) are used. Two motors at the back are used for hip adduction and abduction while flexion and extension are driven by two motors placed on the sides of the pelvis (in parallel to hip joints).

Flexion and extension of the knee joint consist of sliding and rolling, which results in the fact that the axis of rotation is variable. Thus, we assumed that the structure of the exoskeleton should not be a simple kinematic node. In the developed device, the axis of the knee rotation can change by means of air dampers (telescopic elements), see Fig. 4. Namely, the lengths of shanks and thighs change so as to make the axis of rotation of the knee joint of the exoskeleton follow the natural axis of the knee rotation.

As far as the design of a foot segment is concerned, in the distal part, the foot plate has been supplemented by modules enabling the patient to extend and flex their toes during walking (see Fig. 5). The modules are passive and their main purpose is to increase the comfort of the patient and allow for movement as close as possible

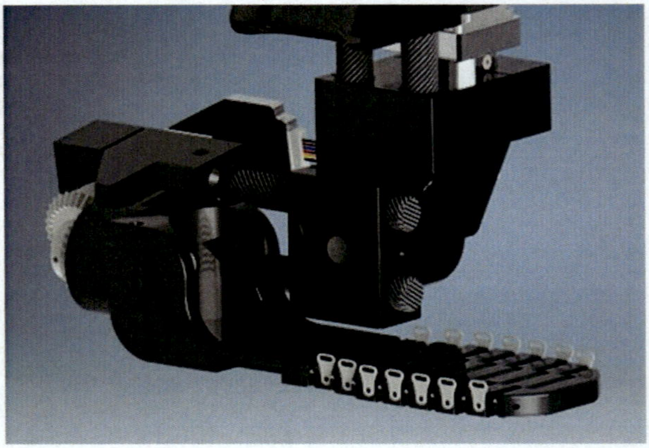

Fig. 5 3D model of the foot segment

Table 3 Literature overview of normalized joint torques during human gait

Flexion [N·m/kg]	Zelik and Kuo [50]	Kirtley [49]	Winter [48]	Linskell [49]	Sloot and van der Krogt [51]
Hip	0.4	0.8	0.55	0.8	0.7
Knee	0.3	0.8	0.65	0.4	0.5
Ankle	1.3	1.45	1.6	1.45	1.4

to the natural one. To adjust the length of the foot segment, some of the above-mentioned modules can be removed. The foot segment is mounted to the patient's shoe by means of Velcro fasteners.

In order to select the appropriate drive for the exoskeleton, the dynamic gait parameters [46, 47] were investigated and compared with the known normal gait patterns given by Winter [48], Kirtley and Linskell [49], and others [50, 51]. As a preliminary condition, it was assumed that the exoskeleton should generate sufficient driving torques to propel the movement of a man weighing at least 85 kg. The reconstruction of correct angular velocities of motion was considered less important in purely rehabilitation applications. Table 3 presents a list of peak values of normalized driving torques for the most loaded joints (rotation in the sagittal plane of the hips, knees, and ankles).

Assuming the maximum values from the sources presented above, it follows that the motors used should generate the minimum holding moments for: hip – 65 N·m; knee – 65 N·m; ankle – 130 N·m.

In order to meet the assumptions mentioned above, it is suggested to use SM 57/76-3008B stepper motors with a holding moment of 1.9 N·m, together with 40:1 planetary gears. The advantage of this solution is the moderate weight of the drive system, and simple control of stepper motors. Bevel gears have been introduced to increase ergonomics of use by means of placing the motors along the main profiles of the device. The main profiles of the K11 exoskeleton are intended to be eventually made of carbon fibre round tubes. Such a solution will ensure more than satisfactory

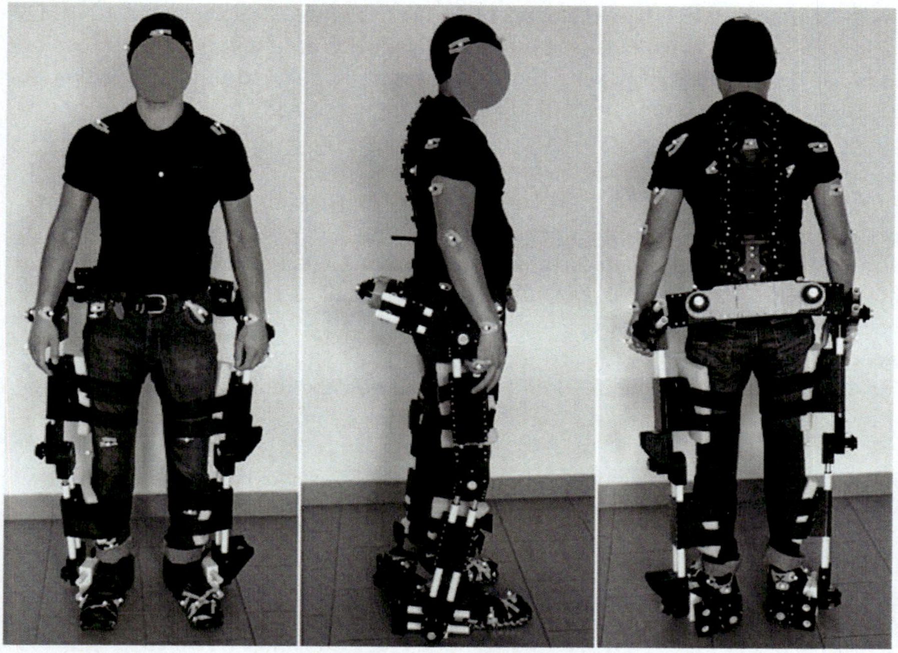

Fig. 6 Passive prototype of the developed exoskeleton. Front view (on the left), side view (centre), rear view (right)

mechanical properties due to high resistance to compression that is likely to occur in the device. In the prototype, PVC (polyvinyl chloride) tubes have been successfully used. The mechanical design of the lower limb exoskeleton developed in the present study is shown in Fig. 6.

4 Discussion

In the present paper, a literature overview concerning active lower limb orthoses (exoskeletons) has been presented. Many academic and commercial constructions have been studied and described with the emphasis put on the number of passive and active DOFs, kinematics, and drive. Also an analysis of torque requirements for selected joints has been conducted. Based on the findings, some guidelines regarding the construction of a lower limb exoskeletons have been presented. They have been used to design a lower limb and spine exoskeleton. The developed exoskeleton has been made for gait rehabilitation of patients suffering from different mobility impairments (paraplegia). The device can be scaled for a specific patient and actuation of particular movements can be adjusted (allowed/blocked), depending on the therapy needed. These features make the use of the exoskeleton possible and affordable for many patients as the personalization is easy and one device can be used for more than one person. The only segment, the length of which cannot be adjusted, is the back module, and thus the future work will cover redesign of this element so as to solve this problem.

Concurrently, the control system is being developed [52], based on the previous studies of the authors [53, 54]. However, further tests and research have to be performed in order to test the operation of the whole exoskeleton and analyse real effects of the introduced concepts. Recently, experimental studies on the influence of the exoskeleton, both passive and active, on the gait of a healthy individual have been conducted with the use of a force plate, electromyography and a motion capture system which has also been used in other studies of authors' department [55, 56]. Furthermore, stability, kinematic and dynamic biocompatibility with a natural limb, as well as material strength should be investigated.

Funding This research was funded by the National Science Centre of Poland, grant OPUS 9 no. 2015/17/B/ST8/01700 for years 2016–2019.

References

1. World Health Organization: World report on disability, 2011; Malta (2011)
2. Villarroel, M., Blackwell, D., Jen, A.: Tables of summary health statistics for U.S. adults: 2018 National Health Interview Survey (2019)
3. Piekarzewska, M., Wieczorkowski, R., Zajenkowska-Kozłowska, A.: Health Status of Population in Poland in 2014. Statistical Publishing Establishment, Warsaw (2016)
4. Bach, J.P., Ziegler, U., Deuschl, G., Dodel, R., Doblhammer-Reiter, G.: Projected numbers of people with movement disorders in the years 2030 and 2050. Mov. Disord. **26**, 2286–2290 (2011)
5. Wenning, G.K., Kiechl, S., Seppi, K., Müller, J., Högl, B., Saletu, M., Rungger, G., Gasperi, A., Willeit, J., Poewe, W.: Prevalence of movement disorders in men and women aged 50–89 years (Bruneck Study cohort): a population-based study. Lancet Neurol. **4**, 815–820 (2005)
6. Mehrholz, J., Thomas, S., Werner, C., Kugler, J., Pohl, M., Elsner, B.: Electromechanical-assisted training for walking after stroke. Stroke. **48**, e188–e189 (2017)
7. Bayón, C., Lerma, S., Ramírez, O., Serrano, J.I.I., Del Castillo, M.D.D., Raya, R., Belda-Lois, J.M.M., Martínez, I., Rocon, E.: Locomotor training through a novel robotic platform for gait rehabilitation in pediatric population: short report. J. Neuroeng. Rehabil. **13**, 1–6 (2016)
8. Federici, S., Meloni, F., Bracalenti, M., De Filippis, M.L.: The effectiveness of powered, active lower limb exoskeletons in neurorehabilitation: a systematic review. Neuro Rehabil. **37**, 321–340 (2015)
9. Farris, R.J., Quintero, H.A., Murray, S.A., Ha, K.H., Hartigan, C., Goldfarb, M.: A preliminary assessment of legged mobility provided by a lower limb exoskeleton for persons with paraplegia. IEEE Trans. Neural Syst. Rehabil. Eng. **22**, 482–490 (2014)
10. Chen, B., Ma, H., Qin, L.L.-Y., Gao, F., Chan, K.-M., Law, S., Qin, L.L.-Y., Liao, W.: Recent developments and challenges of lower extremity exoskeletons. J. Orthop. Transl. **5**, 26–37 (2016)
11. West, D.J., Gardner, D.: Occupational injuries of physiotherapists in North and Central Queensland. Aust. J. Physiother. **47**, 179–186 (2001)
12. Glover, W.: Work-related strain injuries in physiotherapists: prevalence and prevention of musculoskeletal disorders. Physiotherapy. **88**, 364–372 (2002)

13. Rugelj, D.: Low back pain and other work-related musculoskeletal problems among physiotherapists. Appl. Ergon. **34**, 635–639 (2003)
14. Pons, J.L., Torricelli, D., Pajaro, M.: Converging clinical and engineering research on neurorehabilitation, biosystems & biorobotics, vol. 1. Springer, Berlin, Heidelberg (2013)
15. Guo, Z., Yu, H., Yin, Y.H.: Developing a mobile lower limb robotic exoskeleton for gait rehabilitation. J. Med. Device. **8**, 44503 (2014)
16. Gregorczyk, K.N., Adams, A.A., O'Donovan, M.P., Schiffman, J.M., Bensel, C.K., Brown, M.L.: Biomechanical and metabolic implications of wearing a powered exoskeleton to carry a backpack load. In: Proceedings of the American Society of Biomechanics (2012)
17. Banala, S.K., Agrawal, S.K., Scholz, J.P.: Active Leg Exoskeleton (ALEX) for gait rehabilitation of motor-impaired patients. In: Proceedings of the 2007 IEEE 10th International Conference on Rehabilitation Robotics, ICORR'07, pp. 401–407 (2007)
18. Agrawal, S.K., Banala, S.K., Fattah, A., Sangwan, V., Krishnamoorthy, V., Scholz, J.P., Hsu, W.L.: Assessment of motion of a swing leg and gait rehabilitation with a gravity balancing exoskeleton. IEEE Trans. Neural Syst. Rehabil. Eng. **15**, 410–420 (2007)
19. Petrarca, M., Patanè, F., Rossi, S., Carniel, S., Cappa, P., Castelli, E.: A new robotic exoskeleton for gait recovery. Gait Posture. **40**, S26–S27 (2014)
20. Van Asseldonk, E.H.F., Ekkelenkamp, R., Veneman, J.F., Van Der Helm, F.C.T., Van Der Kooij, H.: Selective control of a subtask of walking in a robotic gait trainer (LOPES). IEEE 10th Int. Conf. Rehabil. Robot. ICORR'07 2007, 0, 841–848 (2007)
21. Huijing, P.A.: Muscle, the motor of movement: properties in function, experiment and modelling. J. Electromyogr. Kinesiol. **8**, 61–77 (1998)
22. Ferrari, A., Benedetti, M.G., Pavan, E., Frigo, C., Bettinelli, D., Rabuffetti, M., Crenna, P., Leardini, A.: Quantitative comparison of five current protocols in gait analysis. Gait Posture. **28**, 207–216 (2008)
23. Dong, L., Zhu, F., Jin, X., Suresh, M., Jiang, B., Sevagan, G., Cai, Y., Li, G., Yang, K.H.: Blast effect on the lower extremities and its mitigation: a computational study. J. Mech. Behav. Biomed. Mater. **28**, 111–124 (2013)
24. Clark, T., Hawkins, D.: Are fixed limb inertial models valid for dynamic simulations of human movement? J. Biomech. **43**, 2695–2701 (2010)
25. Zach, L., Konvickova, S., Ruzicka, P.: Investigation of in-vivo hinge knee behavior using a quasi-static finite element model of the lower limb. Proc. IFMBE Proc. **37**, 791–794 (2011)
26. Quintero, H.A., Farris, R.J., Goldfarb, M.: A method for the autonomous control of lower limb exoskeletons for persons with paraplegia. J. Med. Device. **6**, 41003 (2012)
27. Zoss, A., Kazerooni, H., Chu, A.: On the mechanical design of the Berkeley Lower Extremity Exoskeleton (BLEEX) In: Proceedings of the 2005 IEEE/RSJ International Conference on Intelligent Robots and Systems IROS, 3132–3139 (2005)
28. Zoss, A.B., Kazerooni, H., Chu, A.: Biomechanical design of the Berkeley Lower Extremity Exoskeleton (BLEEX). IEEE/ASME Trans. Mechatron. **11**, 128–138 (2006)
29. Lee, S., Sankai, Y.: Power assist control for walking aid with HAL-3 based on EMG and impedance adjustment around knee joint. IEEERSJ Int. Conf. Intell. Robot. Syst. **2**, 1499–1504 (2002)
30. Kawamoto, H., Sankai, Y.: Power assist method based on Phase Sequence and muscle force condition for HAL. Adv. Robot. **19**, 717–734 (2005)
31. Veneman, J.F., Kruidhof, R., Hekman, E.E.G., Ekkelenkamp, R., Van Asseldonk, E.H.F., Van Der Kooij, H.: Design and evaluation of the LOPES exoskeleton robot for interactive gait rehabilitation. IEEE Trans. Neural Syst. Rehabil. Eng. **15**, 379–386 (2007)
32. Kim, W., Lee, H., Kim, D., Han, J., Han, C.: Mechanical design of the Hanyang Exoskeleton Assistive Robot (HEXAR). In: Proceedings of the International Conference on Control, Automation and Systems, pp. 479–484 (2014)
33. Wang, S., Wang, L., Meijneke, C., Van Asseldonk, E., Hoellinger, T., Cheron, G., Ivanenko, Y., La Scaleia, V., Sylos-Labini, F., Molinari, M., et al.: Design and control of the MINDWALKER exoskeleton. IEEE Trans. Neural Syst. Rehabil. Eng. **23**, 277–286 (2015)

34. Wang, L., Wang, S., van Asseldonk, E.H.F.F., van der Kooij, H., Wang, L., Wang, S., van Asseldonk, E.H.F.F., van der Kooij, H.: Actively controlled lateral gait assistance in a lower limb exoskeleton. In: Proceedings of the 2013 IEEE/RSJ International Conference on Intelligent Robots and Systems, pp. 965–970. IEEE (2013)
35. Kwa, H.K., Noorden, J.H., Missel, M., Craig, T., Pratt, J.E., Neuhaus, P.D.: Development of the IHMC mobility assist exoskeleton. In: In: Proceedings of the Proceedings – IEEE International Conference on Robotics and Automation, pp. 2556–2562 (2009)
36. GOGOA Mobility Robots HANK for clinical rehabilitation: http://gogoa.eu/products/robotic-neuro-rehabilitation/hank-for-clinical-rehabilitation/. Last accessed 2019/12/02
37. ReWalkRobotics ReWalkTM Rehabilitation: http://rewalk.com/rewalk-rehabilitation/. Last accessed 2019/12/02
38. Cyberdyne HAL®for Medical Use – Lower Limb Model: https://www.cyberdyne.jp/english/products/LowerLimb_medical.html. Last accessed 2019/12/02
39. Indego Indego Therapy Features: http://www.indego.com/indego/en/Indego-Therapy. Last accessed 2019/12/02
40. EksoBionics EksoGT: https://eksobionics.com/eksohealth/products/. Last accessed 2019/12/02
41. Novacheck, T.F.: The biomechanics of running: review paper. Gait Posture. **7**, 77–95 (1998)
42. Whittle, M.W.: Gait analysis, 4th edn. Butterworth-Heinemann, Edinburgh (2007)
43. Kinoshita, S., Kiyama, R., Yoshimoto, Y.: Effect of handrail height on sit-to-stand movement. PLoS One. **10** (2015)
44. Taal, S.R., Sankai, Y.: Exoskeletal spine and shoulders for full body exoskeletons in health care. Adv. Appl. Sci. Res. **2**, 270–286 (2011)
45. Zhang, H., Kadrolkar, A., Sup, F.C.: Design and preliminary evaluation of a passive spine exoskeleton. J. Med. Device. **10**, 11002 (2015)
46. Wojnicz, W., Zagrodny, B., Ludwicki, M., Syczewska, M., Mrozowski, J., Awrejcewicz, J.: Approach for determination of functioning of lower limb muscles. In: Springer Proceedings in Mathematics and Statistics, Awrejcewicz, J. (ed.), Springer, Vol. 249, pp. 423–438 (2018)
47. Wojnicz, W.: Biomechaniczne modele układu mięśniowo-szkieletowego człowieka (Biomechanical models of the human musculoskeletal system). Gdańsk University of Technology Publishing House, Gdańsk (2018)
48. Winter, D.A.: Kinematic and kinetic patterns in human gait: variability and compensating effects. Hum. Mov. Sci. **3**, 51–76 (1984)
49. Kirtley, C.: Clinical gait analysis normative gait database. http://www.clinicalgaitanalysis.com/data. Last accessed 2019/12/02
50. Zelik, K.E., Kuo, A.D.: Human walking isn't all hard work: evidence of soft tissue contributions to energy dissipation and return. J. Exp. Biol. **213**, 4257–4264 (2010)
51. Sloot, L.H., van der Krogt, M.M.: Interpreting joint moments and powers in gait. In: Handbook of Human Motion, pp. 1–19. Springer (2016)
52. Grzelczyk, D., Szymanowska, O., Awrejcewicz, J.: Gait pattern generator for control of a lower limb exoskeleton. Vib. Phys. Syst. **29**, 10 (2018)
53. Grzelczyk, D., Stańczyk, B., Awrejcewicz, J.: Prototype, control system architecture and controlling of the hexapod legs with nonlinear stick-slip vibrations. Mechatronics. **37**, 63–78 (2015)
54. Grzelczyk, D., Stańczyk, B., Awrejcewicz, J.: Kinematics, dynamics and power consumption analysis of the hexapod robot during walking with tripod gait. Int. J. Struct. Stab. Dyn. **17** (2017)
55. Grzelczyk, D., Biesiacki, P., Mrozowski, J., Awrejcewicz, J.: Dynamic simulation of a novel "broomstick" human forward fall model and finite element analysis of the radius under the impact force during fall. J. Theor. App. Mech. **56**(1), 239–253 (2018)
56. Nigmatullin, R.R., Morozov, A.L., Awrejcewicz, J., Ludwicki, M.: Modeling and experimental validation of walking processes. Biocybern. Biomed. Eng. **40**(1), 200–210 (2020)

Impact Wave Propagation in a Thin Elastic Isotropic Plate

Frantisek Klimenda, Josef Soukup, and Lenka Rychlikova

Abstract The paper deals with the theory of solution of transverse shock wave propagation in thin plane elastic isotropic plate (Al 99,9). The solution is made for various material and geometric models of the plate. The calculation is performed analytically for Kirchhoff and Rayleigh geometric models and Hook 's material model. The plate is fixed around its perimeter. The plate is loaded continuously or by the solitary force acting on the upper facial surface in the perpendicular direction to the midline surface of the unloaded plate. The paper presents the relations and results for transverse displacement, velocity and stress. In the conclusion, the analytical results are compared with the experiment.

Keywords Kirchhoff · Wave propagation · Thin plates

1 Introduction

The problem of plate vibration is an issue more than 200 years old. The first one is devoted to mathematical problem solving was Euler, who in 1776 analysed the free vibration of the plates. German physicist Chladni with his experiment with vibration of fixed horizontal plates proved the existence of various types of free vibration. Chladini applied a thin layer of powder to the plate, which created regular patterns after vibrations were induced. These results of experiments attempted to justify theoretically James II Bernoulli. Its solution was based on Euler 's previous work and resulted in Euler-Bernoulli 's beam bending theory. It represents a Bernoulli plate as a system of belts, wherein each belt is considered beam. In 1850, Kirchhoff published important work on the theory of thin plates, which contributed to the clarification of the theory of plate bending and is widely used in practice. Thin plate

F. Klimenda (✉) · J. Soukup · L. Rychlikova
Faculty of Mechanical Engineering, University of J. E. Purkyne in Usti nad Labem, Usti nad Labem, Czech Republic
e-mail: frantisek.klimenda@ujep.cz

J. Awrejcewicz (ed.), *Perspectives in Dynamical Systems I: Mechatronics and Life Sciences*, Springer Proceedings in Mathematics & Statistics 362,
https://doi.org/10.1007/978-3-030-77306-9_19

theory was further elaborated, for example Rayleigh, Flügge, Timoshenko-Mindlin and others contributed significantly to the solution.

2 Current State

The problem of stress wave propagation in plates has not yet been satisfactorily resolved. At present, research in the world is focused primarily on the solution of shells. This is a complex problem in terms of the interaction of the pressure and tensile phases of the wave generated at the interface of material in homogeneities. The smaller the particles, the greater the number of material interfaces that interact with the traveling wave and the greater the attenuation and dispersion. The material model used has a significant influence on the solution, because materials of diametrically different rheological properties are used, which give the structure a higher strength. Material properties must be described for different velocity of the moving body. These are the modulus of elasticity, Poisson number, thermal expansion of the material, etc. [1].

The solution of deformation and state of the plate under impulse load is realized on the basis of simplifying assumptions, in particular [2].

- *plate geometry*
- *the size and nature of its deformation, bearing and excitation load*
- *the rheological properties of the plate material*
- *simplifying assumptions within the applied solution method* – assumption of small deformations, linearity of determining relations, superposition principle, neglecting the influence of shear, etc.

Important is the definition of geometric assumptions (see Fig. 1), which are based in particular

1. the plate is a body whose one dimension in the vertical z-direction – the thickness h is less than the other two dimensions in the x and y directions – a, b. The median plane of the unloaded plate is parallel to the faces, halved by the plate thickness, which for a prismatic plate is $h = \text{const}$.
2. the origin of the coordinate system is selected in the median plane, the vertical z-axis is perpendicular to it, for rectangular plates the axes x and y are either identical with the axes of symmetry or with the axes of the side walls of the plate
3. perpendicular cuts to the median plane of the undeformed plate remain perpendicular to the curved medial surface of the deformed plate, they do not deform – Kirchhoff hypothesis of preservation of normal
4. the elements of the centring surface do not change their length dimensions, i.e. the deformed plate has no normal stress in the centring surface $\sigma_x = 0$, $\sigma_y = 0$,
5. stress in the direction of normal to the median surface of the deformed plate is negligible in relation to the other two components of the bending stress $\tau_z = 0$,

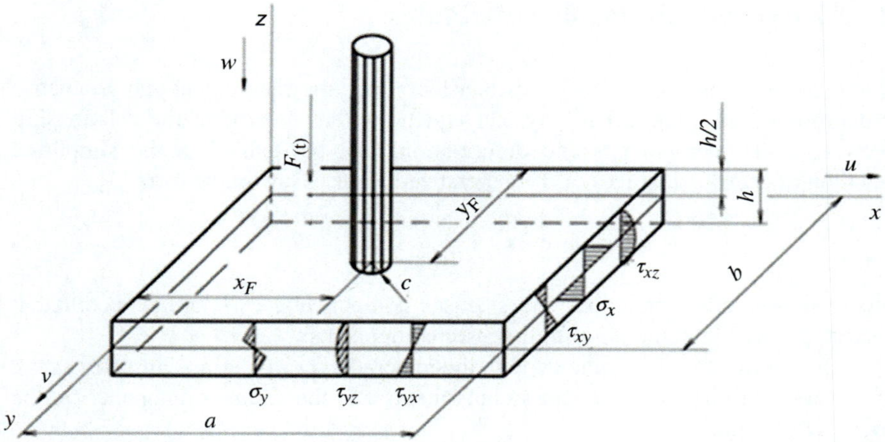

Fig. 1 Plate model [1]
Legend: *a, b, h* – plate dimensions, c – punch radius, $F(t)$ – loading force, x_F, y_F – coordinates of the applied force, *u, v, w* – displacements in axis direction, σ_x, σ_y – bending stresses, τ_{yx}, τ_{xy} – torsional stress, τ_{xz}, τ_{yz} – shear forces

6. the plate is loaded by a continuous load $p(x, y, t)$ in the upper face or in its part, event. by a single force, always in a direction perpendicular to the median plane of the unformed plate.

2.1 Material and Geometric Models of Plates

Material models gradually improved as materials knowledge and behaviour evolved. Models have been created for elastic and viscoelastic material, e.g.: Hook, Voigt-Kelvin, Maxwell and Zener standard body models [3–7].

When solving plates, we start from geometric models, to which we assign material models. The basic simplified geometric model is the Kirchhoff model, which is based on the basic Kirchhoff theory, where only vertical plate displacements and their corresponding inertial effects are considered. Rayleigh 's model extends the basic Kirchhoff theory by the influence of the cross-section rotation and the corresponding inertia effects of the plate. The Flügge model considers the effect of shear on the resulting vertical displacement, the Timoshenko-Mindlin model considers both influences, i.e. the effect of the cross-section rotation (Rayleigh) and the effect of the shear (Flügge). This model is closest to reality.

3 Analytical Solution of Thin Plates

The analytical solution of the plates considers both the rheological and geometric properties of the plates. The physical equations that determine the relationship between stress components and deformations can be defined in the simplified notation of truncated indices of the generalized Hook 's law in the form [1].

$$\boldsymbol{\sigma}_i = \mathbf{c}_{ij}\ \boldsymbol{\varepsilon}_j, i, j = 1, 2, \ldots, 6 \tag{1}$$

where $\boldsymbol{\sigma}_i$ – vector containing stress tensor components, $\mathbf{c}_{ij}$ – elastic coefficient matrix, $\boldsymbol{\varepsilon}_j$ – vector containing strain tensor components.

If the coordinate axes coincide with the material axes (material symmetry axes), the relationship between the stress components and the strain components can be presented in matrix form

$$\begin{bmatrix} \sigma_x \\ \sigma_y \\ \sigma_z \\ \tau_{yz} \\ \tau_{xz} \\ \tau_{xy} \end{bmatrix} = \begin{bmatrix} c_{11} & c_{12} & c_{13} & 0 & 0 & 0 \\ c_{21} & c_{22} & c_{23} & 0 & 0 & 0 \\ c_{31} & c_{32} & c_{33} & 0 & 0 & 0 \\ 0 & 0 & 0 & c_{44} & 0 & 0 \\ 0 & 0 & 0 & 0 & c_{55} & 0 \\ 0 & 0 & 0 & 0 & 0 & c_{66} \end{bmatrix} \cdot \begin{bmatrix} \varepsilon_x \\ \varepsilon_y \\ \varepsilon_z \\ \gamma_{yz} \\ \gamma_{xz} \\ \gamma_{xy} \end{bmatrix} \tag{2}$$

Relations between stress and strain components given by Eqs. (1), resp. (2) can be expressed in inverse form

$$\begin{bmatrix} \varepsilon_x \\ \varepsilon_y \\ \varepsilon_z \\ \gamma_{yz} \\ \gamma_{xz} \\ \gamma_{xy} \end{bmatrix} = \begin{bmatrix} \frac{1}{E_x} & -\frac{\mu_{yz}}{E_y} & -\frac{\mu_{zx}}{E_z} & 0 & 0 & 0 \\ -\frac{\mu_{xy}}{E_x} & \frac{1}{E_y} & -\frac{\mu_{zy}}{E_z} & 0 & 0 & 0 \\ -\frac{\mu_{zx}}{E_x} & -\frac{\mu_{yz}}{E_y} & \frac{1}{E_z} & 0 & 0 & 0 \\ 0 & 0 & 0 & \frac{1}{G_{yz}} & 0 & 0 \\ 0 & 0 & 0 & 0 & \frac{1}{G_{xz}} & 0 \\ 0 & 0 & 0 & 0 & 0 & \frac{1}{G_{xy}} \end{bmatrix} \cdot \begin{bmatrix} \sigma_x \\ \sigma_y \\ \sigma_z \\ \tau_{yz} \\ \tau_{xz} \\ \tau_{xy} \end{bmatrix} \tag{3}$$

Assuming small deformations, i.e. according to geometric linear theory, the deformation tensor components are functions of the displacement vector which can be written in the form of a Cartesian coordinate system

$$\varepsilon_x = \frac{\partial u}{\partial x}, \varepsilon_y = \frac{\partial v}{\partial y}, \varepsilon_z = \frac{\partial w}{\partial z}, \tag{4a}$$

$$\gamma_{xy} = \frac{\partial u}{\partial y} + \frac{\partial v}{\partial x}, \gamma_{xz} = \frac{\partial u}{\partial z} + \frac{\partial w}{\partial x}, \gamma_{yz} = \frac{\partial v}{\partial z} + \frac{\partial w}{\partial y} \tag{4b}$$

The vertical displacement w of the general point of the center plane is not a function of z

$$w = w\,(x, y, t) \tag{5}$$

Horizontal displacements in the x and y directions are a function of z

$$u = u\,(x, y, z, t) = -z\frac{\partial w}{\partial x},\; v = v\,(x, y, z, t) = -z\frac{\partial w}{\partial y} \tag{6}$$

In the theory of thin plates, the equations of motion are usually formulated in an integral form for the element $dx \times dy \times h$, i.e. by means of specific forces (displaceable) and specific moments – bending and torsion, always per unit of length – Fig. 2.

For further solution it is necessary to formulate three equations of motion for forces and moments acting on the slab element $dx \times dy \times h$, which can be expressed in an integral form after adjustment in the form

$$\frac{\partial q_{xz}}{\partial x} + \frac{\partial q_{yz}}{\partial y} + p\,(x, y, t) = \rho h \frac{\partial^2 w}{\partial t^2} \tag{7a}$$

$$q_{xy} = -\frac{\partial m_x}{\partial x} - \frac{\partial m_{xy}}{\partial y} = \rho \frac{h^3}{12}\frac{\partial^2 \varphi_x}{\partial t^2} \tag{7b}$$

$$q_{yz} = -\frac{\partial m_y}{\partial y} - \frac{\partial m_{yx}}{\partial x} = \rho \frac{h^3}{12}\frac{\partial^2 \varphi_y}{\partial t^2} \tag{7c}$$

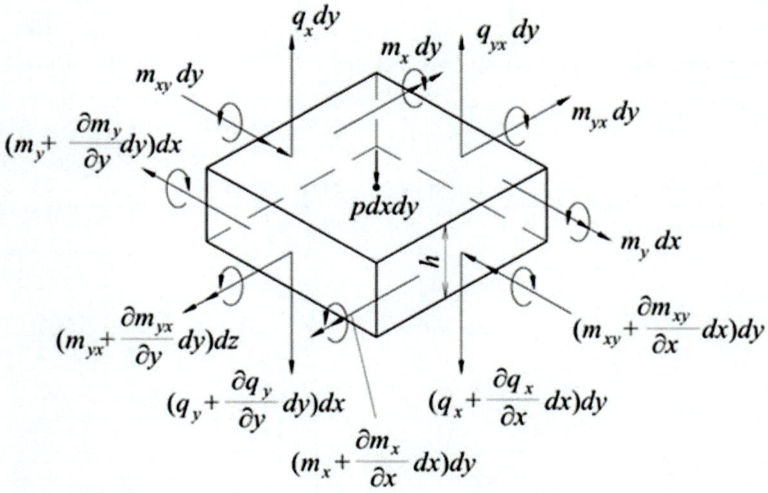

Fig. 2 Shear forces and bending moments acting on a plate element [7]
Legend: q – specific shear forces, m – specific bending moments

After adjusting the equations of motion (7), we obtain the resulting relations for the Kirchhoff and Rayleig model of a thin isotropic plate.

(a) Displacement in the z direction

Kirchhoff

$$w = \frac{16F_0}{abc\rho h}\sum_{m=1}^{\infty}\sum_{n=1}^{\infty}\frac{J_1(\gamma_{mn}c)}{\gamma_{mn}}\frac{\sin(\alpha_n x_F)\cos(\beta_m y_F)}{\omega_{mn}^2}\sin(\alpha_n x)\sin(\beta_m y)\sin^2\frac{\omega_{mn}}{2}t \tag{8a}$$

$$\omega_{mn} = \left(\alpha_n^2+\beta_m^2\right)\sqrt{\frac{D}{\rho h}} = \left(\alpha_n^2+\beta_m^2\right)\frac{ch}{\sqrt{12}} \tag{8b}$$

Rayleigh

$$w = \frac{16F_0}{abc\rho h}\sum_{m=1}^{\infty}\sum_{n=1}^{\infty}\frac{J_1(\gamma_{mn}c)}{\gamma_{mn}}\frac{\sin(\alpha_n x_F)\cos(\beta_m y_F)}{\left[\frac{h^2}{12}\left(\alpha_n^2+\beta_m^2\right)+1\right]\omega_{mn}^2} \times \sin(\alpha_n x)\sin(\beta_m y)\sin^2\frac{\omega_{mn}}{2}t \tag{9a}$$

$$\omega_{mn} = \frac{\left(\alpha_n^2+\beta_m^2\right)}{\sqrt{\frac{h^2}{12}\left(\alpha_n^2+\beta_m^2\right)+1}}\sqrt{\frac{D}{\rho h}} = \frac{\alpha_n^2+\beta_m^2}{\sqrt{\frac{h^2}{12}\left(\alpha_n^2+\beta_m^2\right)+1}}\frac{ch}{\sqrt{12}} \tag{9b}$$

$$D = \frac{Eh^3}{12\left(1-\mu^2\right)}\ \gamma_{mn} = \sqrt{\alpha_n^2+\beta_m^2}\ \alpha_n = n\frac{\pi}{a}\ \beta_m = m\frac{\pi}{b} \tag{10}$$

(b) Velocity of displacement in the z direction

Kirchhoff

$$\dot{w} = -\frac{8zF_0}{abc\rho h}\sum_{m=1}^{\infty}\sum_{n=1}^{\infty}\frac{J_1(\gamma_{mn}c)}{\gamma_{mn}}\frac{\sin(\alpha_n x_F)\cos(\beta_m y_F)}{\omega_{mn}}\sin(\alpha_n x)\sin(\beta_m y)\cdot \tag{11a}$$

$$\cdot\sin(\omega_{mn})\ t \tag{11b}$$

Rayleigh

$$\dot{w} = -\frac{8zF_0}{abc\rho h}\sum_{m=1}^{\infty}\sum_{n=1}^{\infty}\frac{J_1(\gamma_{mn}c)}{\gamma_{mn}}\frac{\sin(\alpha_n x_F)\cos(\beta_m y_F)}{\left[\frac{h^2}{12}\left(\alpha_n^2+\beta_m^2\right)+1\right]\omega_{mn}}\sin(\alpha_n x)\sin(\beta_m y)\cdot \tag{12a}$$

$$\cdot\sin(\omega_{mn})\ t \tag{12b}$$

(c) Bending stresses in x-direction (σ_x) and y-direction (σ_y)

Kirchhoff

$$\sigma_x = \frac{E_z 16F_0}{\left(1-\mu^2\right)abc\rho h}\sum_{m=1}^{\infty}\sum_{n=1}^{\infty}\frac{J_1(\gamma_{mn}c)}{\gamma_{mn}}\frac{\alpha^2+\mu\beta^2}{\omega_{mn}^2} \tag{13a}$$
$$\times\sin(\alpha_n x_F)\sin(\beta_m y_F)\sin(\alpha_n x)\cdot$$

$$\cdot\sin(\beta_m y)\sin^2\left(\frac{\omega_{mn}}{2}\right)\ t \tag{13b}$$

$$\sigma_y = \frac{E_z 16F_0}{\left(1-\mu^2\right)abc\rho h}\sum_{m=1}^{\infty}\sum_{n=1}^{\infty}\frac{J_1(\gamma_{mn}c)}{\gamma_{mn}}\frac{\beta^2+\mu\alpha^2}{\omega_{mn}^2} \tag{13c}$$
$$\times\sin(\alpha_n x_F)\sin(\beta_m y_F)\sin(\alpha_n x)\cdot$$

$$\cdot\sin(\beta_m y)\sin^2\left(\frac{\omega_{mn}}{2}\right)\ t \tag{13d}$$

Rayleigh

$$\sigma_x = \frac{E_z 16F_0}{\left(1-\mu^2\right)abc\rho h}\sum_{m=1}^{\infty}\sum_{n=1}^{\infty}\frac{J_1(\gamma_{mn}c)}{\gamma_{mn}}\frac{\alpha^2+\mu\beta^2}{\left[\frac{h^2}{12}\left(\alpha_n^2+\beta_m^2\right)+1\right]\omega_{mn}^2} \tag{14a}$$
$$\times\sin(\alpha_n x_F)\sin(\beta_m y_F)\cdot$$

$$\cdot\sin(\alpha_n x)\sin(\beta_m y)\sin^2\left(\frac{\omega_{mn}}{2}\right)\ t \tag{14b}$$

$$\sigma_y = \frac{E_z 16F_0}{\left(1-\mu^2\right)abc\rho h}\sum_{m=1}^{\infty}\sum_{n=1}^{\infty}\frac{J_1(\gamma_{mn}c)}{\gamma_{mn}}\frac{\beta^2+\mu\alpha^2}{\left[\frac{h^2}{12}\left(\alpha_n^2+\beta_m^2\right)+1\right]\omega_{mn}^2} \tag{14c}$$
$$\times\sin(\alpha_n x_F)\sin(\beta_m y_F)\cdot$$

$$\cdot\sin(\alpha_n x)\sin(\beta_m y)\sin^2\left(\frac{\omega_{mn}}{2}\right)\ t \tag{14d}$$

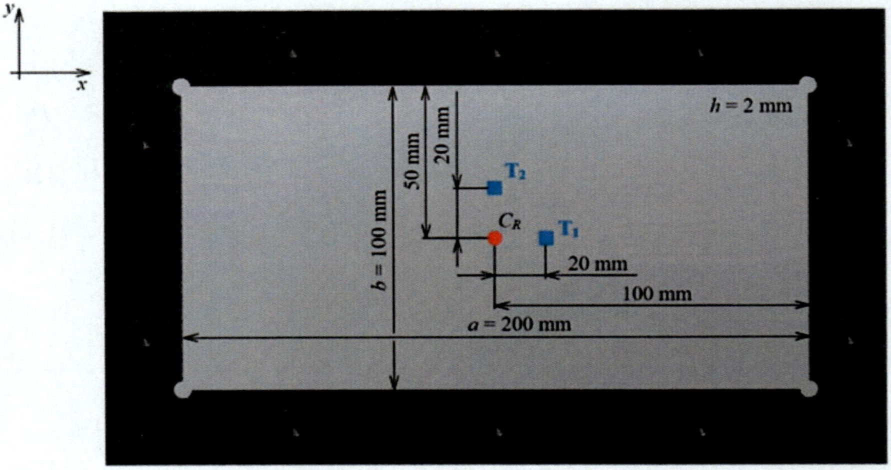

Fig. 3 Control points and center of impact C_RLegend: C_R – center of impact, $T_{1,2}$ – control points

Table 1 Material properties of Al plate [4]

Name	Mark	Unit	Value
Tensile modulus	E	Pa	7.1×10^{10}
Poisson number	μ	–	0.34
Density	ρ	$Kg{\cdot}m^{-3}$	2699

where *a, b, h* – plate dimensions, *c* – radius of load circle, ρ – density of plate material, F_0 – loading force, J_1 $(\gamma_{mn}c)$ – Bessel function of first kind, first order for argument γmnc, α, β – constants, m, n – matrix elements, x_F, y_F – circular load coordinates, ω_{mn} – natural frequency, *t* – time, *D* – stiffness modulus, *E* – tensile modulus, *G* – shear modulus, μ – material viscosity coefficient.

The analytical solution is performed in the MATLAB program for isotropic aluminum plate Al 99.9. The plate has dimensioned $a \times b \times h$, where $a = 200$ mm, $b = 100$ mm and $h = 2$ mm. The plate is fixed around its perimeter, its geometrical dimensions, the place of loading and the location of control points are shown in Fig. 3.

The rheological properties of the plate are given in Table 1.

To assess the wave propagation of the first wave sled before the chalk reflection from the point of attachment is displayed (or the edges of the boards with loosely signed).

The waveform of the spreading shock wave was determined according to formulas (8a) and (14c) for both models at points T_1 and T_2. The vertical displacements and velocities at the individual points in the z-axis direction and the normal stresses in the *x*-*y*-axis direction were determined.

To determine the time at which the waveform reaches the local extreme, times t_1, t_2 and t_3 were determined.

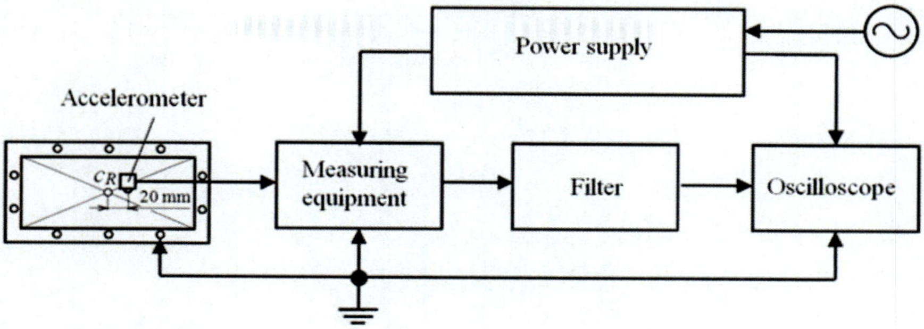

Fig. 4 Measuring chain

4 Experimental Solutions

The velocity of the spreading transverse wave was experimentally measured using a piezoelectric accelerometer. The measurement was performed at two measuring points (T_1 and T_2) perpendicular to each other at a distance of 20 mm from the center of impact C_R in the x and y axes (Fig. 3). The impact force of 1 N was caused by the impact of a cylindrical impact of Ø 5 mm with a spherical end to the center of the impact C_R, which is in the center of the plate. The measuring chain is shown on Fig. 4.

5 Solution of Results

On Figs. 5, 6, 7 and 8 shows a comparison of individual theories. The displacements, velocities, and bending stresses at the control point T_1 ($x = 20$ mm, $y = 0$ mm from the center of impact C_R) are compared. The local extremes at time t_1, t_2 and t_3 are determined for the individual functions (see Tables 2, 3, 4, and 5). Experimentally measured was only velocity w (z direction).

At first glance they look during movement of both models in Fig. 5 almost identically. The greatest local extreme difference at time t1 is approximately 58%. This is due to the fact that the Rayleig model is enhanced by the effect of the cross-section rotation and its corresponding inertial effects. Shockwaves therefore arrive at the point T_1 later than wave at Kirchhoff model and thus gives rise to a large deviation.

On Fig. 6 is a comparison of velocity graphs w at point T_1. The Kirchhoff and Reyleigho model is compared with the experiment. At first glance, it is evident that both models differ only minimally in local extremes t_2 and t_3. The course of the experiment differs with both models. This is because the experiment includes all corrections of the basic Kirchhoff model, so it is a Timoshenko-Mindlin model. Normal stresses are almost identical in both directions; the greatest deviation is

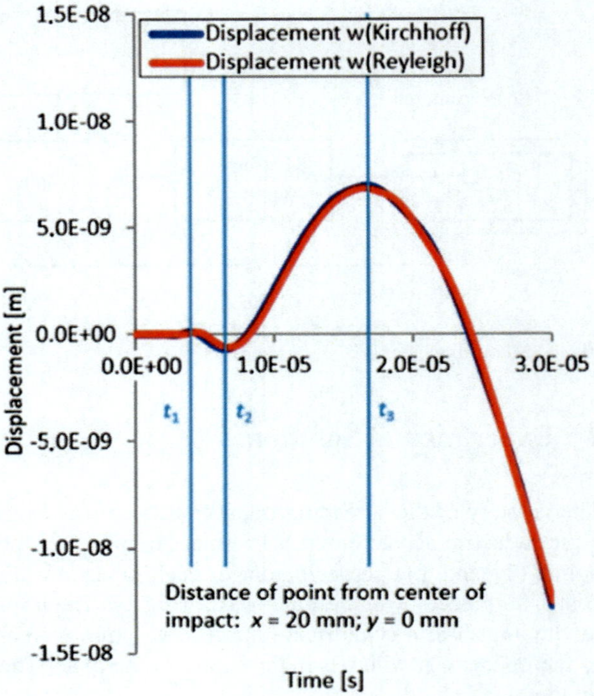

Fig. 5 Wave displacement after shock (w) at point T_1

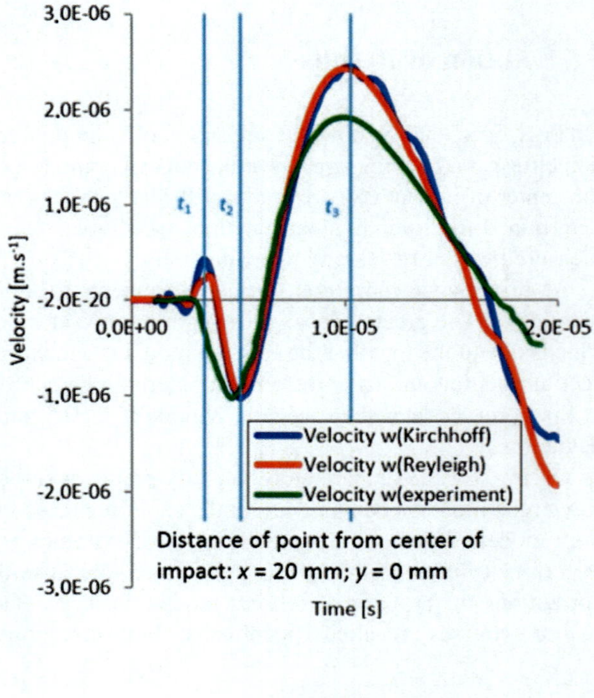

Fig. 6 Wave velocity after shock (w) at point T_1

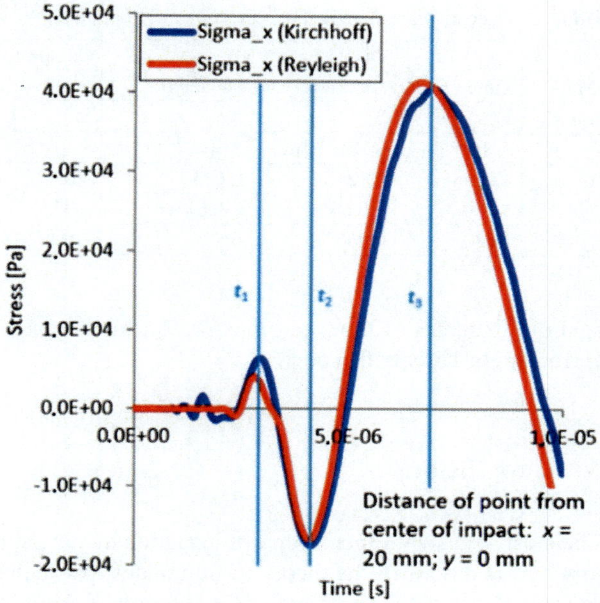

Fig. 7 Normal stress in the x-axis (σ_x) at pint T_1

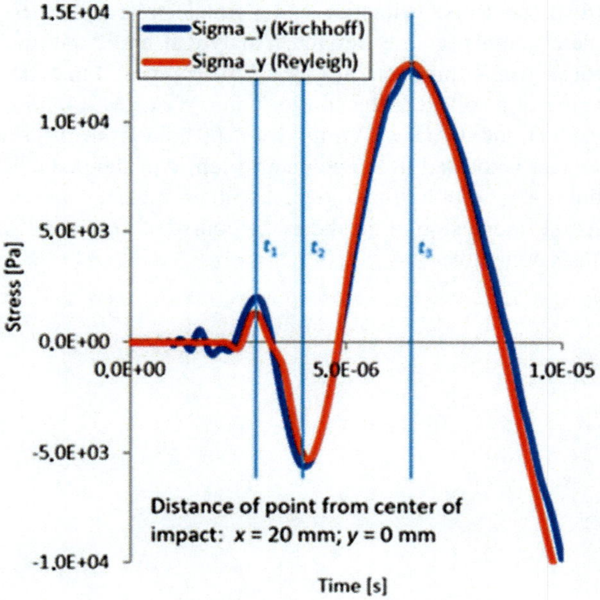

Fig. 8 Normal stress in the x-axis (σ_y) at pint T_1

Table 2 Wave displacement in the z-direction (w) at point T_1

Mark	Time	Displacement w		Comparison	
		Kirchhoff	Rayleigh	col. 4 – col. 3	(col. 5/coll. 3) × 100
1	2	3	4	5	6
	$\times 10^{-6}$ [s]	$\times 10^{-10}$ [m]	$\times 10^{-10}$ [m]	$\times 10^{-10}$ [m]	[%]
t_1	4.000	0.128	0.053	−0.075	58.6
t_2	6.500	−0.686	−0.613	0.073	10.6
t_3	1.680	6.966	6.825	−0.141	2.0

again at extremes at time t_1 – max. 43%, this is given by the already mentioned correction of Kirchhoff model.

6 Conclusion

The thesis deals with shock wave propagation in thin isotropic plate. In the introductory part is described the theory of thin plates, particular concrete assumptions for solving plates and derived analytical solution of thin plates. In the next part of the thesis, the basic equations of motion are derived, from which the basic relations (displacements, velocities and normal stresses) for Kirchhoff 's and Rayleig 's plate geometries are derived. Analytical and experimental solution is performed for isotropic thin plate of given dimensions and material properties. The individual values are solved at the control points T_1 ($x = 200$ mm, $y = 0$ mm from the shock center) and T_2 ($x = 0$ mm, $y = 200$ mm from the shock center). The center of impact is located in the geometric center of the plate. The shockwave is caused by the impact of a ball-end impact with an impact force of 1 N from a certain height. At the conclusion of the thesis the course of individual variables in the control point T_1 is compared.

Table 3 Wave velocity in the z-direction (w) at point T_1

	Velocity w				Comparison					
	Time	Kirchhoff	Reyleigh	Experiment	col. 4 – col. 3	col. 5 – col. 3	(col. 5 – col. 4)	(col. 6/col. 3) ×100	(col. 7/col. 3) ×100	(col. 8/col. 4) ×100
1	2	3	4	5	6	7	8	9	10	11
	$\times 10^{-5}$ [m.s^{-1}]	$\times 10^{-6}$ [m.s^{-1}]	$\times 10^{-6}$ [m.s^{-1}]	$\times 10^{-6}$ [m.s^{-1}]	$\times 10^{-6}$ [m.s^{-1}]	$\times 10^{-6}$ [m.s^{-1}]	$\times 10^{-6}$ [m.s^{-1}]	[%]	[%]	[%]
t_1	0.304	0.429	0.208	−0.069	−0.221	−0.498	−0.277	51.5	116.1	133.2
t_2	0.510	−0.103	−0.097	−0.111	0.006	−0.008	−0.014	5.8	7.8	14.4
t_3	1.030	2.452	2.409	2.157	−0.043	−0.295	−0.252	1.8	12.0	10.5

Table 4 Normal stress in the *x*-axis (σ_x) at pint T_1

Mark	Time	Normal stress		Comparison	
		Kirchhoff	Reyleigh	col. 4 – col. 3	(col. 5/col. 3) × 100
1	2	3	4	5	6
	$\times 10^{-6}$ [s]	$\times 10^4$ [Pa]	$\times 10^4$ [Pa]	$\times 10^4$ [Pa]	[%]
t_1	2.900	0.643	0.365	−0.278	43.2
t_2	4.100	−1.745	−1.680	0.065	3.7
t_3	6.900	4.071	4.101	0.030	0.7

Table 5 Normal stress in the *x*-axis (σ_y) at pint T_1

Mark	Time	Normal stress		Comparison	
		Kirchhoff	Reyleigh	col. 4 – col. 3	(col. 5/col. 3) × 100
1	2	3	4	5	6
	$\times 10^{-6}$ [s]	$\times 10^4$ [Pa]	$\times 10^4$ [Pa]	$\times 10^4$ [Pa]	[%]
t_1	2.900	0.203	0.134	−0.069	34.0
t_2	4.100	−0.558	−0.506	0.052	9.3
t_3	6.500	1.245	1.268	0.023	1.8

Acknowledgments This work was supported by internal grant of Jan Evangelista Purkyně University no. UJEP SGS-2018-48-002-2.

References

1. Brepta, R.: Rázy a vlny napětí v pevných elastických tělesech. Vydavetelství ČVUT, Praha (1997)
2. Sokolnikoff, I.S.: Mathematical Theory of Elasticity. McGrow-Hill (1956)
3. Huffington, N.J., Hoppmann, W.J.: On the transverse vibrations of rectangular orthotropic bases. J. Appl. Mech. **25**, 389–395 (1958)
4. Volek, J., Červ, F., Valeš, J.: Vlny v tenké ortotropní desce (Model Kirchhoffův a Rayleigho) generované příčným pulsním zatížrním. In: Výpočtová mechanika 2003, 19. konference s mezinárodní účastí, Nečtiny (2003)
5. Kolsky, H.: The propagation of stress waves in viscoelastic solids. Appl. Mech. Rev. **11**(9), 465–468 (1958)
6. Leitmann, M.J.: The Linear Theory of Viscoelasticity. Encyklopedia of Physics, Berlin (1973)
7. Soukup, J., Volek, J.: A thin rectangular viscoelastic orthotropic plate under transverse impuls loading. In: 9th Conference on Dynamical Systems Theory and Applications, Poland (2007)

Optimal Rendezvous with Proportional Navigation Unmanned Aerial Vehicle

Oleg Cherkasov and Elina Makieva

Abstract Two-dimensional optimal rendezvous problem with proportional navigation unmanned aerial vehicle is analyzed using a non-linear model. The velocities of both players have a constant modulus, but vary in direction. The problem is to minimize the final distance between the pursuer and the drone in the transition from the given initial conditions. The angle between line-of-sight and pursuer velocity vector is considered as a control variable. The process time is given beforehand and is fixed. Application may be interesting for the rendezvous problem tanker-drone. The Maximum Principle procedure allows to reduce optimal control problem to the boundary-value problem for two nonlinear differential equations. The qualitative analysis of the equations of the boundary-value problem allows to determine the characteristic features of the extremal trajectories. These analysis is based on the consideration of the phase portrait. Simulation results are presented to illustrate the analytical conclusions. A numerical comparison of extreme solutions is made with the case when the pursuer applies the pure pursuit method.

Keywords Optimal rendezvous · Proportional navigation · Phase portrait

1 Introduction

Pursuit-evasion problems have been traditionally classified among the classical examples of differential game theory. An overview of the key works of this kind is given in the article [1]. In the last decades another approach has been applied to these problems, namely, to fix the pursuer's strategy and to form a one-sided optimal control problem for the evader. This approach, being conceptually simpler than the former, enables more realistic models to be applied for the dynamics of the opponents. In general, the fixed pursuer's strategy has been taken as constant gain proportional navigation which, under some formulations, is an optimal strategy

O. Cherkasov (✉) · E. Makieva
Lomonosov Moscow State, Moscow, Russia

J. Awrejcewicz (ed.), *Perspectives in Dynamical Systems I: Mechatronics and Life Sciences*, Springer Proceedings in Mathematics & Statistics 362,
https://doi.org/10.1007/978-3-030-77306-9_20

for the pursuer [2]. In most works on optimal pursuit-evasion, where the results could be obtained in a closed form, it was assumed that the relative trajectory can be linearized in the neighborhood of the initial line of sight (LOS). This assumption is acceptable when the final stage of the rendezvous is analyzed. In this case, it is reasonable to take into account the time delay in the navigation system of the pursuer. If the time of the process is long enough, and LOS can be rotated at large angles, non-linear kinematics should be applied. In this case, the delay in the pursuer guidance system can be neglected. In the paper [3] using the exact nonlinear equations of motion optimal guidance law for a vehicle pursuing maneuvering target was derived. Complete knowledge of the evader's motion was assumed available to the pursuer. The same approach to the planar interception was considered in the paper [4]. The problem of the optimal evasion from a pursuer employing proportional navigation guidance is considered in paper [5]. The problem of maximizing the time-to-capture was considered under assumption that the problem parameters, namely, the speed ratio and the proportional navigation constant, are such that capturability of the evader is guaranteed. Qualitative analysis of the optimal planar evasion against proportional navigation pursuer is presented in the paper [6].

This work is devoted to the problem of minimizing the final distance between the players for a fixed time. Instead of evasion problem against proportional navigation pursuer we consider the optimal rendezvous with proportional navigation unmanned aerial vehicle. Such a problem may be relevant when planning the approach trajectories of a tanker aircraft with an unmanned aerial vehicle and also in case of interception of the attacking unmanned aerial vehicle by the simulator of the target launched from the real target. The research method is based on the techniques proposed in [6].

2 Problem Formulation

The equations of relative motion of the players are:

$$\dot{r} = cos\alpha - bcos\beta, \tag{1}$$

$$\dot{\beta} = a\,(sin\alpha - bsin\beta)\,/r, \tag{2}$$

where r – dimensionless distance between the players, β – the angle between the velocity vector of the Player 2 (P2) and the LOS, α – the angle between the velocity vector of the Player 1 (P1) and the LOS, considered as a control variable, a is a constant reflecting the fixed strategy of the P2, a is linked with the proportional navigation gain k by the formula $a = k - 1$, $a > 0$, b is a positive constant representing the ratio of the P2 velocity magnitude to the P1 velocity magnitude (see Fig. 1). The final time of the process is fixed.

The initial conditions for the Eqs. (1 and 2) have the form:

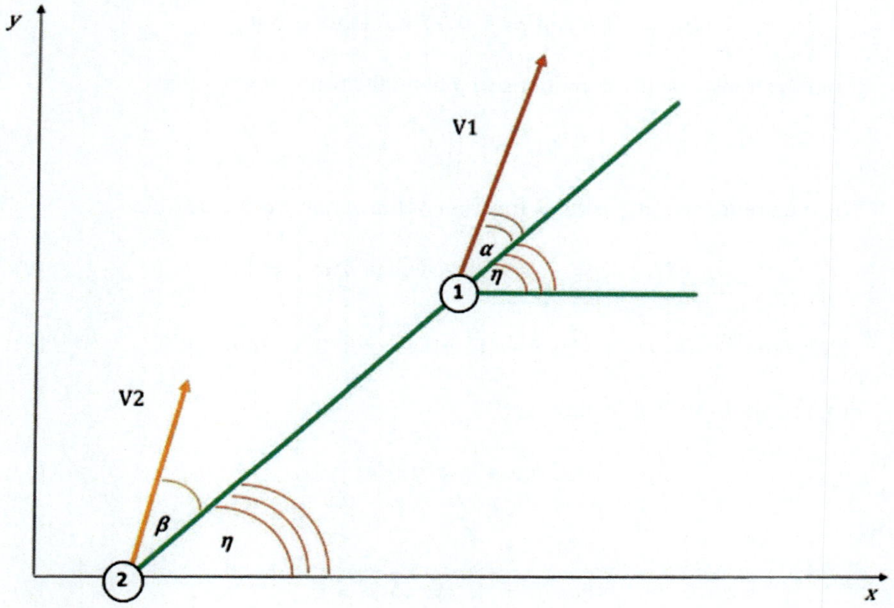

Fig. 1 The geometry of rendezvous

$$r(0) = r_0, \beta(0) = \beta_0, \tag{3}$$

The final conditions are free. The purpose of control is to minimize the functional:

$$J = r(T). \tag{4}$$

3 Problem Analysis

Pontryagin's function [7] for the problem (1)–(4) is:

$$\mathrm{H} = \psi_r\,(cos\alpha - bcos\beta) + \psi_\beta\,(a\,(sin\alpha - bsin\beta)\,/r) = \mathrm{C}, \tag{5}$$

where C is a constant.

The equations for the conjugate variables are:

$$\dot{\psi}_r = -\partial H/\partial r = \psi_\beta \cdot a\,(sin\alpha - bsin\beta)\,/r^2, \tag{6}$$

$$\dot{\psi}_\beta = -\partial H/\partial\beta = -bsin\beta\psi_r + abcos\beta\psi_\beta/r \tag{7}$$

From the transversality conditions we obtain the following relations:

$$\psi_{r_T} = -1, \psi_{\beta_T} = 0 \tag{8}$$

The maximum conditions of H function of the variable α are as follows:

$$\partial H/\partial\alpha = -\psi_r sin\alpha + \psi_\beta \cdot acos\alpha/r = 0, \tag{9}$$

$$\partial^2 \mathrm{H}/\partial\alpha^2 \leq 0 \Rightarrow -\psi_r cos\alpha - \psi_\beta \cdot asin\alpha/r \leq 0, \psi_r/cosa \geq 0. \tag{10}$$

From relations (9, 10) one can get:

$$\psi_\beta = \psi_r rtg\alpha/a, \tag{11}$$

and

$$tg\alpha(T) = 0, \quad \psi_r \cos\alpha > 0. \tag{12}$$

Therefore,

$$\alpha(T) = \pi. \tag{13}$$

By differentiating relation (11) with respect to t according to (1, 2, 6, 7), the equation for the control variable could be obtained:

$$\dot{\alpha} = b\,(a+1)\,\mathrm{cos}\alpha \cdot \sin\,(\alpha - \beta)\,/r - (sin\alpha - bsin\beta)\,/r. \tag{14}$$

Thus, the optimal control problem (1)–(4) is reduced to the following boundary value problem for a system of three differential equations:

$$\dot{r} = cos\alpha - bcos\beta, r(0) = r_0 \tag{15a}$$

$$\dot{\beta} = a\,(sin\alpha - bsin\beta)\,/r, \beta(0) = \beta_0, \tag{15b}$$

$$\dot{\alpha} = b\,(a+1)\,\mathrm{cos}\alpha \cdot \sin\,(\alpha - \beta)\,/r - (sin\alpha - bsin\beta)\,/r, \alpha(T) = \pi. \tag{15c}$$

For the analysis of the boundary value problem (15) let us consider a section of the phase space of a dynamical system by plane $r = const$. Further confine ourselves to the analysis of the system (2, 14), taking into account, that the phase portrait of this system, types and location of equilibrium points do not depend on r, its value determines only the speed of movement along phase trajectories.

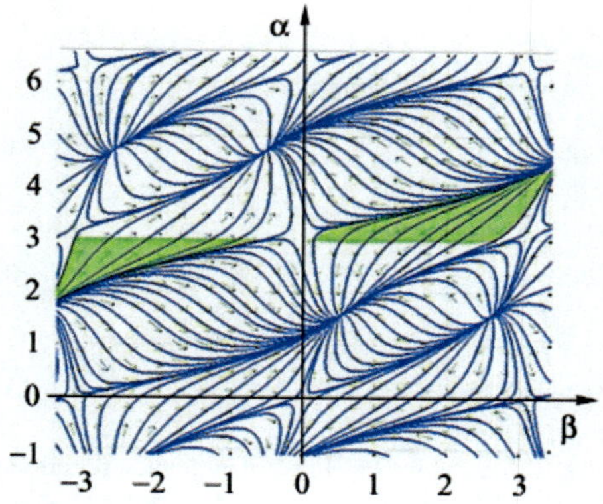

Fig. 2 The phase portrait of the system (2), (14) for $a = 1$, $b = 2$

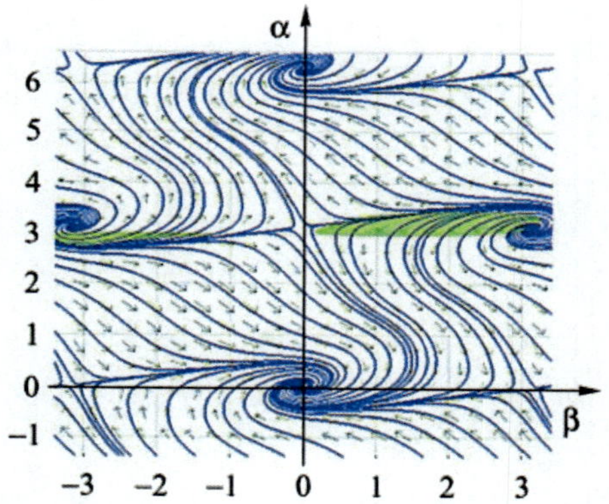

Fig. 3 The phase portrait of the system (2), (14) for $a = 1$, $b = 1/2$

The system (2), (14) considered on the torus has eight equilibrium points. Expanding the torus on a plane, we obtain in the square $[-\pi;\pi] \times [-\pi;\pi]$ thirteen equilibrium points:

(a) $(0;0)$, $(\pi;\pi)$, $(-\pi;\pi)$, $(\pi;-\pi)$, $(-\pi;-\pi)$, $(0;\pi)$, $(\pi;0)$, $(-\pi;0)$, $(0;-\pi)$,
(b) $(\pm\, arcsin\ 1/b;\ \pm\,\pi/2)$,
(c) $(\pm(\pi\ -\ arcsin\ 1/b;\ \pm\,\pi/2)$.

Analysis of the equations, linearized in the neighborhood of the equilibrium points, allows to determine the type of the stationary solutions. If $b > 1$, solutions (a) are the saddle type points; solutions (b) are the stable node type; solutions (c) are the unstable node type. The phase portrait of the system (2, 14) is shown in Fig. 2. The analysis of phase portrait allows to determine qualitative properties of extreme trajectories.

The phase portrait of the system (2), (14) for $0 < b < 1$ is shown in Fig. 3.

From the analysis of the phase portrait it follows that in the case when $\beta_0 > 0$, $\alpha_0 > \pi$ and when $\beta_0 < 0$, $\alpha_0 < \pi$. Also, according to the phase portrait, α_0 and β_0 lie in different half-planes. If the time of the process is sufficient large, the main contribution to the functional is made by the motion in the neighborhood of the saddle type point. It is easy to observe the monotone behavior of the $\alpha(t)$.

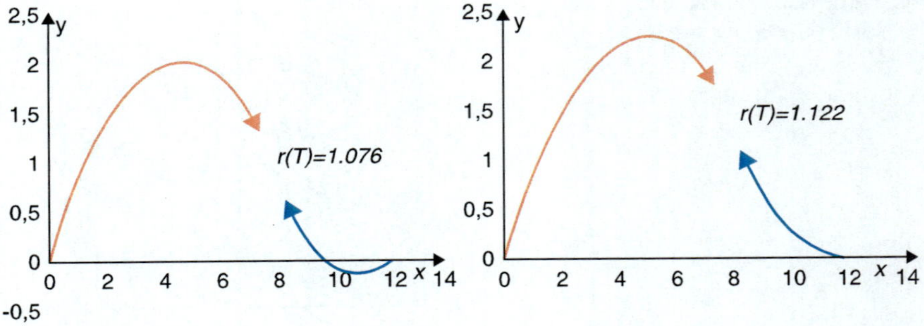

Fig. 4 Trajectories of Players in the horizontal plane at $\beta_0 = \pi/4$, $r_0 = 12$, $T = 4$; (**a**) P1 applies the law (14), (**b**) P1 applies the pure pursuit method

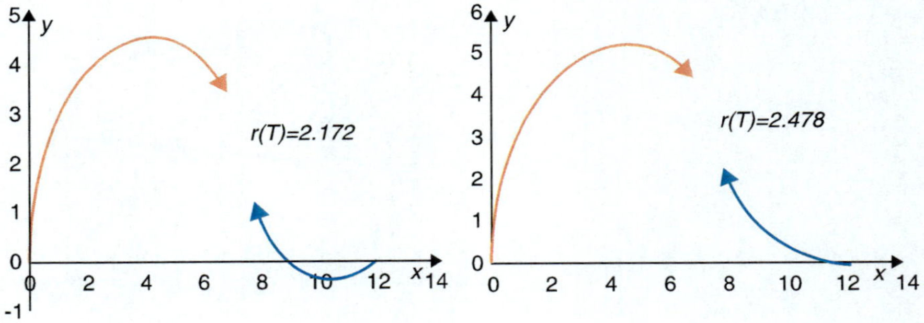

Fig. 5 Trajectories of Players in the horizontal plane at $\beta_0 = \pi/2$, $r_0 = 12$, $T = 5$; (**a**) P1 applies the law (14), (**b**) P1 applies the pure pursuit method

4 Numerical Solution

The numerical solution of the boundary value problem (15) was carried out for the values $a = 1$, $b = 2$, the calculation results are shown in Figs. 4, 5, 6 and 7.

The boundary value problem for system (2, 14) was solved by the shooting and dichotomy method; the corresponding Cauchy problems were solved in Matlab ODE 45 (fourth-order Runge-Kutta method with automatic selection of the integration step).

The simulation results illustrate the conclusions made based on the analysis of the phase portrait. Nevertheless, the direction of the Player's velocity vectors at the initial moment of time on either side of the LOS seems unexpected. For this reason, it was decided to carry out the simulation for an alternative strategy for P1. As an alternative strategy, the pure pursuit method when the velocity vector of P1 is aimed exactly at P2 at each the moment of time was chosen (the final condition on the position of the velocity vector of P1 is also satisfied).

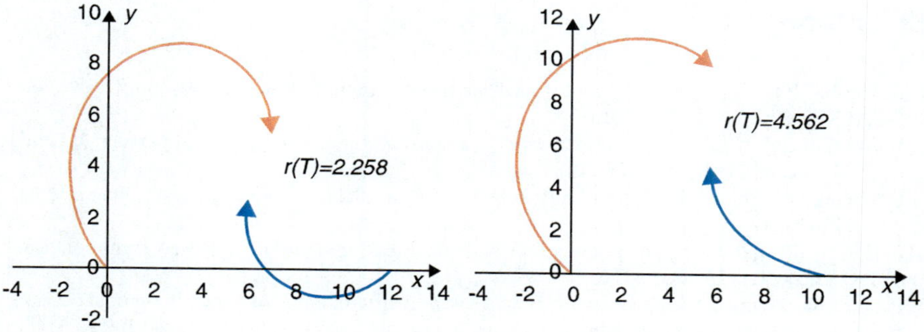

Fig. 6 Trajectories of Players in the horizontal plane at $\beta_0 = 3\pi/4$, $r_0 = 12$, $T = 9$; (**a**) P1applies the law (14), (**b**) P1 applies the pure pursuit method

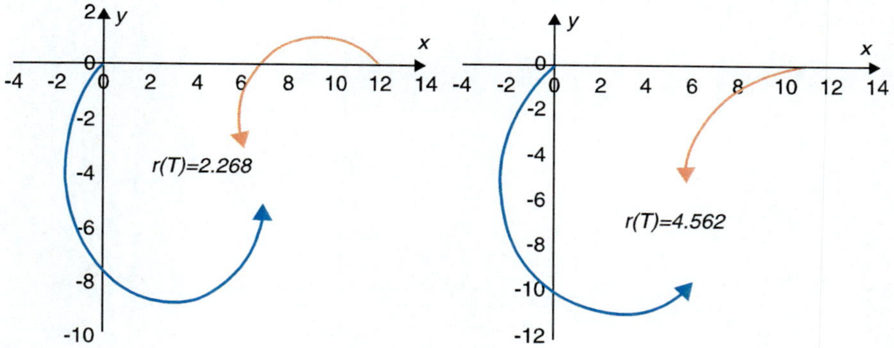

Fig. 7 Trajectories of Players in the horizontal plane at $\beta_0 = 5\pi/4$, $r_0 = 12$, $T = 9$; (**a**) P1 applies the law (14), (**b**) P1 applies the pure pursuit method

The corresponding comparison paths are shown in Figs. 4b, 5b, 6b, and 7b. It can be seen from the figures that the distance between the players at the final moment of time, if P1 uses the pursuit method, is greater. Corresponding final distance is indicated in Figs. 4, 5, 6 and 7.

5 Conclusions

The rendezvous problem of moving objects with a fixed guidance strategy of one player and the optimal one of the other is investigated. Application of the Pontryagin maximum principle allows us to reduce the optimal control problem to a boundary value problem for the initial variables. A qualitative analysis of this system is carried out, characteristic properties are illustrated by computer modeling.

References

1. Kabamba, P.T., Girard, A.R.: Proportional navigation: optimal homing and optimal evasion. SIAM Rev. **57**(4), 611–624 (2015)
2. Ben-Asher, J.Z., Cliff, E.M.: Optimal evasion against a proportionally guided pursuer. J. Guid. Control Dynam. **12**(4), 598–600 (1989)
3. Guelman, M., Shinar, J.: Optimal guidance law in the plane. J. Guid. Control Dynam. **7**(4), 471–476 (1984)
4. Glizer, V.Y.: Optimal planar interception with fixed end conditions: a closed form solution. JOTA. **88**(3), 503–539 (1996)
5. Pachter, M., Yavin, Y.: Simple-motion pursuit-evasion differential games, part 2: optimal evasion from proportional navigation guidance in the deterministic and stochastic cases. JOTA. **51**(1), 129–159 (1986)
6. Cherkasov, O.Y., Yakushev, A.G.: Singular arcs in the optimal evasion against a proportional navigation vehicle. JOTA. **113**(2), 211–226 (2002)
7. Pontryagin, L.S., et al.: The Mathematical Theory of Optimal Processes. Interscience Publishers, New York (1962)

Optimization of the Geometry of Aeroelastic Energy Harvester

Filip Sarbinowski and Roman Starosta

Abstract Geometry optimization was performed using a genetic algorithm (GA) that processes data from computational fluid dynamics (CFD) calculations. This algorithm generated a random population of twenty-arm geometrical figures. Each geometry was subjected to a numerical experiment during which the resultant force acting on the body was calculated. The calculations were repeated for angular orientation of the object varying from 0 to 180 degrees, at 5 degrees step, to obtain a complete characteristic of aerodynamical forces acting on the body related to its angular orientation. For each of the obtained functions, the satisfaction of Den Hartog's criterion is examined, which is the basis for geometry evaluation. To accelerate the calculations, classical GA has been modified by implementing a chaotic crossover process based on a logistic map. The numerical calculations were performed using the Method of Fundamental Solutions.

Keywords Energy harvesting · Optimization · Galloping

1 Introduction

The aeroelastic energy harvester is a device that allows energy recovery from vibrations induced by the flow. Man has used the flow energy since the dawn of time, but only the recent development of electronics and autonomous devices has contributed to the need for small-scale systems that are capable of continuous low power generation. Devices of this type use the phenomenon of galloping caused by negative aerodynamic damping, which – assuming the quasi-stationarity of the phenomenon – was first described by Den Hartog [1] and extended in [2].

In these works, a body with one degree of freedom (translation parallel to axis Z), mounted on a damped spring system, subjected to flow in a direction parallel to

F. Sarbinowski (✉) · R. Starosta
Poznan University of Technology, Poznan, Poland
e-mail: filip.j.sarbinowski@doctorate.put.poznan.pl

J. Awrejcewicz (ed.), *Perspectives in Dynamical Systems I: Mechatronics and Life Sciences*, Springer Proceedings in Mathematics & Statistics 362,
https://doi.org/10.1007/978-3-030-77306-9_21

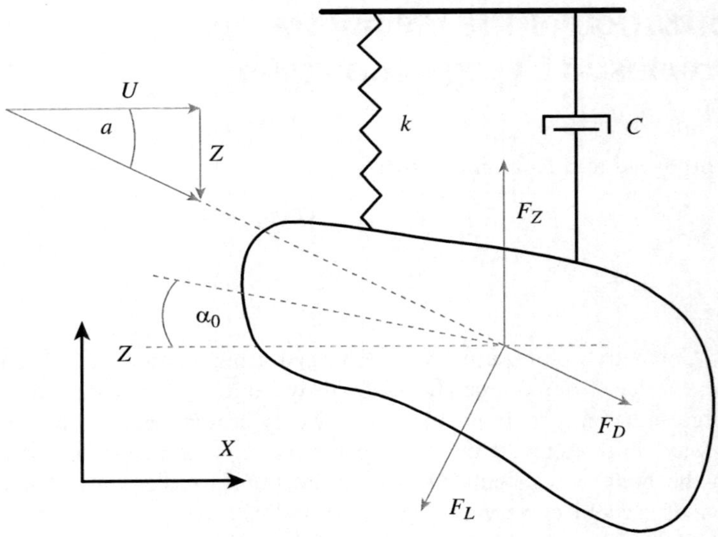

Fig. 1 Physical model of the aeroelastic energy harvester. F_L – lift force, F_D – drag force

the X-axis was analyzed (see Fig. 1). The dynamics of the system is described by the equation:

$$m\,\ddot{z}(t) + c\,\dot{z}(t) + k\,z(t) = F_z\,(\alpha) = \frac{1}{2}\,b\,\rho\,U^2\,C_z\,(\alpha)\,, \tag{1}$$

where: m – the mass of the body, c –damping coefficient, k –stiffness coefficient, $z(t)$ – displacement in the Z direction, $\dot{()}$ and $\ddot{()}$ – first and second differential with respect to time, $C_z(\alpha)$ – coefficient of aerodynamic force acting in the Z direction at α angular orientation of the body, F_Z – aerodynamic force component acting in the Z direction, b – characteristic length of the body, ρ – fluid density, U – flow velocity.

Galloping occurs for $(c - C_z(\alpha_0)) < 0$.

In Fig. 1 it can be seen that for small enough α it is true that:

$$F_z\,(\alpha) = -F_D \sin(\alpha) - F_L \cos(\alpha) \approx -F_L - F_D\,\alpha, \tag{2}$$

and after expanding in Taylor's series:

$$-F_L - F_D\,\alpha \approx -F_L - \left(\frac{dF_L}{d\alpha} + F_D\right)\alpha - \frac{1}{2}\left(\frac{d^2F_L}{d\alpha^2} + \frac{dF_D}{d\alpha}\right)\alpha^2$$
$$-\frac{1}{6}\left(\frac{d^3F_L}{d\alpha^3} + \frac{d^2F_D}{d\alpha^2}\right)\alpha^3 - \ldots \tag{3}$$

Since the constant force component F_L does not affect the dynamics of the system, it will be neglected in further considerations. The only factors affecting the aerodynamic force depending on the α angle are aerodynamics coefficients, hence:

$$C_z(\alpha) \approx \left(\frac{dC_L}{d\alpha} + C_D\right)\alpha + \frac{1}{2}\left(\frac{d^2C_L}{d\alpha^2} + \frac{dC_D}{d\alpha}\right)\alpha^2 + \frac{1}{6}\left(\frac{d^3C_L}{d\alpha^3} + \frac{d^2C_D}{d\alpha^2}\right)\alpha^3 + \ldots, \tag{4}$$

As reported in [3, 4] the approximation of the $C_z(\alpha)$ function with a third-order polynomial is sufficient for energy harvesting purposes:

$$\begin{cases} C_z(\alpha) \approx a_1\,\alpha + a_2\,\alpha^2 + a_3\,\alpha^3 \\ a_1 = \left(\frac{dC_L}{d\alpha} + C_D\right) \\ a_2 = \frac{1}{2}\left(\frac{d^2C_L}{d\alpha^2} + \frac{dC_D}{d\alpha}\right) \\ a_3 = \frac{1}{6}\left(\frac{d^3C_L}{d\alpha^3} + \frac{d^2C_D}{d\alpha^2}\right) \end{cases} \tag{5}$$

The mathematical model of the phenomenon thus takes the form:

$$m\ddot{z}(\mathrm{t}) + c\dot{z}(\mathrm{t}) + kz(\mathrm{t}) = -\frac{1}{2}\rho Ub\left(a_1\frac{\dot{z}(t)}{U} + a_2\left(\frac{\dot{z}(t)}{U}\right)^2 + a_3\left(\frac{\dot{z}(t)}{U}\right)^3\right) \tag{6}$$

Conversion of mechanical energy into electricity can be realized by utilizing a variety of transducers, however, the most commonly used are electrostatic, electromagnetic and piezoelectric. Electrostatic may be the cheapest solution for large scale production [5, 6]. Electromagnetic [7, 8], due to their complicated construction turn out to be the most expensive but at the same time characterized by high efficiency.

The multitude of possible transducer designs of this type increases their versatility, allowing their use in both small devices [9] and massive hydro or wind power plants. Prototypes, however, most often are consist of piezoelectric transducers [10–12], which is justified by great simplicity in their implementation while maintaining high efficiency. It will also be used in this work. The piezoelectric vibration energy harvester (PVEH) mathematical model takes the form [13]:

$$\begin{cases} m\,\ddot{z}(t) + c\,\dot{z}(t) + k\,z(t) + \theta\,v(t) = \\ = -\frac{1}{2}\rho Ub\left(a_1\,\frac{\dot{z}(t)}{U} + a_2\left(\frac{\dot{z}(t)}{U}\right)^2 + a_3\left(\frac{\dot{z}(t)}{U}\right)^3\right) \\ Cp\,\dot{v}(t) + \frac{v(t)}{R} - \theta\,\dot{z}(t) = 0, \end{cases} \tag{7}$$

where: $v(t)$ – generated voltage, θ – electromechanical coupling, R – circuit resistance, Cp – circuit equivalent capacity.

Mathematical model Eq. (7) can be rewritten in nondimensional form by introducing nondimensional parameters:

$$\begin{cases} \ddot{\eta}(\tau) + c\dot{\eta}(\tau) + \eta(\tau) + \kappa\, \nu(\tau) = a_1\, aD\, \mathcal{U}\, \dot{\eta}(\tau) + a_2\, aD\, \dot{\eta}(\tau)^2 + a_3\, aD \frac{\dot{\eta}(\tau)^3}{\mathcal{U}} \\ \dot{\nu}(\tau) + \nu(\tau)\, r^{-1} - \dot{\eta}(\tau) = 0, \end{cases} \tag{8}$$

where: $\eta(\tau) = \frac{z(t)}{h}$, $\nu(\tau) = \frac{v(t)}{\theta\, Cp} b$, $= \frac{c}{m\, \omega n}$, $\kappa = \frac{\theta^2}{Cp\, m\, \omega n^2}$, $aD = \frac{h^2\, b\, \rho}{2m}$, $r = Cp\, \omega n\, R$, $\mathcal{U} = \frac{U}{h\, \omega n}$, $\Omega = \frac{\omega}{\omega n}$, $\tau = \omega n\, t$.

The efficiency of PVEH is significantly affected by its mechanical structure. The typical one degree of freedom beam devices [14–16] seem to give way in this respect to more complex systems with many degrees of freedom [17, 18]. It is worth noting that devices showing also torsional vibrations should not be modeled using the Den Hartog's hypothesis – for torsional vibrations the quasi-stationarity condition is never satisfied. The non-stationary flow model was used, among others in works [19, 20]. Researchers also signal the possibility of achieving higher efficiency by constructing devices with non-linear dynamic properties whose motion is regular [21] or chaotic [22].

From the point of view of device efficiency, of course, the geometry of the flowing body is of key importance. In [23], elliptical cross-sections with different ratios between the length of the semi-minor axis and the semi-major axis were examined. A substantial set of aerodynamic coefficients of various typical sections is included in [24]. The maximum efficiency of PVEH depending on the shape of the flowing body was analyzed in [25]. The work [26] is devoted to the analysis of the impact of trapezoid arm inclination on its aerodynamic coefficients. To the knowledge of the authors, so far no research would include analysis of any, unusual and irregular geometry. The purpose of this work is to fill this gap.

2 Impact of Structure Parameters on Device Performance

The optimization goal function was the maximization of the electrical power generated by the system. The analytical form of the expression for the device's power as a function of the coefficients a_1, a_2 and a_3 can be obtained by assuming that both the vibration amplitude and the voltage have harmonic solutions:

$$\eta(\tau) = \eta A\, \mathrm{Cos}(\Omega\, \tau), \tag{9}$$

$$\nu(\tau) = \nu A\, \mathrm{Cos}(\Omega\, \tau + \varphi), \tag{10}$$

where: ηA and νA – dimensionless amplitudes of vibrations and voltage respectively, Ω – dimensionless natural frequency, φ – phase shift.

The phase shift φ functions was determined by solving the second equation of the model Eq. (10) into which the solutions of the assumed form Eqs. (9 and 10) were substituted and adopting $\tau = 0$:

$$\sin(\varphi) = \frac{1}{\sqrt{(r\ \Omega)^2 + 1}}, \tag{11}$$

$$\cos\ (\varphi) = \frac{r\ \Omega}{\sqrt{(r\ \Omega)^2 + 1}}. \tag{12}$$

The relation between dimensionless voltage and dimensionless displacement amplitudes was determined by integrating the second equation Eq. (8) in terms of dimensionless time in half-period boundary and substituting Eqs. (9 and 10):

$$\nu A = \frac{r\ \eta A\ \Omega}{\sqrt{1 + r^2\ \Omega^2}}. \tag{13}$$

Then, the energy balance was performed based on a system of equations Eq. (8), which by taking into account equations Eqs. (10, 11, 12, 13 and 14) led to an expression for dimensionless vibration amplitude:

$$\eta A = \frac{\mathcal{U}\left(-32\ \mathrm{a2\ aD}\ \Omega + \sqrt{1024\ \mathrm{a2}^2\ \mathrm{aD}^2\ \Omega^2 + 36\ \mathrm{a3\ aD}\ \pi\ \Omega^2\left(-12\ \mathrm{a1\ aD}\ \pi\ \mathcal{U} + 12\ \pi + \frac{12\ \pi\ r}{1+r^2\Omega^2}\right)\mathcal{U}^{-1}}\right)}{18\ \mathrm{a3\ aD}\ \pi\ \Omega^2} \tag{14}$$

By substituting Eqs. (10, 11, 12, 13, 14 and 15) to the first equation of model Eq. (9) and examining it for $\tau = 0$ one can get an explicit expression for the dimensionless frequency:

$$\Omega = \frac{1}{\sqrt{2}}\sqrt{1 - \frac{1}{r^2} + \kappa + \frac{\sqrt{4\,r^2 + \left(1 - r^2 - r^2\,\kappa\right)^2}}{r^2}} \tag{15}$$

The dimensionless power of the system is therefore given by the formula:

$$\mathcal{P} = \frac{\kappa\ \nu A^2}{r} = \frac{r\mathcal{U}^2\kappa\left(-32\ \mathrm{a}_2\ \mathrm{aD}\ \Omega + \sqrt{1024\ \mathrm{a}_2{}^2\ \mathrm{aD}^2\ \Omega^2 + 36\ \mathrm{a}_3\ \mathrm{aD}\ \pi\ \Omega^2\left(-12\ \mathrm{a}_1\ \mathrm{aD}\ \pi\ \mathcal{U} + 12\pi + \frac{12\pi r}{1+r^2\Omega^2}\right)\mathcal{U}^{-1}}\right)^2}{324\ \mathrm{a}_3{}^2\ \mathrm{aD}^2\ \pi^2\ \Omega^2\left(1+r^2\Omega^2\right)} \tag{16}$$

3 Optimization

As can be seen in chapter "On the Vibrational Analysis for the Motion of a Rotating Cylinder", to determine the body's galloping potential, it is necessary to have data on $C_L(\alpha_0)$ and $C_D(\alpha_0)$ functions determined by the geometry of the resonator. Thus, they will store information about the extent that the geometry affects the power generated by the device. It is therefore reasonable to develop – through optimization – a geometry that allows the most efficient energy generation. The optimization process was performed using a genetic algorithm, while the aerodynamic characteristics were obtained by simulating flow using the Method of Fundamental Solutions.

3.1 Genetic Algorithm

This chapter is written with the assumption that the reader has elementary knowledge of the essence of genetic algorithms (GA), which, if necessary, can be taken from one of the numerous monographs, e.g [27, 28].

At the first stage of the GA, 200 random, having a horizontal axis of symmetry geometries were generated. They were defined as a closed broken line described on points, each of which lay on one (and only one) of 20 uniformly distributed radial axes (see Fig. 2). Then each of them was evaluated according to the procedure described in Sect. 2. To improve calculations, the classic crossover process has been replaced here by the chaotic process proposed in [29], based on the logistic map, given by the formula:

$$z_{n+1} = \lambda\, z_n\, (1 - z_n) \tag{17}$$

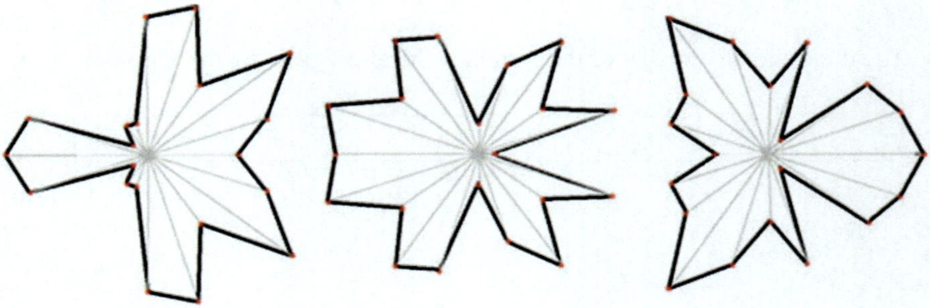

Fig. 2 Examples of body geometries

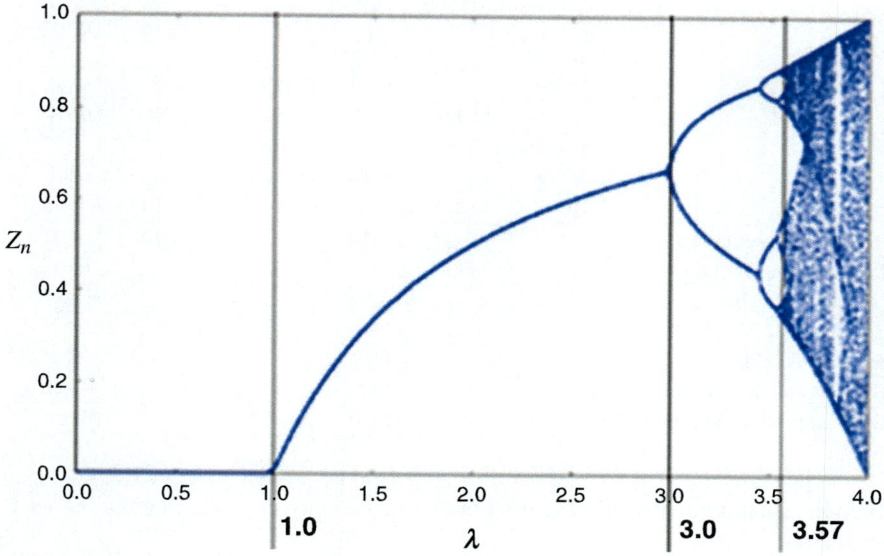

Fig. 3 Logistic map

Function Eq. (17) describes the iterative process of generating the value of z_{n+1} based on any of the values chosen from the range < 0; 1>. Then the generated value is adopted as the new value of z_n and the process is repeated. Depending on the arbitrarily chosen parameter λ, the process can be convergent for $\lambda \in (1; 3>$, periodic for $\lambda \in (3; 3.57)$, or chaotic for λ >3.57 (includes stable manifolds) (Fig. 3).

The implementation of the chaotic crossover model consists of extending the standard genome containing only the solution by two additional sections: information about the randomly assigned value of the parameter λ and the random crossover mask (which should be encoded in the Gray code). The crossover mask stores the information about which bits the offspring will inherit from which parent – if the value 1 appears in the first place of the mask, the offspring inherits the gene first from the first parent, if 0 appears in the second place of the mask, the offspring inherits the second gene the second parent, etc.

The mask of each parent is used to generate one child. The value of λ is inherited without any change and is used to generate the offspring mask. This process starts from decoding the parent mask, normalizing its value to the range <0; 1>, and then processing this value by a logistic map with the assigned parameter λ. Processed and re-encoded in Gray code value is assigned as the offspring mask. The example of the chaotic crossover procedure is presented in Table 1.

As reported in [29], λ values that appear in the initial population have a significant impact on the algorithm performance. Individuals having this parameter within the convergent and chaotic range are desirable, and parameters causing periodic

Table 1 Chaotic crossover algorithm

	Parent 1			**Parent 2**		
	solution	λ	mask	solution	λ	mask
	0110	11.111	1011	1001	10.100	0101
	Offspring 1			**Offspring 2**		
	solution	λ	mask	solution	λ	mask
	0010	11.111		1000	10.100	
Binary representation		11.111	1101		10.100	0110
Decimal representation		3.88	13		2.5	6
Normalized value (z_n)			0.87			0.40
$z_n + 1$			0.45			0.60
Expanded value			6.72			9.00
Binary representation			0111			1001
Gray code representation			0101			1101

solutions should be avoided. The distribution of parameters in the population does not seem to matter.

The crossover process was performed with the strategy of elitism, assuming the probability of its occurrence at the level of 0.8. Besides, the solution may have been subjected to a flip bit mutation with a probability of 0.05. The condition for the algorithm to stop was the invariability of the best solution within a 100 interactions.

3.2 *Determination of Aerodynamic Characteristics*

The aerodynamic coefficients C_L and C_D necessary to determine the parameters a_1, a_2 and a_3 of the body were obtained by numerical simulation of the vibration of the body in a channel filled with flowing fluid (air). The physical model was prepared according to the benchmark [30], with the difference that the cylinder was replaced by the tested geometry. The simulation was repeated for angular orientations of the object from 0 to 180 degrees at 5 degrees step to obtain full characteristics describing the aerodynamics of the body $C_L(\alpha_0)$ and $C_D(\alpha_0)$. For each simulation constant Reynolds number Re $= 10^5$ was maintained. Based on the received characteristics and according to Eq. (6), the functions $a_1(\alpha_0)$, $a_2(\alpha_0)$, $a_3(\alpha_0)$ were calculated, these in turn, after substituting them into Eq. (17), were used to characterize the power $\mathcal{P}(\alpha_0)$ generated by the device equipped with the resonator of the studied geometry. The maximum value of $\mathcal{P}(\alpha_0)$ was the basis for the evaluation of geometry by the genetic algorithm.

4 Results and Discussion

Figure 4 shows the optimized geometry, while Fig. 5 shows its aerodynamic characteristics. Individual axes lengths are shown in Table 2. The blunt side orientation in the normal direction to the wind flow corresponds to the 90-degree orientation and for this orientation, the body will show the greatest aerodynamic instability.

Table 3 lists the coefficients a_1, a_2 and a_3 calculated for optimized geometry and for standard geometries that are commonly studied. Based on these coefficients and using the formula Eq. (17), the potential power $\mathcal{P}$ generated by a device equipped

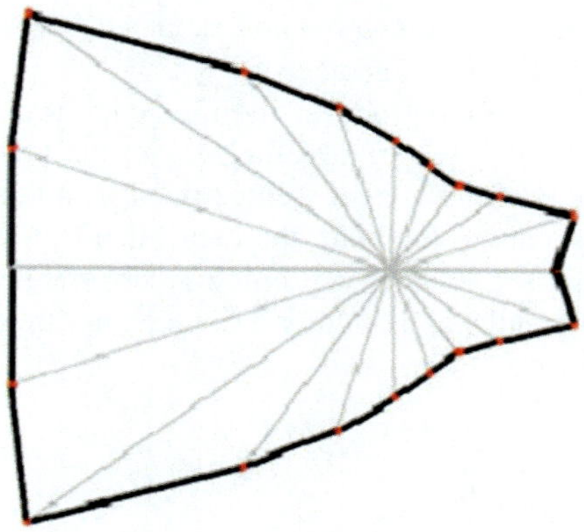

Fig. 4 Optimized geometry

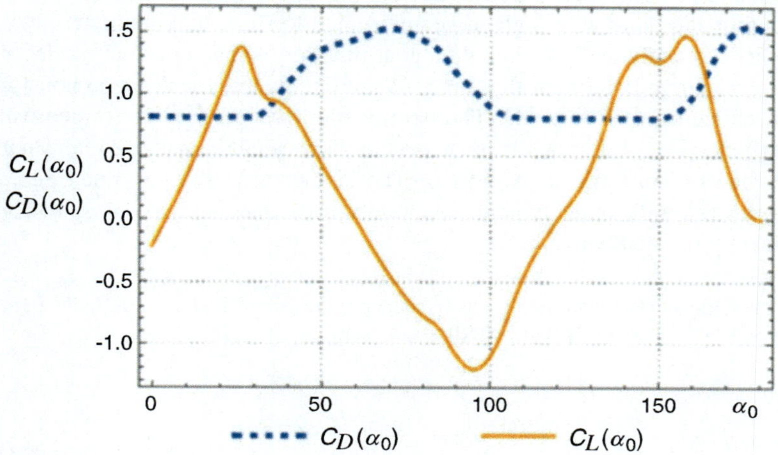

Fig. 5 Aerodynamic characteristic of the geometry

Table 2 Length of individual arms of optimized geometry [mm]

1	2	3	4	5	6	7	8	9	10	11
30.20	35.50	24.40	20.95	21.55	25.40	34.00	48.70	85.10	75.60	71.00

Table 3 Summary of obtained results

	Optimized geometry	D-section	Equilateral triangle	Square
a_1	3.56	0.097	1.87	2.69
a_2	0.28	4.25	5.11	0
a_3	−9.74	−28.83	−1418	−0.0068
$\mathcal{P}/\mathcal{P}_{opt}$	1	0.77	0.87	0.59

with a resonator of a given geometry was calculated and related to the power generated by a device with optimal geometry $\mathcal{P}_{opt}$.

As can be seen from the table above, the efficiency provided by the optimal resonator with geometry is closest to the efficiency of the system with a triangular body (13% difference). Both figures are geometrically similar, however, the shape of the optimized one seems to facilitate the detachment of the fluid stream from the surface. More complex geometry promotes performance, although, can be problematic from a technological point of view, which in turn can be a barrier when using it in prototype devices.

5 Conclusions

The geometry of the resonator implemented in the aeroelastic energy harvester was optimized. The optimization goal function was to maximize the efficiency of the device, thus the body with a given geometry had to show maximum aerodynamic instability. To analytically formulate a goal function, a mathematical model of the device was derived and then through its solution, an analytical expression for the system efficiency was obtained. The optimization process was performed using a chaotic genetic algorithm, while the necessary flow calculations were realized using the Method of Fundamental Solutions. The efficiency of the generated geometry was compared with other typical, commonly used ones and it was shown to be significantly more effective.

Acknowledgments The work is financially supported by grant 02/21/DS-MK/3529/2018 from the Ministry of Science and Higher Education in Poland.

References

1. Den Hartog, J.P.: Mechanical Vibrations, 4th edn. McGraw-Hill, New York (1956)
2. Novak, M.: Aeroelastic galloping of prismatic bodies. J .Eng. Mech. Div. **96**, 115–142 (1969)

3. Ng, Y.T., Luo, S.C., Chew, Y.T.: On using high-order polynomial curve fits in the quasi-steady theory for square-cylinder galloping. J. Fluids Struct. **20**(1), 141–146 (2005). https://doi.org/10.1016/j.jfluidstructs.2004.02.008
4. Barrero-Gil, A., Alonso, G., Sanz-Andrés, A.: Energy harvesting from transverse galloping. J. Sound Vib. **329**(14), 2873–2883 (2010). https://doi.org/10.1016/j.jsv.2010.01.028
5. Boisseau, S., Despesse, G., Seddik, A.B.: Electrostatic Energy Harvesting Systems. Small-Scale Energy Harvesting, Intech (2012)
6. Kumar, A., Balpande, S.S., Anjankar, S.C.: Electromagnetic energy harvester for low frequency vibrations using MEMS. Procedia Comput. Sci. **79**, 785–792 (2016). https://doi.org/10.1109/JMEMS.2011.2162488
7. Dai, H.L., Abdelkefi, A., Javed, U., Wang, L.: Modeling and performance of electromagnetic energy harvesting from galloping oscillations. Smart Mater. Struct. **24**(4) (2015). https://doi.org/10.1088/0964-1726/24/4/045012
8. Haroun, A., Yamada, I., Warisawa, S.: Investigation of kinetic energy harvesting from human body motion activities using free/impact based micro electromagnetic generator. J. Diabetes Cholesterol Metab. **1**, 12–16 (2016)
9. Abdelkefi, A., Nayfeh, A., Hajj, M.R.: Enhancement of power harvesting from piezoaeroelastic systems. Nonlinear Dynam. **68**(4) (2016). https://doi.org/10.1007/s11071-011-0234-9
10. Abdelkefi, A., Nayfeh, A., Hajj, M.R.: Modeling and analysis of piezoaeroelastic energy harvesters. Nonlinear Dynam. **67**(2), 925–939 (2011). https://doi.org/10.1007/s11071-011-0035-1
11. Tao, J.X., Nguyen, V., Carpinteri, A., Wang, Q.: Energy harvesting from wind by a piezoelectric harvester. Eng. Struct. **133**, 74–80 (2017). https://doi.org/10.1016/j.engstruct.2016.12.021
12. Howells, C.: Piezoelectric energy harvesting. Energy Convers. Manag. **50**, 1847–1850 (2009). https://doi.org/10.1016/j.enconman.2009.02.020
13. De Marqui, C., Erturk, A., Inman, D.J.: An electromechanical finite element model for piezoelectric energy harvester plates. J. Sound Vib. **327**, 9–25 (2019). https://doi.org/10.1016/j.jsv.2009.05.015
14. Jayant, S., Rohan, M.: Harvesting wind energy using a galloping piezoelectric beam. J. Vib. Acoust. **134**(1) (2012). https://doi.org/10.1115/SMASIS2009-1479
15. Wu, Y., Li, D., Xiang, J.: Dimensionless modeling and nonlinear analysis of a coupled pitch–plunge–leadlag airfoil-based piezoaeroelastic energy harvesting system. Nonlinear Dynam. **92**(2), 153–167 (2018). https://doi.org/10.1007/s11071-017-3954-7
16. Abdelkefi, A., Hajj, M.R., Nayfeh, A.: Power harvesting from transverse galloping of square cylinder. Nonlinear Dynam. **70** (2012). https://doi.org/10.1007/s11071-012-0538-4
17. Hwang, M., Arrieta, A.: Input-independent energy harvesting in bistable lattices from transition waves. Sci. Rep. **8**(1) (2018). https://doi.org/10.1038/s41598-018-22003-7
18. Vasconcellos, R., Abdelkefi, A.: Nonlinear dynamical analysis of an aeroelastic system withmulti-segmented moment in the pitch degree-of-freedom. Commun. Nonlinear Sci. Numer. Simul. **20**(1), 324–334 (2015). https://doi.org/10.1016/j.cnsns.2014.05.017
19. Wu, Y., Li, D., Xiang, J., Da Ronch, A.: A modified airfoil-based piezoaeroelastic energy harvester with double plunge degrees of freedom. Theor. Appl. Mech. Lett. **6**, 244–247 (2016). https://doi.org/10.1016/j.taml.2016.08.009
20. Abdelkefi, A., Ghommem, M., Nuhait, A., Hajj, M.R.: Nonlinear analysis and enhancement of wing-based piezoaeroelastic energy harvesters. J. Sound Vib. **333**, 166–177 (2014). https://doi.org/10.1016/j.jsv.2013.08.032
21. Kumar, A., Shaikh, F.A., Arockiarajan, A.: Enhanced energy harvesting from nonlinear oscillators via chaos control. IFAC-PapersOnLine. **49**, 35–40 (2016). https://doi.org/10.1016/j.ifacol.2016.03.025
22. Litak, G., Friswell, M., Adhikari, S.: Regular and chaotic vibration in a piezoelectric energy harvester. Meccanica. **51**(5), 1017–1025 (2016). https://doi.org/10.1140/epjp/i2015-15103-8
23. Alonso, G., Meseguer, J., Sanz-Andres, A., Valero, E.: On the galloping instability of two-dimensional bodies having elliptical cross-sections. J. Wind Eng. Ind. Aerodyn. **98**, 438–448 (2010). https://doi.org/10.1016/j.jweia.2010.02.002

24. Nikitas, N., Macdonald, J.H.G.: Misconceptions and generalizations of the Den Hartog galloping criterion. J. Eng. Mech. **140** (2014)
25. Meseguer, J., Sanz-Andres, A., Alonso, G.: Determination of maximum mechanical energy efficiency in energy galloping systems. J. Eng. Mech. **141** (2015). https://doi.org/10.1061/(ASCE)EM.1943-7889.0000817
26. Kluger, J., Moon, F.C., Rand, E.: Shape optimization of a blunt body Vibro-wind galloping oscillator. J. Fluids Struct. **40**, 185–200 (2013). https://doi.org/10.1016/j.jfluidstructs.2013.03.014
27. Goldberg, D.E.: Genetic Algorithms in Search, Optimization and Machine Learning. Addison-Wesley Longman Publishing Co. Inc. USA (1989)
28. Fogel, L.J., Owens, A.J., Walsh, M.J.: Artificial Intelligence Through Simulated Evolution. Wiley, Chichester (1966)
29. Snaselova, P., Zboril, F.: Genetic algorithm using theory of Chaos. Procedia Comput. Sci. **51**(1), 316–325 (2015). https://doi.org/10.1016/j.procs.2015.05.248
30. Schäfer, M., Turek, S., Durst, F., Krause, E., Rannacher, R.: Benchmark computations of laminar flow around a cylinder. Flow Simul. High-Perform. Comput. II. **52**, 547–566 (1996)

Rolling Heavy Ball Over the Surface with Arbitrary Shape in Real Rn3 Space

Katica R. (Stevanović) Hedrih

Abstract The research results of the rolling, without slipping, of a homogeneous heavy ball over the surface with arbitrary shape, in the real Rn 3 space, are presented. The system is holonomic stationary, since the ball is subjected to geometric constraints, and has three degrees of freedom of movement. Two orthogonal unit vectors, in the tangent plane in their contact point, to the surface of ball and surface, along which ball rolls, are determined. The unit vector of the normal to the surface of the ball and the surface along which ball rolls without slipping, through the current contact point of ball and surface, and passes through the center of the ball, is determined. At each moment, for the current position of the point of the contact between the ball and the surface, the position vector of the center of the ball is determined. Also, the corresponding vector of velocity of the center of the ball is determined. Using the velocity vector of the center of the ball, the current angular velocity of rolling the ball over the surface in the function of generalized coordinates is determined, as well as their direction. The direction of the elementary arch of the curvilinear trace through current contact point, of rolling the ball over the surface, as well as the direction of the momentary axis around which the ball is rolling without slipping, are determined. All presented ideas and results are new original generalized approach.

Keywords Rolling heavy ball · Momentary axis of rolling · Momentary angular velocity of rolling Arbitrary surface · Nonlinear differential equation · Geometrical constraints

K. R. (Stevanović) Hedrih (✉)
Department of Mechanics, Mathematical Institute of Serbian Academy of Science and Arts, Belgrade, Serbia

Faculty of Mechanical Engineering, University of Nis, Nis, Serbia
e-mail: katicah@mi.sanu.ac.rs; khedrih@eunet.rs; khedrih@sbb.rs

J. Awrejcewicz (ed.), *Perspectives in Dynamical Systems I: Mechatronics and Life Sciences*, Springer Proceedings in Mathematics & Statistics 362,
https://doi.org/10.1007/978-3-030-77306-9_22

1 Introduction

In the last series of the author's papers [1–13], nonlinear dynamics of different generalized rolling pendulums along different curvilinear lines is investigated and presented as mechanical systems with one degree of freedom and with holonomic, pure geometrical scleronomic system dynamics. In the last author's paper [12] rolling of a heavy ball over the sphere surface is described in curvilinear sphere coordinates using meridian and circular angle coordinates. Rolling of a ball is decomposed into two components of the rolling, one along meridians and second along comparators of the spherical curvilinear coordinate lines. Investigation shown that constraints are pure geometrical and stationary, and that system is holonomic and scleronomic. In the congress presentation, on the basis of previous results, a natural approach for investigation rolling of a heavy ball over the curved coordinate surfaces and corresponding parallel surfaces in different orthogonal curvilinear coordinate system is presented. Rolling ball motion is decomposed, into two components of rolling along orthogonal coordinate lines of the curved coordinate surface.

2 Description of the Model of a Rolling Heavy Ball Over the Arbitrary Surface

We study the rolling of a heavy rigid homogeneous ball having mass $\mathbf{M}$, radius r, and axial moments $\mathbf{J}_P^{(y,z)} = \mathbf{J}_C^{(y,z)} + \mathbf{M}r^2$ and $\mathbf{J}_P^{(x,z)} = \mathbf{J}_C^{(x,z)} + \mathbf{M}r^2$ of mass inertia for the corresponding instantaneous rolling axes of the ball, orthogonal to the coordinate plane (x, z) and (y, z) plane. Instantaneous axes of the rolling of the heavy ball are orthogonal to the instantaneous component rolling velocities of the ball's mass center and also orthogonal to the curve lines in the plane of rolling, which is defined in the three-dimensional coordinate system (x, y, z) by the following equation $z = f(x, y)$. Let's assume that the ball began to roll from the position $P_0\,(x_0, y_0, z_0 = f(x_0, y_0))$ in which the point of the contact of the ball and the surface of the rolling and ball was in, and that the center of the ball was at the point $C_0\,(x_{C0}, y_{C0}, z_{C0} = f_C(x_{C0}, y_{C0}))$.

The ball as a heavy rigid (brittle) homogeneous body in the general case, when it is not supported by the constraints, has six degrees of freedom of movement, and when rolling without slipping, it is subjected by the constraints, which are three, and has three degrees of freedom of movement. The constraints to which the ball is subjected to rolling without slipping over the defined surface are as follows:

1* The center of the mass (weight) $C(x, y, z = f_C(x, y))$ of the rolling ball is always at a orthogonal distance r measured from the point $P(x, y, z = f(x, y))$ of the count between rolling ball and surface of the rolling, which gives one geometric connection, which is both mutually retaining.

2* The condition that rolling the ball around the surface is without slipping, gives another geometric relation to the equality of the rolling arc - a set of points of contact of ball in rolling and surfaces that are described on the ball and rolling trace over the surface are of the same length. The trace in considered surface of rolling ball is two dimensional line, then follows that two additional constraints appear.

Trace of the rolling ball in surface is defined by the direction on the ball surface and on the surface of the rolling, this gives another two geometric constraints, and which are both sides mutually retaining.

3* When ball is rolling, the ball cannot be translate moved in the direction of the trace, because it rolls without slipping, and this is, also, explication of the third geometric constraints.

This means that the rolling of the heavy rigid ball on the surface of an arbitrary shape has three degrees of freedom, two component rolling by orthogonal elementary arches and one rotation around its own axis of self-rotation, which is in the orthogonal direction to the surface upon which it rolls, and at the point of the momentary contact of the ball and surface of rolling.

As we have shown by the analysis that there are three geometric, stationary constraints, it follows that the rolling of the ball, without slipping, on the arbitrary surface is a mechanical system with three degrees of freedom of movement, in rolling without slipping, and that it represents the holonomic stationary system, because all the constraints that act on the system, are holonomic, geometric and stationary. For the generalized coordinates, we will adopt the coordinates x and y of the point $P(x, y, z = f(x, y))$ of the contact of the ball and $P_s(x, y, z = f(x, y))$ of the surface on which it is rolling, without slipping (see Fig. 2).

These points $P(x, y, z = f(x, y))$ and $P_s(x, y, z = f(x, y))$ of the contact are two sets, one on the ball surface and other in the form of trace of ball's rolling, without slipping in a considered surface over which ball is in the rolling motion.

The third independent generalized coordinate is the angular coordinate - of its own self-rotatiots around an axis orthogonal to the surface of rolling at the point $P(x, y, z = f(x, y)) = P_s(x, y, z = f(x, y))$ of contact of the ball and surface it, which passed through the center of the ball's mass $C(x_C, y_C, z_C = f_C(x, y))$ at each momentary position, and the point of the contact $P(x, y, z = f(x, y))$. Current angular velocity of rolling of the heavy rigid ball over the arbitrary surface without slipping is:

$$\omega_P\left(x, \dot{x}, y, \dot{y}\right) = \frac{1}{r} v_C\left(x, \dot{x}, y, \dot{y}\right), \tag{1}$$

along the curvilinear path (trace) in that surface, and can be explained in each time of the moment by two components – the angular velocity of the rotation by two orthogonal elementary components of the rolling of the ball in two orthogonal directions along two orthogonal elementary arch of traces in considered surface of the rolling, as it is shown in Fig. 1.a * and 1.b *, and also in Fig. 2. With this in mind, we can display the rolling of the ball on a curved surface without slipping with the

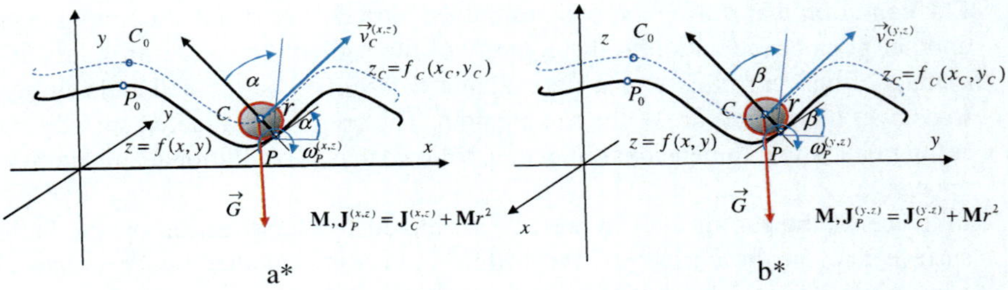

Fig. 1 Presentations of the decomposition of the dynamics of a rolling heavy ball into component rolls along orthogonal curvilinear line traces in arbitrary surface: a* in coordinate surface (x, z) and v* in coordinate surface (y, z)

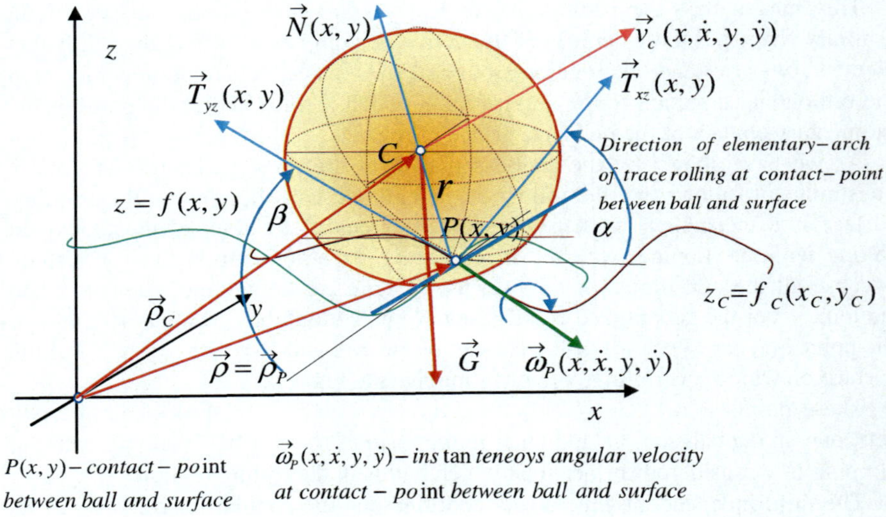

$$\mathbf{M}, \mathbf{J}_p = \mathbf{J}_c + \mathbf{M}r^2$$

$$\vec{\rho}_C(x, y, z = f(x,)) = \vec{\rho}(x, y, z = f(x,)) + r\vec{N}(x, y, z = f(x,))$$

$$\vec{N}(x, y, z = f(x, y)) = \frac{1}{\left|[\vec{T}_{xz}, \vec{T}_{yz}]\right|}[\vec{T}_{xz}, \vec{T}_{yz}]$$

Fig. 2 Model of a rolling heavy homogeneous ball over the arbitrary surface in three dimensional Rn3 real space with notation of corresponding geometrical and kinetic parameters

system of elementary components of the rolling of the ball by the elementary arcs of the coordinate lines in the curved surface, which are defined by the cross-sections of the parallel coordinate plane, and the resulting rolling as the momentary sum of two elementary rolling by the orthogonal coordinate lines of the surface.Based on this, it follows:

In Fig. 1.a *, with the angle α of inclination of the tangent to the curvilinear elementary component trace in surface $z = f(x, y)$ at the level of the parallel plane (x, z) for the fixed coordinate plane (O, x, z), and at the point $P(x, y, z = f(x, y))$ of contact of the ball in rolling, without slipping, and the tangent of the component trace in considered surface and plane for the fixed y, it is indicated that:

$$tg\ \alpha\,(x, y) = \frac{\partial z}{\partial x} = \frac{\partial f(x,y)}{\partial x},\quad \sin\,\alpha\,(x, y) = \frac{\frac{\partial z}{\partial x}}{\sqrt{1+\left[\frac{\partial z}{\partial x}\right]^2}},\quad \cos\,\alpha\,(x, y) = \frac{1}{\sqrt{1+\left[\frac{\partial z}{\partial x}\right]^2}} \tag{2}$$

In Fig. 1.b *, the angle β of inclination of the tangent to the curvilinear elementary component trace is fixed at the level of the parallel plane (y, z) for the fixed x, and at the point of contact $P(x, y, z = f(x, y))$ of the ball in rolling, without slipping, and the fixing x in $P(x, y, z = f(x, y))$, it follows that:

$$tg\ \beta\,(x \cdot y) = \frac{\partial z}{\partial y} = \frac{\partial f(x,y)}{\partial y},\quad \sin\,\beta\,(x, y) = \frac{\frac{\partial z}{\partial y}}{\sqrt{1+\left[\frac{\partial z}{\partial y}\right]^2}},\quad \cos\,\beta\,(x, y) = \frac{1}{\sqrt{1+\left[\frac{\partial z}{\partial y}\right]^2}} \tag{3}$$

Now, at the point $P(x, y, z = f(x, y))$ of the contact in the surface, and between the ball and curved surface, by which the ball is rolling, we can set up two single orthogonal unit vectors, $\overrightarrow{T}_{xz}$ and $\overrightarrow{T}_{yz}$, which touch the surface of the ball and the surface of the rolling, at the point $P(x, y, z = f(x, y))$ of their contact:

$$\overrightarrow{T}_{xz} = \overrightarrow{i}\ \cos\,\alpha\,(x, y) + \overrightarrow{k}\ \sin\,\alpha\ (x, y) = \overrightarrow{i}\,\frac{1}{\sqrt{1+\left[\frac{\partial z}{\partial x}\right]^2}} + \overrightarrow{k}\,\frac{\frac{\partial z}{\partial x}}{\sqrt{1+\left[\frac{\partial z}{\partial x}\right]^2}}, \tag{4}$$

$$\overrightarrow{T}_{yz} = \overrightarrow{j}\ \cos\,\beta\,(x, y) + \overrightarrow{k}\ \sin\,\beta\ (x, y) = \overrightarrow{j}\,\frac{1}{\sqrt{1+\left[\frac{\partial z}{\partial y}\right]^2}} + \overrightarrow{k}\,\frac{\frac{\partial z}{\partial y}}{\sqrt{1+\left[\frac{\partial z}{\partial y}\right]^2}} \tag{5}$$

The unit vector $\overrightarrow{N}\ (x, y, z = f\ (x, y))$ is normal on the surface of the ball rolling and on the surface of the corresponding surface of the ball's rolling at the point.

$P(x, y, z = f(x, y))$ of their contact

$$\overrightarrow{N}(x, y, z = f(x,)) = \frac{1}{\sqrt{1 + \left[\frac{\partial z}{\partial x}\right]^2 + \left[\frac{\partial z}{\partial y}\right]^2}} \left\langle -\overrightarrow{i}\frac{\partial z}{\partial x} - \overrightarrow{j}\frac{\partial z}{\partial y} + \overrightarrow{k} \right\rangle \tag{6}$$

This unit vector $\overrightarrow{N}(x, y, z = f(x, y))$ of the normal at contact point, is in the direction of the gradient on the surface of the ball rolling:

$$\overrightarrow{N}(x, y, z = f(x,)) = \frac{grad z(x,y)}{|grad z(x,y)|} = \frac{\nabla z(x,y)}{|\nabla z(x,y)|} = \frac{1}{\sqrt{1+\left[\frac{\partial z}{\partial x}\right]^2+\left[\frac{\partial z}{\partial y}\right]^2}} \left\langle -\overrightarrow{i}\frac{\partial z}{\partial x} - \overrightarrow{j}\frac{\partial z}{\partial y} + \overrightarrow{k} \right\rangle \tag{7}$$

The center $C(x_C, y_C, z_C = f_C(x_C, y_C))$ of the heavy rigid ball's mass, in rolling, without slipping, along the curved surface of the equation $z = f(x, y)$, when the constraint is both sides keeping, it is always is in the orthogonal direction of which is determined by the unit vector $\overrightarrow{N}(x, y, z = f(x,))$, on the surface of the ball and the surface of the ball's rolling, and at a normal distance r from the contact point $P(x, y, z = f(x, y))$. Based on this, we can write:

$$\overrightarrow{\rho}_C(x, y, z = f(x,)) = \overrightarrow{\rho}(x, y, z = f(x,)) + r\overrightarrow{N}(x, y, z = f(x,)),$$
$$\overrightarrow{\rho}_C(x, y, z = f(x,)) = x_C\overrightarrow{i} + y_C\overrightarrow{j} + z_C\overrightarrow{k} = x\overrightarrow{i} + y\overrightarrow{j} + z\overrightarrow{k} + r\frac{\left\langle -\overrightarrow{i}\frac{\partial z}{\partial x} - \overrightarrow{j}\frac{\partial z}{\partial y} + \overrightarrow{k} \right\rangle}{\sqrt{1+\left[\frac{\partial z}{\partial x}\right]^2+\left[\frac{\partial z}{\partial y}\right]^2}} \tag{8}$$

In the scalar form, we can write the coordinates $x_C(x, y, z = f(x, y))$, $y_C(x, y, z = f(x, y))$ and $z_C(x, y, z = f(x, y))$ of the center $C(x_C, y_C, z_C = f_C(x_C, y_C))$ of the rolling ball mass, in the following form:

$$x_C(x, y, z = f(x, y)) = x - r\frac{\frac{\partial z}{\partial x}}{\sqrt{1 + \left[\frac{\partial z}{\partial x}\right]^2 + \left[\frac{\partial z}{\partial y}\right]^2}}, \tag{9}$$

$$y_C(x, y, z = f(x, y)) = y - r\frac{\frac{\partial z}{\partial y}}{\sqrt{1 + \left[\frac{\partial z}{\partial x}\right]^2 + \left[\frac{\partial z}{\partial y}\right]^2}}, \tag{10}$$

$$z_C(x, y, z = f(x, y)) = z + r\frac{1}{\sqrt{1 + \left[\frac{\partial z}{\partial x}\right]^2 + \left[\frac{\partial z}{\partial y}\right]^2}}. \tag{11}$$

3 Velocity of the Centre of a Rolling Heavy Rigid Ball

The components $\dot{x}_C(x, y, z = f(x, y), \dot{x}, \dot{y})$, $\dot{y}_C(x, y, z = f(x, y), \dot{x}, \dot{y})$ and $\dot{z}_C(x, y, z = f(x, y), \dot{x}, \dot{y})$ of the velocity $\overrightarrow{v}_C(\dot{x}_C, \dot{y}_C \cdot \dot{z}_C) = \overrightarrow{v}_{CC}(x, y, \dot{x}, \dot{y})$ of the center $C(x_C, y_C, z_C = f_C(x_C, y_C))$ of the mass of the heavy rigid ball are determined by differentiating the coordinates $x_C(x, y, z = f(x, y))$, $y_C(x, y, z = f(x, y))$ and $z_C(x, y, z = f(x, y))$ of the centre by the time:

$$\dot{x}_C(x, y, z = f(x, y), \dot{x}, \dot{y}) = \dot{x} - r\dot{x}\frac{\frac{\partial^2 z}{\partial x^2}\left(1+\left[\frac{\partial z}{\partial x}\right]^2+\left[\frac{\partial z}{\partial y}\right]^2+\frac{\partial z}{\partial x}\left(\frac{\partial z}{\partial x}\frac{\partial^2 z}{\partial x^2}+\frac{\partial z}{\partial y}\frac{\partial^2 z}{\partial x \partial y}\right)\right)}{\left(1+\left[\frac{\partial z}{\partial x}\right]^2+\left[\frac{\partial z}{\partial y}\right]^2\right)\sqrt{1+\left[\frac{\partial z}{\partial x}\right]^2+\left[\frac{\partial z}{\partial y}\right]^2}} - \\ - r\dot{y}\frac{\frac{\partial^2 z}{\partial x \partial y}\left(1+\left[\frac{\partial z}{\partial x}\right]^2+\left[\frac{\partial z}{\partial y}\right]^2-\frac{\partial z}{\partial x}\left(\frac{\partial z}{\partial x}\frac{\partial^2 z}{\partial x \partial y}+\frac{\partial z}{\partial y}\frac{\partial^2 z}{\partial y^2}\right)\right)}{\left(1+\left[\frac{\partial z}{\partial x}\right]^2+\left[\frac{\partial z}{\partial y}\right]^2\right)\sqrt{1+\left[\frac{\partial z}{\partial x}\right]^2+\left[\frac{\partial z}{\partial y}\right]^2}} \tag{12}$$

$$\dot{y}_C(x, y, z = f(x, y), \dot{x}, \dot{y}) = \dot{y} - r\dot{x}\frac{\frac{\partial^2 z}{\partial y \partial x}\left(1+\left[\frac{\partial z}{\partial x}\right]^2+\left[\frac{\partial z}{\partial y}\right]^2-\frac{\partial z}{\partial y}\left(\frac{\partial z}{\partial x}\frac{\partial^2 z}{\partial x^2}+\frac{\partial z}{\partial y}\frac{\partial^2 z}{\partial x \partial y}\right)\right)}{\left(1+\left[\frac{\partial z}{\partial x}\right]^2+\left[\frac{\partial z}{\partial y}\right]^2\right)\sqrt{1+\left[\frac{\partial z}{\partial x}\right]^2+\left[\frac{\partial z}{\partial y}\right]^2}} - \\ - r\dot{y}\frac{\frac{\partial^2 z}{\partial y^2}\left(1+\left[\frac{\partial z}{\partial x}\right]^2+\left[\frac{\partial z}{\partial y}\right]^2-\frac{\partial z}{\partial y}\left(\frac{\partial^2 z}{\partial x \partial y}\frac{\partial z}{\partial x}+\frac{\partial z}{\partial y}\frac{\partial^2 z}{\partial y^2}\right)\right)}{\left(1+\left[\frac{\partial z}{\partial x}\right]^2+\left[\frac{\partial z}{\partial y}\right]^2\right)\sqrt{1+\left[\frac{\partial z}{\partial x}\right]^2+\left[\frac{\partial z}{\partial y}\right]^2}} \tag{13}$$

$$\dot{z}_C(x, y, z = f(x, y), \dot{x}, \dot{y}) = \dot{z} - r\dot{x}\frac{\frac{\partial z}{\partial x}\frac{\partial^2 z}{\partial x^2}+\frac{\partial z}{\partial y}\frac{\partial^2 z}{\partial x \partial y}}{\sqrt{1+\left[\frac{\partial z}{\partial x}\right]^2+\left[\frac{\partial z}{\partial y}\right]^2}} - r\dot{y}\frac{\frac{\partial z}{\partial x}\frac{\partial^2 z}{\partial x \partial y}+\frac{\partial z}{\partial y}\frac{\partial^2 z}{\partial y^2}}{\sqrt{1+\left[\frac{\partial z}{\partial x}\right]^2+\left[\frac{\partial z}{\partial y}\right]^2}} \tag{14}$$

The velocity $\overrightarrow{v}_P(x, y, \dot{x}, \dot{y})$ of displacement of the point $P(x, y, z = f(x, y))$ of contact of the ball in the rolling, without slipping, over the surface of the arbitrary shape $z = f(x, y)$ given in the analytical form is in the following form:

$$\overrightarrow{v}_P(x, y, \dot{x}, \dot{y}) = \dot{x}\,\overrightarrow{i} + \dot{y}\,\overrightarrow{j} + \left(\dot{x}\frac{\partial z}{\partial x} + \dot{y}\frac{\partial z}{\partial x}\right)\overrightarrow{k}. \tag{15}$$

The intensity of this velocity $\overrightarrow{v}_P(x, y, \dot{x}, \dot{y})$ of displacement of the point $P(x, y, z = f(x, y))$ of the instantaneous contact of the ball in the rolling, without slipping, over the surface, is in the following form:

$$\left|\vec{v}_P\left(x,y,\dot{x},\dot{y}\right)\right|=\sqrt{\dot{x}^2\left(1+\left(\frac{\partial z}{\partial x}\right)^2\right)+\dot{y}\left(1+\left(\frac{\partial z}{\partial y}\right)^2\right)+2\dot{x}\dot{y}\frac{\partial z}{\partial x}\frac{\partial z}{\partial y}}. \tag{16}$$

We introduce the following notations:

$$F\left(x,y,f_x',f_y'\right)=\left(1+\left[\frac{\partial z}{\partial x}\right]^2+\left[\frac{\partial z}{\partial y}\right]^2\right). \tag{17}$$

We introduce, also, new notations $F_{11}(x,y)$, $F_{12}(x,y)$., $F_{21}(x,y)$., $F_{21}(x,y)$., $F_{31}(x,y)$. and $F_{32}(x,y)$. for short writing of the expressions and terms in following denotations:

$$F_{11}\left(x,y\right)=\frac{\frac{\partial^2 z}{\partial x^2}F\left(x,y,f_x',f_y'\right)-\frac{\partial z}{\partial x}\left(\frac{\partial z}{\partial x}\frac{\partial^2 z}{\partial x^2}+\frac{\partial z}{\partial y}\frac{\partial^2 z}{\partial x\partial y}\right)}{F\left(x,y,f_x',f_y'\right)\sqrt{F\left(x,y,f_x',f_y'\right)}}, \tag{18}$$

$$F_{12}\left(x,y\right)=\frac{\frac{\partial^2 z}{\partial x\partial y}F\left(x,y,f_x',f_y'\right)-\frac{\partial z}{\partial x}\left(\frac{\partial z}{\partial x}\frac{\partial^2 z}{\partial x\partial y}+\frac{\partial z}{\partial y}\frac{\partial^2 z}{\partial y^2}\right)}{F\left(x,y,f_x',f_y'\right)\sqrt{F\left(x,y,f_x',f_y'\right)}}, \tag{19}$$

$$F_{21}\left(x,y\right)=\frac{\frac{\partial^2 z}{\partial y\partial x}F\left(x,y,f_x',f_y'\right)-\frac{\partial z}{\partial y}\left(\frac{\partial z}{\partial x}\frac{\partial^2 z}{\partial x^2}+\frac{\partial z}{\partial y}\frac{\partial^2 z}{\partial x\partial y}\right)}{F\left(x,y,f_x',f_y'\right)\sqrt{F\left(x,y,f_x',f_y'\right)}}, \tag{20}$$

$$F_{22}\left(x,y\right)=\frac{\frac{\partial^2 z}{\partial y^2}F\left(x,y,f_x',f_y'\right)-\frac{\partial z}{\partial y}\left(\frac{\partial^2 z}{\partial x\partial y}\frac{\partial z}{\partial x}+\frac{\partial z}{\partial y}\frac{\partial^2 z}{\partial y^2}\right)}{F\left(x,y,f_x',f_y'\right)\sqrt{F\left(x,y,f_x',f_y'\right)}}, \tag{21}$$

$$F_{31}\left(x,y\right)=\frac{\left(\frac{\partial z}{\partial x}\frac{\partial^2 z}{\partial x^2}+\frac{\partial z}{\partial y}\frac{\partial^2 z}{\partial x\partial y}\right)}{\sqrt{F\left(x,y,f_x',f_y'\right)}},\quad F_{32}\left(x,y\right)=\frac{\left(\frac{\partial z}{\partial x}\frac{\partial^2 z}{\partial x\partial y}+\frac{\partial z}{\partial y}\frac{\partial^2 z}{\partial y^2}\right)}{\sqrt{F\left(x,y,f_x',f_y'\right)}} \tag{22}$$

components $\dot{x}_C\left(x,y,z=f\left(x,y\right),\dot{x},\dot{y}\right)$, $\dot{y}_C\left(x,y,z=f\left(x,y\right),\dot{x},\dot{y}\right)$ and $\dot{z}_C\left(x,y,z=f\left(x,y\right),\dot{x},\dot{y}\right)$ of the vector $\vec{v}_C\left(\dot{x}_C,\dot{y}_C\cdot\dot{z}_C\right)=\vec{v}_C\left(x,y,\dot{x},\dot{y}\right)$ of the velocity of the center $C(x_C,y_C,z_C=f_C(x_C,y_C))$ of mass of rolling ball over the considered surface, are now:

$$\dot{x}_C(x, y, z = f(x, y), \dot{x}, \dot{y}) = \dot{x} - r\dot{x}F_{11}(x, y) - r\dot{y}F_{12}(x, y), \tag{23}$$

$$\dot{y}_C(x, y, z = f(x, y), \dot{x}, \dot{y}) = \dot{y} - r\dot{x}F_{21}(x, y) - r\dot{y}F_{22}(x, y), \tag{24}$$

$$\dot{z}_C(x, y, z = f(x, y), \dot{x}, \dot{y}) = \dot{x}f_x{}'(x, y) + \dot{y}f_y{}'(x, y) - r\dot{x}F_{31}(x, y) - r\dot{y}F_{32}(x, y) \tag{25}$$

and square of the intensity of the vector $\vec{v}_C(\dot{x}_C, \dot{y}_C \cdot \dot{z}_C) = \vec{v}_C(x, y, \dot{x}, \dot{y})$ of the velocity is:

$$\begin{aligned}[v_C(x, y, \dot{x}, \dot{y})]^2 = [\dot{x} - r\dot{x}F_{11}(x, y) - r\dot{y}F_{12}(x, y)]^2 + \left[\dot{y} - r\dot{x}F_{212}(x, y) - r\dot{y}F_{22}(x, y)\right]^2 + \\ + \left[\dot{x}f_x{}'(x, y) + \dot{y}f_y{}'(x, y) - r\dot{x}F_{31}(x, y) - r\dot{y}F_{32}(x, y)\right]^2\end{aligned} \tag{26}$$

The direction of the velocity $\vec{v}_C(\dot{x}_C, \dot{y}_C \cdot \dot{z}_C) = \vec{v}_C(x, y, \dot{x}, \dot{y})$ of the center $C(x_C, y_C, z_C = f_C(x_C, y_C))$ of mass movement determines the current direction of the rolling heavy rigid ball along the surface and the current axis of rolling the ball over the surface, which is the direction directed by orthogonal direction to the velocity $\vec{v}_C(\dot{x}_C, \dot{y}_C \cdot \dot{z}_C) = \vec{v}_C(x, y, \dot{x}, \dot{y})$ of the center $C(x_C, y_C, z_C = f_C(x_C, y_C))$ of mass movement of the ball. Also, with this velocity $\vec{v}_C(\dot{x}_C, \dot{y}_C \cdot \dot{z}_C) = \vec{v}_C(x, y, \dot{x}, \dot{y})$, we can easily determine the current angular velocity of the ball's rolling over the determined surface.

4 Kinetic and Potential Energies of a Heavy Rolling Ball Over the Arbitrary Surface and Two Nonlinear Differential Equations

It is, now easy, to determine expressions of the kinetic and potential energies of the rolling ball, without slipping, and write a system of two nonlinear differential equations by use Lagrange equations of the second kind by the chosen independent generalized coordinates x and y.

Bearing in mind that the heavy rigid homogeneous ball is centrally symmetrical in relation to the center of the mass, and that the axial moment of inertia of its mass for any of the rolling axles, because it is tangent to the surface of the ball, it is the same and does not change and it is:

Then, we introduce the following denotations of the parameters (see Reference [12]):

$$\frac{\mathbf{J}_{Pv}}{\mathbf{M}r^2} = \frac{\mathrm{i}_{Pv}^2}{r^2} \frac{\frac{2}{5}\mathbf{M}r^2 + \mathbf{M}r^2}{\mathbf{M}r^2} = \frac{7}{5} \text{ and } \frac{\mathbf{J}_{Pc}}{\mathbf{M}r^2} = \frac{\mathrm{i}_{Pc}^2}{r^2} = \frac{\frac{2}{5}\mathbf{M}r^2 + \mathbf{M}r^2}{\mathbf{M}r^2} = \frac{7}{5}; \tag{27}$$

* and reduced length of the generalized rolling pendulum of a rigid ball

$$\lambda = \frac{\mathbf{i}_{PC}^2}{r^2}(R \pm r) = \left(\frac{\mathbf{i}_C^2}{r^2} + 1\right)(R \pm r) = \lambda = \kappa\,(R \pm r) = \frac{7}{5}(R \pm r); \tag{28}$$

* coefficient of rolling of a rigid ball in rolling

$$\kappa = \left(\frac{\mathbf{i}_C^2}{r^2} + 1\right) = \frac{\frac{2}{5}\mathbf{M}r^2 + \mathbf{M}r^2}{\mathbf{M}r^2} = \frac{7}{5}, \tag{29}$$

where $\frac{\mathbf{J}_{Pv}}{\mathbf{M}} = \mathbf{i}_{Pv}^2$ is the square of radius of axial mass inertia moment for momentary axis of ball rolling along coordinate line of a arbitrary surface. Also, $\mathbf{i}_{Pv}^2$ is square of radius of axial mass inertia moment for momentary axis of the rolling tangent to the coordinate line of a arbitrary surface $\frac{\mathbf{J}_{Pv}}{\mathbf{M}} = \mathbf{i}_{Pv}^2$, and $\mathbf{i}_{PC}^2$ is square of radius of axial mass inertia moment for momentary axis of rolling tangent to the corresponding coordinate line of a arbitrary surface $\frac{\mathbf{J}_{Pv}}{\mathbf{M}} = \mathbf{i}_{Pc}^2$, and for rolling homogeneous heavy rigid ball is: $\mathbf{i}_{Pv}^2 = \mathbf{i}_{Pc}^2$.

The kinetic energy $\mathbf{E}_k$ of the ball in rolling by the given surface is now:

$$\mathbf{E}_k = \frac{1}{2}\mathbf{J}_P\omega_P^2 = \frac{1}{2}\mathbf{J}_C\omega_C^2 + \frac{1}{2}Mv_C^2 = \frac{1}{2}\left(\mathbf{J}_C + Mr^2\right)\omega_P^2; \tag{30}$$

where $v_C = (r\omega_C)$; $\omega_C = \omega_P$ and $\mathbf{J}_P = \mathbf{J}_C + Mr^2$.

If the ball, also, gets its own self rotation, determined by angle coordinate γ with angular velocity $\dot{\gamma}$, around its own axis in the direction of the normal $\overrightarrow{N}\,(x, y, z = f\,(x,))$ to the surface at the point $P(x, y, z = f(x, y))$ of the contact between ball and surface, passing through center of mass $C(x_C, y_C, z_C = f_C(x_C, y_C))$, then for expression of the kinetic energy $\mathbf{E}_k$ we can write the following"

$$\mathbf{E}_k = \frac{1}{2}\mathbf{J}_C\dot{\gamma}^2 + \frac{1}{2}\mathbf{J}_P\omega_P^2. \tag{31}$$

If we assume that there is no own self rotation of the ball in rolling, but only rolling, then kinetic energy is determined by following expression:

$$\mathbf{E}_k = \frac{1}{2}\mathbf{J}_P\omega_P^2 + \frac{1}{2r^2}\mathbf{J}_P[v_C(x, y, \dot{x}, \dot{y})]^2, . \tag{32}$$

$$\begin{aligned}\mathbf{E}_k = {} & \tfrac{1}{2r^2}\mathbf{J}_P[\dot{x} - r\dot{x}F_{11}(x, y) - r\dot{y}F_{12}(x, y)]^2 + \tfrac{1}{2r^2}\mathbf{J}_P\left[\dot{y} - r\dot{x}F_{212}(x, y) - r\dot{y}F_{22}(x, y)\right]^2 + \\ & + \tfrac{1}{2r^2}\mathbf{J}_P\left[\dot{x}f_x'(x, y) + \dot{y}f_y'(x, y) - r\dot{x}F_{31}(x, y) - ryF_{32}(x, y)\right]^2\end{aligned} \tag{33}$$

Ball is heavy, and force of Earth attraction of the ball through mass center is $\vec{G} = -Mg\vec{k}$. The expression for potential energy $\mathbf{E_p}$ of the ball in rolling is in the form:

$$\mathbf{E_p} = Mg(z_C - z_{C0}) = Mg(f_C(x, y) - f_C(x_0, y_0)),$$

$$\mathbf{E_p} = Mg\left\langle (f(x, y) - f(x_0, y_0)) + r\left(\frac{1}{\sqrt{1+\left[\frac{\partial z}{\partial x}\right]^2+\left[\frac{\partial z}{\partial y}\right]^2}} - \frac{1}{\sqrt{1+\left[\frac{\partial z}{\partial x}\right]_0^2+\left[\frac{\partial z}{\partial y}\right]_0^2}}\right)\right\rangle, \tag{34}$$

Since, we adopted the generalized coordinates x and y, by which we have expressed kinetic and potential energies, we, now, write by the Lagrange differential equations of the second kind for independent general coordinates x and y, in the following forms:

$$\frac{d}{dt}\frac{\partial \mathbf{E}_k}{\partial \dot{x}} - \frac{\partial \mathbf{E}_k}{\partial x} + \frac{\partial \mathbf{E}_p}{\partial x} = 0 \text{ and } \frac{d}{dt}\frac{\partial \mathbf{E}_k}{\partial \dot{y}} - \frac{\partial \mathbf{E}_k}{\partial y} + \frac{\partial \mathbf{E}_p}{\partial y} = 0, \tag{35}$$

follows a system of two non-linear differential equation of rolling the ball along surface is now in the following form:

$$\begin{aligned}&\tfrac{d}{dt}\left\langle \tfrac{1}{r^2}\mathbf{J}_P[\dot{x} - r\dot{x}F_{11}(x, y) - r\dot{y}F_{12}(x, y)](1 - rF_{11}(x, y)) - \tfrac{1}{r^2}\mathbf{J}_P \right.\\ &\quad \left. \times\left[\dot{y} - r\dot{x}F_{212}(x, y) - r\dot{y}F_{22}(x, y)\right] rF_{212}(x, y)\right\rangle + \\ &\quad + \tfrac{d}{dt}\left\langle \tfrac{1}{r^2}\mathbf{J}_P\left[\dot{x}f_x'(x, y) + \dot{y}f_y'(x, y) - r\dot{x}F_{31}(x, y) - ryF_{32}(x, y)\right]\right.\\ &\quad \left. \times\left\langle f_x'(x, y) - rF_{31}(x, y)\right\rangle\right\rangle - \tfrac{\partial}{\partial x}\left\langle \tfrac{1}{2r^2}\mathbf{J}_P[\dot{x} - r\dot{x}F_{11}(x, y) - r\dot{y}F_{12}(x, y)]^2\right.\\ &\quad \left. + \tfrac{1}{2r^2}\mathbf{J}_P\left[\dot{y} - r\dot{x}F_{212}(x, y) - r\dot{y}F_{22}(x, y)\right]^2 - \right\rangle\\ &\quad - \tfrac{\partial}{\partial x}\left\langle \tfrac{1}{2r^2}\mathbf{J}_P\left[\dot{x}f_x'(x, y) + \dot{y}f_y'(x, y) - r\dot{x}F_{31}(x, y) - r\dot{y}F_{32}(x, y)\right]^2\right\rangle\\ &\quad + \tfrac{\partial}{\partial x}\langle Mg(f_C(x, y) - f_C(x_0, y_0))\rangle = 0\end{aligned} \tag{36}$$

$$
\begin{aligned}
&\frac{d}{dt}\left\langle \frac{1}{r^2}\mathbf{J}_P\left[\dot{x} - r\dot{x}F_{11}(x,y) - r\dot{y}F_{12}(x,y)\right] rF_{12}(x,y) - \frac{1}{r^2}\mathbf{J}_P \right.\\
&\quad \times \left[\dot{y} - r\dot{x}F_{212}(x,y) - r\dot{y}F_{22}(x,y)\right]\left\langle 1 - rF_{22}(x,y)\right\rangle\Big\rangle +\\
&\quad + \frac{d}{dt}\left\langle \frac{1}{r^2}\mathbf{J}_P\left[\dot{x}f'_x(x,y) + \dot{y}f'_y(x,y) - r\dot{x}F_{31}(x,y) - ryF_{32}(x,y)\right]\right.\\
&\quad \times \left\langle f'_y(x,y) - rF_{32}(x,y)\right\rangle\Big\rangle - \frac{\partial}{\partial y}\left\langle \frac{1}{2r^2}\mathbf{J}_P\left[\dot{x} - r\dot{x}F_{11}(x,y) - r\dot{y}F_{12}(x,y)\right]^2\right.\\
&\quad + \frac{1}{2r^2}\mathbf{J}_P\left[\dot{y} - r\dot{x}F_{212}(x,y) - r\dot{y}F_{22}(x,y)\right]^2\Big\rangle -\\
&\quad - \frac{\partial}{\partial y}\left\langle \frac{1}{2r^2}\mathbf{J}_P\left[\dot{x}f'_x(x,y) + \dot{y}f'_y(x,y) - r\dot{x}F_{31}(x,y) - r\dot{y}F_{32}(x,y)\right]^2\right\rangle\\
&\quad + \frac{\partial}{\partial y}\left\langle Mg\left(f_C(x,y) - f_C(x_0,y_0)\right)\right\rangle = 0
\end{aligned}
\tag{37}
$$

Where we introduced the coefficient κ of rolling of a ball over the arbitrary surface in three dimensional space and if we make new labels, $G_{11}(x, y, \dot{x}, \dot{y}, \ddot{x}, \ddot{y})$, $G_{12}(x, y, \dot{x}, \dot{y})$, $G_{22}(x, y, \dot{x}, \dot{y}, \ddot{x}, \ddot{y})$ and $G_{21}(x, y, \dot{x}, \dot{y})$ the previous system of two nonlinear differential equations can be written in the following simpler form, from which we can more clearly see the structure if two differential equations:

$$
\begin{aligned}
&G_{11}(x,y,\dot{x},\dot{y},\ddot{x},\ddot{y}) = \frac{d}{dt}\Big\langle \left[\dot{x} - r\dot{x}F_{11}(x,y) - r\dot{y}F_{12}(x,y)\right]\left(1 - rF_{11}(x,y)\right)\\
&\quad - \left[\dot{y} - r\dot{x}F_{212}(x,y) - r\dot{y}F_{22}(x,y)\right] rF_{212}(x,y)\Big\rangle +\\
&\quad + \frac{d}{dt}\left\langle\left[\dot{x}f'_x(x,y) + \dot{y}f'_y(x,y) - r\dot{x}F_{31}(x,y) - r\dot{y}F_{32}(x,y)\right]\left\langle f'_x(x,y) - rF_{31}(x,y)\right\rangle\right\rangle,\\
&G_{12}(x,y,\dot{x},\dot{y}) = \left\langle\left[\dot{x} - r\dot{x}F_{11}(x,y) - r\dot{y}F_{12}(x,y)\right]\left[-r\dot{x}F'_{11x}(x,y) - r\dot{y}F'_{12x}(x,y)\right]\right\rangle\\
&\quad + \left\langle\left[\dot{y} - r\dot{x}F_{212}(x,y) - r\dot{y}F_{22}(x,y)\right]\left[-r\dot{x}F'_{21x}(x,y) - r\dot{y}F'_{22x}(x,y)\right]\right\rangle +\\
&\quad + \left\langle\left[\dot{x}f'_x(x,y) + \dot{y}f'_y(x,y) - r\dot{x}F_{31}(x,y) - r\dot{y}F_{32}(x,y)\right]\right\rangle\\
&\quad \times \left[\dot{x}f''_x(x,y) + \dot{y}f''_{xy}(x,y) - r\dot{x}F'_{3x1}(x,y) - r\dot{y}F'_{32x}(x,y)\right],\\
&G_{22}(x,y,\dot{x},\dot{y},\ddot{x},\ddot{y}) = \frac{d}{dt}\Big\langle \left[\dot{x} - r\dot{x}F_{11}(x,y) - r\dot{y}F_{12}(x,y)\right] rF_{12}(x,y)\\
&\quad - \left[\dot{y} - r\dot{x}F_{212}(x,y) - r\dot{y}F_{22}(x,y)\right]\left\langle 1 - rF_{22}(x,y)\right\rangle\Big\rangle +\\
&\quad + \frac{d}{dt}\left\langle\left[\dot{x}f'_x(x,y) + \dot{y}f'_y(x,y) - r\dot{x}F_{31}(x,y) - r\dot{y}F_{32}(x,y)\right]\left\langle f'_y(x,y) - rF_{32}(x,y)\right\rangle\right\rangle,\\
&G_{21}(x,y,\dot{x},\dot{y}) = \left\langle\left[\dot{x} - r\dot{x}F_{11}(x,y) - r\dot{y}F_{12}(x,y)\right]\left[-r\dot{x}F'_{11y}(x,y) - r\dot{y}F'_{12y}(x,y)\right]\right\rangle\\
&\quad + \left\langle\left[\dot{y} - r\dot{x}F_{212}(x,y) - r\dot{y}F_{22}(x,y)\right]\left[-r\dot{x}F'_{21y}(x,y) - r\dot{y}F'_{22y}(x,y)\right]\right\rangle +\\
&\quad + \Big\langle\left[\dot{x}f'_x(x,y) + \dot{y}f'_y(x,y) - r\dot{x}F_{31}(x,y) - r\dot{y}F_{32}(x,y)\right]\\
&\quad \times \left[\dot{x}f''_{xy}(x,y) + \dot{y}f''_y(x,y) - r\dot{x}F'_{31y}(x,y) - r\dot{y}F'_{32y}(x,y)\right]\Big\rangle
\end{aligned}
\tag{38}
$$

This system of the non-linear differential equations of the nonlinear dynamics describing the rolling ball are, in general, not solvable, so that the system of nonlinear differential equations must be solved for specific cases by the given analytical expression of the surface and the properties of the rolling dynamics of the ball by the same from the examination of the nonlinear phenomena in singularity environments.

$$G_{11}(x, y, \dot{x}, \dot{y}, \ddot{x}, \ddot{y}) - G_{12}(x, y, \dot{x}, \dot{y}) + \frac{g}{\kappa} f'_{Cx}(x, y) = 0, \tag{39}$$

$$G_{22}(x, y, \dot{x}, \dot{y}, \ddot{x}, \ddot{y}) - G_{21}(x, y, \dot{x}, \dot{y}) + \frac{g}{\kappa} f'_{Cy}(x, y) = 0. \tag{40}$$

5 Concluding Remarks

The system of two ordinary nonlinear differential equations of dynamics of a rolling heavy rigid ball, expressed by independent generalized coordinates, is derived, and is new research generalized result. Main problem for slowing series of numerous particular tasks is in slowing these coupled two nonlinear differential equations. In the case that is possible to solve these two nonlinear differential equations and find solutions of the two generalized coordinates in the functions of time, then it is easier to determine momentary angular velocity of the ball rolling, and translator velocity of the ball centre of mass. Also, at each moment, is easier to determine corresponding instantaneous axis of ball rolling and corresponding momentary direction of rolling as well as trace of rolling. In numerous special cases is possible to obtain first integrals along generalized coordinates and determine previous listed kinetic parameters.. Special case of the considered dynamics is rolling heavy ball over the sphere surface, considered in sphere curilinear coordinates, and published previously in reference [12]. Then, presented results are generalization of author previously published results.

Acknowledgement Parts of this research were supported by the Ministry of Sciences of Republic Serbia through Mathematical Institute SANU Belgrade Grants OI 174001" Dynamics of hybrid systems with complex structures. Mechanics of materials.".

References

1. Hedrih (Stevanović), K.R.: Non-linear phenomena in vibro-impact dynamics: Central collision and energy jumps between two rolling bodies, Dedicated to memory of Professor and important scientist Ali Nayfeh (December 21, 1933-March 27, 2017). Nonlinear Dyn. **91**(3), 1885–1907 (2018). https://doi.org/10.1007/s11071-017-3988-x
2. Hedrih (Stevanović) KR.: Vibro-impact dynamics in systems with trigger of coupled three singular points: Collision of two rolling bodies, The 24th International Congress of Theoretical and Applied Mechanics (IUTAM ICTAM 2016), Montreal, Canada, 21–26 August, 2016, Book of Papers, pp. 212–213. IUTAM permanent site. (2016). ISBN: NR16-127/2016E-EPUB; Catalogue Number: 978-0-660-05459-9
3. Hedrih (Stevanović) KR: Generalized rolling pendulum along curvilinear trace: Phase portrait, singular points and total mechanical energy surface. In: Prokopenya, A., Gil-Swiderska, A.

(eds.) Computer Algebra Systems in Teaching and Research, vol. VI, p. 204216. Publisher Siedlce University of Natural Sciences and Humanities, Siedlce (2017). ISSN 2300-7397. http://www.castr.uph.edu.pl

4. Hedrih (Stevanović) KR: Tangent space extension of the position vectors of a discrete rheonomic mechanical system, Professor N. R. Sen Memorial Lecture. Bull. Calcutta Math. Soc. **104**(2), 81–102 (2012)
5. Hedrih (Stevanović) KR: Angular velocityand intensity under change of basic vectors of position vector of tangent space of a meterial system kinetic point –Consideration of the difference betwenn linear and nonlinear transformations, To memory of academician Vladimir Metodievich Matrosov (May 8, 1932-April 17,2011) President of Academy of nonlinear Sciences. *Tensor*. **75**(1), 71–93 (2014). Tensor Society (Tokyo), c/o Kawaguchi Inst. of Math. Soc. Japan
6. Hedrih (Stevanović) RK: Central collision of two rolling balls: theory and examples. Adv. Theor. Appl. Mech. **10**(1), 33–79 (2017). https://doi.org/10.12988/atam.2017.765
7. Hedrih (Stevanović) KR: Chapter 13: Dynamics Of Impacts And Collisions Of The Rolling Balls, dynamical systems: theoretical and experimental analysis. In: Springer Proceedings in Mathematics & Statistics, vol. 182, pp. 157–168. © Springer, Part of Springer Science+Business (2017). ISBN 978-3-319-42407-1. ISSN 2194-1009 ISSN 2194-1017 (electronic)
8. Hedrih, (Stevanović) KR., ENERGY ANALYSUS AND METHODOLOGY OF VIBRO-IMPACT DYNAMICS INVESTIGATION OF SYSTEM WITH ROLLING BODIES (ЕНЕРГЕТИЧНИЙ АНАЛІЗ ТА МЕТОДОЛОГІЯ ДОСЛІДЖЕННЯ ДИНАМІКИ ВІБРОУДАРНО СИСТЕМИ З ТІЛАМИ КОЧЕННЯ), Proceedings, 14-th International Symposium of UkrainianMechanical Engineers in Lviv, pp. 51-52; Національний університет «Львівська політехніка», автори, 2019 р.© Оформлення ТзОВ «КІНПАТРІ ЛТД», 2019 р. Свідоцтво про внесення до Державного реєстру суб'єктів видавничо справи ДК №884 від 04.04.2002 р.
9. Hedrih, (Stevanović) KR., ROLLING A HEAVY BALL OVER THE CURVE COORDINATE SURFACES OF ORTHOGONAL CURVILINEAR COORDINATE SYSTEMS, In Memory of scientists and academician RAS V.V. Rumyantsev and V.M. Matrosov, 19-24 августа 2019 г. Уфа, Республика Башкортостан, Россия, СБОРНИК ТРУДОВ, в 4 томах, ТОМ 1, Общая и прикладная механика, , pp. 295-297. Уфа, РИЦ БашГУ 2019, ISBN 978-5-7477-4951-1. DOI: 10.22226/2410-3535-2019-congress-v1. XII Всероссийский съезд по фундаментальным проблемам теоретической и прикладной механики: сборник трудов в 4 томах. Д23 Т. 1: Общая и прикладная механика.— Уфа: РИЦ БашГУ, 2019.—780 с. http://ruscongrmech2019.bashedu.ru/ru/trudy-sezda
10. Hedrih (Stevanović) KR.: Predavanja: Mehanika III – Dinamika školska godina 2006/2007 (Lectures: Mechanics III- Dynamics University year 2006/2007), Faculty of Mechanical Engineering at University of Nish (2006/2007). Electronic version at link: http://www.hm.co.rs/mehanika/
11. Hedrih (Stevanović) KR.: Vibro-impact dynamics of two rolling heavy thin disks along rotate curvilinear line and energy analysis. In: Nonlinear Dynamics. Springer. Submission NODY-D-18-02888R2; (M21a=10) (2019). https://doi.org/10.1007/s11071-019-04988-6
12. Hedrih (Stevanović) KR: Rolling heavy ball over the sphere in real Rn3 space. Nonlinear Dyn. **97**, 63–82; 1–20, Springer (2019). https://doi.org/10.1007/s11071-019-04947-1
13. Hedrih (Stevanović) KR: Nonlinear phenomena in the dynamics of a class of rolling pendulums: a trigger of coupled singularities, plenary lecture, book of abstracts CHAOS, pp. 6–7, (2021). http://www.cmsim.org/chaos2021.html

Explicit Model for Surface Waves on an Elastic Half-Space Coated by a Thin Vertically Inhomogeneous Layer

Ali Mubaraki, Danila Prikazchikov, and Askar Kudaibergenov

Abstract The study is focussed on surface waves propagating in an isotropic elastic half-space coated with a thin, vertically inhomogeneous layer, subject to action of a prescribed normal surface stress. The effective boundary conditions modelling an inhomogeneous coating are derived in the long-wave limit, generalising the those for a thin homogeneous isotropic layer. A singularly perturbed hyperbolic equation on the interface is then deduced, governing surface wave propagation. The effect of the perturbative pseudo-differential operator including the structure of the quasi-front emerging for a point impulse loading, is analysed.

Keywords Surface waves · Thin coating · Inhomogeneous

1 Introduction

Thin films and coatings have numerous applications in engineering and biological sciences, see e.g. [1–6], to name a few. In addition, a number of technological developments are associated with related multi-layered structures, see e.g. [7] and references therein.

Often the effect of a thin coating on the half-space is modelled by means of the so-called effective boundary conditions, starting from the original work [8], and still popular, see e.g. [9, 10] and references therein.

A. Mubaraki
Keele University, Keele, UK

D. Prikazchikov (✉)
Keele University, Keele, UK

Institute for Problems in Mechanical Engineering, St. Petersburg, Russia
e-mail: d.prikazchikov@keele.ac.uk

A. Kudaibergenov
Al-Farabi Kazakh National University, Almaty, Kazakhstan

J. Awrejcewicz (ed.), *Perspectives in Dynamical Systems I: Mechatronics and Life Sciences*, Springer Proceedings in Mathematics & Statistics 362,
https://doi.org/10.1007/978-3-030-77306-9_23

The method of effective boundary conditions was also implemented for analysis of surface wave field in a coated half-space, within the framework of hyperbolic-elliptic models for the Rayleigh wave induced by a prescribed surface load, see [11, 12] for more detail. As a result, the contribution of surface wave to the overall dynamic response in the long wave limit is described by elliptic equations over the interior associated with decay away from the surface, and a singularly perturbed wave equation on the boundary governing surface wave propagation.

In this paper, we extend these results for a thin vertically inhomogeneous coating layer, with density and material parameters being depth-dependent. First, we derive the effective boundary conditions by employing a standard long wave asymptotic procedure, well established for thin structures, see e.g. [13, 14]. Then, we follow a slow-time perturbation scheme proposed in [11], with the small parameter corresponding to the proximity of the wave phase velocity to that of the Rayleigh wave. As a result, we obtain a wave equation for the longitudinal elastic potential, which is singularly perturbed by a pseudo-differential operator. The amplitude of the perturbation depends on the combination of the material parameters of both coating and the substrate. As observed earlier in [11] for the case of a homogeneous coating layer, the sign of this coefficient plays a crucial role, distinguishing between the case of a local maximum/minimum of the phase speed at the Rayleigh wave speed in the long wave limit. Finally, we illustrate the developments by considering a model example of a concentrated vertical impulse loading applied on the surface of a two-layered coating.

2 Basic Equations

Consider an elastic layer of thickness h, occupying the domain $0 \leq x_3 \leq h$, coating a homogeneous half-space $x_3 \geq h$, see Fig. 1.

The layer is assumed to be vertically inhomogeneous, with the constitutive relations given by

$$\sigma_{ij} = \lambda_c \left(u_{1,1} + u_{2,2} + u_{3,3}\right) \delta_{ij} + \mu_c \left(u_{i,j} + u_{j,i}\right), \tag{1}$$

where σ_{ij}, $i, j = 1, 2, 3$, are the Cauchy stress tensor components, u_i are displacement components, $\lambda_c = \lambda(x_3)$ and $\mu_c = \mu(x_3)$ are the Lamé elastic moduli, and

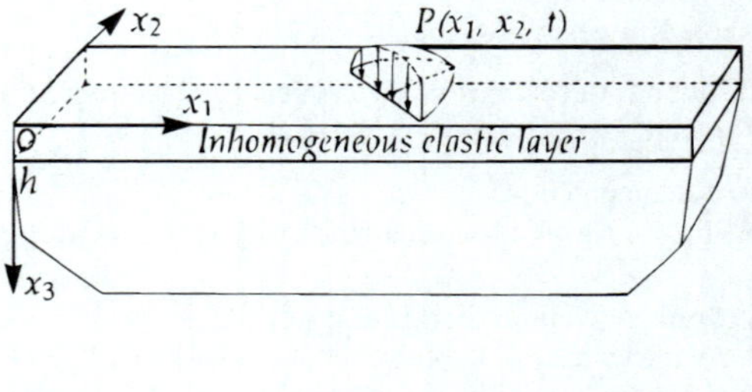

Fig. 1 An inhomogeneous layer by a coated half-space

δ_{ij} is the Kronecker delta. Here and below a comma denotes differentiation with respect to the corresponding variable. The governing equations of motion in the 3D elasticity are taken as (see e.g. [15])

$$\sigma_{i1,1} + \sigma_{i2,2} + \sigma_{i3,3} = \rho_c \, u_{i,tt}, \tag{2}$$

where $\rho_c = \rho\,(x_3)$ is volume mass density. The longitudinal and transverse wave speeds are introduced as

$$c_1\,(x_3) = \sqrt{\frac{\lambda_c + 2\mu_c}{\rho_c}}, \qquad \text{and} \qquad c_2\,(x_3) = \sqrt{\frac{\mu_c}{\rho_c}}, \tag{3}$$

respectively. The boundary conditions at the surface $x_3 = 0$ are taken in the form

$$\sigma_{3m} = 0, \qquad \text{and} \qquad \sigma_{33} = -P, \qquad m = 1, 2, \tag{4}$$

where $P = P(x_1, x_2, t)$ is a prescribed vertical load, with the continuity conditions at the interface assumed as

$$u_i = v_i \qquad \text{at} \qquad x_3 = h, \tag{5}$$

where $v_i = v_i(x_1, x_2, t)$, $i = 1, 2, 3$ are displacements on the surface of the substrate.

3 Effective Boundary Conditions

First, we derive the effective boundary conditions, accounting for the effect of the thin coating layer. Below we implement the direct asymptotic integration of the equations in elasticity, see e.g. [11]. A small parameter ϵ, associated with the long-wave limit, is specified as

$$\epsilon = \frac{h}{L} \ll 1, \tag{6}$$

where L is the typical wave length. We introduce the scaling

$$\xi_m = \frac{x_m}{L}, \qquad \eta = \frac{x_3}{h}, \qquad \tau = \frac{t\,c_h}{L}, \tag{7}$$

with

$$u_i^* = \frac{u_i}{L}, \quad v_i^* = \frac{v_i}{L}, \quad \sigma_{mn}^* = \frac{\sigma_{mn}}{\mu_h}, \quad \sigma_{3i}^* = \frac{\sigma_{3i}}{\epsilon\,\mu_h}, \quad p^* = \frac{P}{\epsilon\,\mu_h}, \tag{8}$$

where $c_h = c_2(h)$, $\mu_h = \mu_c(h)$, $\rho_h = \rho_c(h)$, $m, n = 1, 2$ and all quantities with the asterisk are assumed to be of the same asymptotic order. Then the equation of motion (2) and the constitutive relations (1) can be written explicitly as

$$\begin{aligned} \sigma^*_{mm,\xi_m} + \sigma^*_{mn,\xi_n} + \sigma^*_{m3,\eta} &= \rho_*\, u^*_{m,\tau\tau}, \\ \sigma^*_{33,\eta} + \epsilon\left(\sigma^*_{3m,\xi_m} + \sigma^*_{3n,\xi_n}\right) &= \rho_*\, u^*_{3,\tau\tau}, \end{aligned} \tag{9}$$

and

$$\begin{aligned} \sigma^*_{mn} &= \kappa_2^2\left(u^*_{m,\xi_n} + u^*_{n,\xi_m}\right), \\ \epsilon\,\sigma^*_{mm} &= \left(\kappa_1^2 - 2\kappa_2^2\right) u^*_{3,\eta} + \epsilon\left(\kappa_1^2\, u^*_{m,\xi_m} + \left(\kappa_1^2 - 2\kappa_2^2\right) u^*_{n,\xi_n}\right), \\ \epsilon^2\,\sigma^*_{m3} &= \kappa_2^2\left(u^*_{m,\eta} + \epsilon\, u^*_{3,\xi_m}\right), \\ \epsilon^2\,\sigma^*_{33} &= \kappa_1^2\, u^*_{3,\eta} + \epsilon\left(\kappa_1^2 - 2\kappa_2^2\right)\left(u^*_{m,\xi_m} + u^*_{n,\xi_n}\right), \end{aligned} \tag{10}$$

where $\rho_*(\eta) = \rho_c/\rho_h$, $\kappa_1^2 = (\lambda_c + 2\mu_c)/\mu_h$, $\kappa_2^2 = \mu_c/\mu_h$ and $\kappa_c^2 = \kappa_1^2/\kappa_2^2$, with $1 \le m \ne n \le 2$. On substituting $u^*_{3,\eta}$ from $(10)_4$ into $(10)_2$, we get

$$\sigma^*_{mm} = 4\kappa_2^2\left(1 - \kappa_c^{-2}\right) u^*_{m,\xi_m} + \left(1 - 2\kappa_c^{-2}\right)\left(2\kappa_2^2\, u^*_{n,\xi_n} + \epsilon\,\sigma^*_{33}\right). \tag{11}$$

The conditions (4) and (5) become

$$\begin{aligned} \sigma^*_{3m} = 0, \qquad \sigma^*_{33} &= -p^* \qquad \text{at} \qquad \eta = 0, \\ \text{and} \qquad u^*_i &= v^*_i, \qquad \text{at} \qquad \eta = 1. \end{aligned} \tag{12}$$

Next, expand the displacements and stresses as asymptotic series

$$\begin{pmatrix} u^*_i \\ \sigma^*_{mm} \\ \sigma^*_{mn} \\ \sigma^*_{3i} \end{pmatrix} = \begin{pmatrix} u^{(0)}_i \\ \sigma^{(0)}_{mm} \\ \sigma^{(0)}_{mn} \\ \sigma^{(0)}_{3i} \end{pmatrix} + \epsilon \begin{pmatrix} u^{(1)}_i \\ \sigma^{(1)}_{mm} \\ \sigma^{(1)}_{mn} \\ \sigma^{(1)}_{3i} \end{pmatrix} + \dots . \tag{13}$$

Then, at leading order, we have

$$\begin{aligned} \sigma^{(0)}_{mm,\xi_m} + \sigma^{(0)}_{mn,\xi_n} + \sigma^{(0)}_{m3,\eta} &= \rho_*\, u^{(0)}_{m,\tau\tau}, \\ \sigma^{(0)}_{33,\eta} &= \rho_*\, u^{(0)}_{3,\tau\tau}, \\ \sigma^{(0)}_{mn} &= \kappa_2^2\left(u^{(0)}_{m,\xi_n} + u^{(0)}_{n,\xi_m}\right), \\ \sigma^{(0)}_{mm} &= 4\kappa_2^2\left(1 - \kappa_c^{-2}\right) u^{(0)}_{m,\xi_m} + 2\kappa_2^2\left(1 - 2\kappa_c^{-2}\right) u^{(0)}_{n,\xi_n}, \\ u^{(0)}_{i,\eta} &= 0, \end{aligned} \tag{14}$$

subject to

$$\begin{aligned} \sigma_{3m}^{(0)} = 0\,, \qquad \sigma_{33}^{(0)} &= -p^* \qquad \text{at} \qquad \eta = 0, \\ \text{and} \qquad u_i^{(0)} &= v_i^*, \qquad \text{at} \qquad \eta = 1. \end{aligned} \tag{15}$$

Equations $(14)_5$ with boundary conditions $(15)_2$ imply

$$u_i^{(0)} = v_i^*, \quad i = 1, 2, 3. \tag{16}$$

Therefore, from $(14)_2$ and $(15)_1$ we have

$$\sigma_{33}^{(0)} = v_{3,\tau\tau}^* \int_0^\eta \rho_*(z)\, dz - p^*. \tag{17}$$

Hence, $(14)_1$, $(14)_4$, (16) and $(15)_1$ yield

$$\begin{aligned} \sigma_{3m}^{(0)} = v_{m,\tau\tau}^* \left(\int_0^\eta \rho_*(z)\, dz \right) - 4v_{m,\xi_m\xi_m}^* \left(\int_0^\eta \kappa_2^2(z) \left(1 - \kappa_c^{-2}(z)\right) dz \right) \\ - v_{m,\xi_n\xi_n}^* \left(\int_0^\eta \kappa_2^2(z)\, dz \right) - v_{n,\xi_m\xi_n}^* \left(\int_0^\eta \kappa_2^2(z) \left(3 - 4\kappa_c^{-2}(z)\right) dz \right). \end{aligned} \tag{18}$$

Finally, the effective boundary conditions on the interface $x_3 = h$ may be expressed in terms of the original variables as

$$\begin{aligned} \sigma_{3m} &= h\left(\tilde{\rho}\, u_{m,tt} - \tilde{\gamma}\, u_{m,mm} - \tilde{\mu}\, u_{m,nn} - (\tilde{\gamma} - \tilde{\mu})\, u_{n,mn}\right), \\ \sigma_{33} &= h\tilde{\rho}\, u_{3,tt} - P, \end{aligned} \tag{19}$$

where $\gamma(x_3) = 4\mu_c(x_3)\left(1 - \kappa_c^{-2}(x_3)\right)$ and a tilde over a quantity denotes its mean value over the thickness of the layer

$$\tilde{f} = \frac{1}{h} \int_0^h f(x_3) dx_3.$$

Note that in case of a homogeneous isotropic layer the derived effective boundary conditions (19) reduce to the well-known ones first obtained in [8], see also [11], cf. (3.17).

4 Asymptotic Model for Surface Wave

With the effective boundary conditions (19) derived, an asymptotic model for surface wave may now be constructed, generalising the previous results in [11] to a coating with vertically inhomogeneous material properties. We arrive at the following boundary value problem for a homogeneous isotropic substrate, containing the conventional Navier equations of motion

$$(\lambda + \mu)\,\mathrm{grad}\,\mathrm{div}\,\mathbf{u} + \mu\Delta\mathbf{u} = \rho\mathbf{u}_{,tt}, \tag{20}$$

subject to $(x_3 = h)$

$$\begin{aligned} \mu\left(u_{1,3} + u_{3,1}\right) &= h\left(\tilde{\rho}\,u_{1,tt} - \tilde{\gamma}\,u_{1,11} - \tilde{\mu}\,u_{1,22} - (\tilde{\gamma} - \tilde{\mu})\,u_{2,12}\right), \\ \mu\left(u_{2,3} + u_{3,2}\right) &= h\left(\tilde{\rho}\,u_{2,tt} - \tilde{\gamma}\,u_{2,22} - \tilde{\mu}\,u_{2,11} - (\tilde{\gamma} - \tilde{\mu})\,u_{1,12}\right), \\ \lambda(u_{1,1} + u_{2,2}) + (\lambda + 2\mu)u_{3,3} &= h\tilde{\rho}\,u_{3,tt} - P. \end{aligned} \tag{21}$$

In above $\mathbf{u} = (u_1, u_2, u_3)$ is the displacement vector, Δ is a 3D Laplace operator in spatial coordinates, λ and μ are the constant Lamé parameters of the substrate, and ρ is its volume mass density.

Following the procedure in [11], the Radon integral transform is applied to (20) and (21), resulting in a reduction to a 2D formulation. Then, a slow-time perturbation scheme may be established, revealing the free Rayleigh wave at leading order, with the perturbed wave equation following from the analysis of correction terms. The resulting explicit formulation for surface wave field is expressed in terms of for the longitudinal Lamé potential ϕ, and two non-zero components of the vector shear potential, ψ_1 and ψ_2, with the displacement field expressed using the Helmholtz theorem

$$\mathbf{u} = \mathrm{grad}\,\phi + \mathrm{curl}\,\boldsymbol{\psi}, \tag{22}$$

with $\boldsymbol{\psi} = (-\psi_2, \psi_1, 0)$, for more details see [12]. The behaviour over the interior of the half-space is governed by pseudo-static elliptic equations

$$\phi_{,33} + \alpha_R^2\,\Delta_2\phi = 0, \qquad \psi_{m,33} + \beta_R^2\,\Delta_2\psi_m = 0, \quad m = 1, 2, \tag{23}$$

where $\Delta_2 = \partial_{11} + \partial_{22}$ is the 2D Laplacian in x_1 and x_2 and

$$\alpha_R = \sqrt{1 - \frac{c_R^2}{c_1^2}}, \qquad \beta_R = \sqrt{1 - \frac{c_R^2}{c_2^2}}, \qquad c_1^2 = \frac{\lambda + 2\mu}{\rho}, \qquad c_2^2 = \frac{\mu}{\rho},$$

with c_1, c_2, and c_R conventionally denoting the longitudinal, transverse, and Rayleigh wave speeds. The boundary condition for $(23)_1$ is given by a singularly perturbed wave equation

$$\Delta_2\phi - \frac{1}{c_R^2}\phi_{,tt} - bh\sqrt{-\Delta_2}\,(\Delta_2\theta) = -\frac{1 + \beta_R^2}{2\mu B}\,P, \tag{24}$$

with

$$B = \frac{1-\alpha_R^2}{\alpha_R}\beta_R + \frac{1-\beta_R^2}{\beta_R}\alpha_R - 1 + \beta_R^4,$$

and the constant b inheriting properties of both coating and substrate

$$b = \frac{1-\beta_R^2}{2\mu B}\left(\tilde{\rho}c_R^2\left(\alpha_R+\beta_R\right) - \tilde{\gamma}\beta_R\right). \tag{25}$$

It can be easily verified that in case of a homogeneous isotropic coating layer the latter reduces to earlier results (cf. (4.23) in [11]). The differential relations between the potentials on the boundary $x_3 = h$ are

$$\phi_{,3} = -\frac{1+\beta_R^2}{2}\left(\psi_{1,1}+\psi_{2,2}\right), \qquad \phi_{,m} = \frac{2}{1+\beta_R^2}\psi_{m,3}, \quad m = 1, 2. \tag{26}$$

5 Illustrative Example

In order to illustrate the derived formulation, let us restrict ourselves to a the plane-strain problem for a concentrated impact force $P(x_1, t) = P_0\delta(x_1)\delta(t)$, acting on the surface of a two-layered coating, with the material and geometrical parameters of the layers denoted with subscripts 1 and 2. The wave equation (24) may be rewritten in the form

$$\theta_{,ss} - \frac{1}{c_R^2}\theta_{,\tau_R\tau_R} - h_L \operatorname{sgn} b\sqrt{-\partial_{ss}}\left(\theta_{,ss}\right) = -\delta(s)\delta(\tau_R), \tag{27}$$

where $s = x_1/L$, $\tau_R = tc_R/L$ are the dimensionless coordinates, and

$$\theta = -\frac{4\mu B}{(1+\beta_R^2)c_R P_0}\phi\big|_{x_2=h_1+h_2}, \qquad h_L = \frac{(h_1+h_2)|b|}{L} \ll 1, \tag{28}$$

with the constant b defined according to (25) with

$$\tilde{\rho} = \frac{\rho_1 h_1 + \rho_2 h_2}{h_1+h_2}, \qquad \tilde{\gamma} = \frac{4\mu_1 h_1(1-\kappa_{c1}^{-2}) + 4\mu_2 h_2(1-\kappa_{c2}^{-2})}{h_1+h_2}. \tag{29}$$

Equation (27) may be solved by asymptotic matching, see [11], resulting in

$$\theta = \frac{1}{2}\left[1 - \operatorname{sgn}(b)\left(\frac{1}{2} + \operatorname{sgn}(\chi)\big(C(\chi)+S(\chi)\big) - C^2(\chi) - S^2(\chi)\right)\right], \tag{30}$$

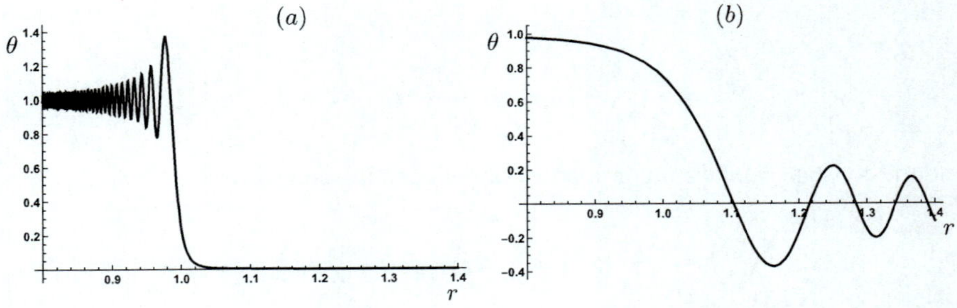

Fig. 2 Quasi-front type behaviour for a two-layered coating: (**a**) rubber-nylon coating on polysterene substrate; (**b**) nylon-polysterene coating on a rubber substrate

where $\chi = (s - \tau_R)\text{sgn}\, b/\sqrt{2h_L\tau_R}$ and $C(x)$ and $S(x)$ denote the Fresnel integrals. Illustrations of the solution (30) is presented below in Fig. 2, showing dependence of θ on s, with $t_R = 1$, $h_1 = 0.1$, $h_2 = 0.2$. The material properties are taken as follows: for rubber the Young's modulus $E = 0.1\,GPa$, volume mass density $\rho = 930\,\text{kg/m}^3$, Poisson ratio $\nu = 0.49$, for nylon $E = 2.95\,\text{GPa}$, $\rho = 1130\,\text{kg/m}^3$, $\nu = 0.39$, for polystyrene $E = 3.1\,\text{GPa}$, $\rho = 1040\,\text{kg/m}^3$, $\nu = 0.35$. As may be seen from the graphs, there are possibilities of receding and advancing quasi-fronts, as noticed previously in [11], associated with the local min/max of the phase velocity at the Rayleigh wave speed in the long-wave limit. Moreover, the velocity of oscillations could also differ on the material parameters. In case of the coating involving soft rubber layer (with contrast in stiffness between rubber and polystyrene exceeding 30), the oscillations of the quasi-front are rapid, whereas in case of a soft rubber substrate, the oscillations are relatively slow.

6 Concluding Remarks

The methodology of hyperbolic-elliptic models for surface wave field has been extended to the case of a half-space coated by a vertically inhomogeneous layer. Further developments may include analysis of other types of boundary conditions [16], near-resonant regimes of moving loads [17], anisotropy [18], as well as a more general treatment of a vertically inhomogeneous half-space, see [19].

Acknowledgments Support by the Ministry of Education and Science of the Republic of Kazakhstan, Grant IRN AP08857255 is acknowledged. For Section 4 DP was supported from the Russian Science Foundation, grant number 20-11-20133. AM acknowledges support by Taif University. The authors are grateful to J. Kaplunov for fruitful discussions.

References

1. Chattopadhyay, D.K., Raju, K.: Structural engineering of polyurethane coatings for high performance applications. Prog. Polym. Sci. **32**(3), 352–418 (2007). https://doi.org/10.1016/j.progpolymsci.2006.05.003
2. Hauert, R.: A review of modified DLC coatings for biological applications. Diam. Relat. Mater. **12**(3–7), 583–589 (2003). https://doi.org/10.1016/S0925-9635(03)00081-5
3. Pompe, W., Worch, H., Epple, M., Friess, W., Gelinsky, M., Greil, P., Hempel, U., Scharnweber, D., Schulte, K.: Functionally graded materials for biomedical applications, Mater. Sci. Eng. A **362**, 40–60 (2003). https://doi.org/10.1016/S0921-5093(03)00580-X
4. Argatov, I., Mishuris, G.: Contact Mechanics of Articular Cartilage Layers. In: Asymptotic Models. Springer, Berlin (2016)
5. Borodich, F.M.: The Hertz-type and adhesive contact problems for depth-sensing indentation. Adv. Appl. Mech. **47**, 225–366 (2014). https://doi.org/10.1016/B978-0-12-800130-1.00003-5
6. Veprek, S., Veprek-Heijman, M.J.: Industrial applications of superhard nanocomposite coatings. Surf. Coat. Tech. **202**(21), 5063–5073 (2008). https://doi.org/10.1016/j.surfcoat.2008.05.038
7. Asmus, M., Nordmann, J., Naumenko, K., Altenbach, H.: A homogeneous substitute material for the core layer of photovoltaic composite structures. Comp. B: Eng. **112**, 353–372 (2017). https://doi.org/10.1016/j.compositesb.2016.12.042
8. Tiersten, H.: Elastic surface waves guided by thin films. J. Appl. Phys. **40**(2), 770–789 (1969). https://doi.org/10.1063/1.1657463
9. Pham, C.V., Vu, A.: Effective boundary condition method and approximate secular equations of Rayleigh waves in orthotropic half-spaces coated by a thin layer. J. Mech. Mater. Struct. **11**(3), 259–277 (2016). https://doi.org/10.2140/jomms.2016.11.259
10. Kaplunov, J., Prikazchikov, D.A., Sultanova L.: On higher order effective boundary conditions for a coated elastic half-space. In: Andrianov, I.V. et al. (eds.) Advanced Structured Materials, vol. 94, pp. 449–462. Springer, Cham (2019). https://doi.org/10.1007/978-3-319-92234-8_25
11. Dai, H.H., Kaplunov, J., Prikazchikov, D.A.: A long-wave model for the surface elastic wave in a coated half-space. Proc. Roy. Soc. A. **466**(2122), 3097–3116 (2010). https://doi.org/10.1098/rspa.2010.0125
12. Kaplunov, J., Prikazchikov, D.A.: Asymptotic theory for Rayleigh and Rayleigh-type waves. Adv. Appl. Mech. **50**, 1–106 (2017). https://doi.org/10.1016/bs.aams.2017.01.001
13. Aghalovyan, L.: Asymptotic Theory of Anisotropic Plates and Shells. World Scientific, New Jersey (2015)
14. Andrianov, I.V., Awrejcewicz, J., Manevitch, L.I.: Asymptotical Mechanics of Thin-Walled Structures. Springer, Berlin (2013)
15. Graff, K.F.: Wave Motion in Elastic Solids. Dover, New York (1975)
16. Kaplunov, J., Prikazchikov, D., Sultanova, L.: Rayleigh-type waves on a coated elastic half-space with a clamped surface. Phil. Trans. Roy. Soc. A **377**(2156), 20190111 (2019). https://doi.org/10.1098/rsta.2019.0111
17. Erbaş, B., Kaplunov, J., Prikazchikov, D.A., Şahin, O.: The near-resonant regimes of a moving load in a three-dimensional problem for a coated elastic half-space. Math. Mech. Solids **22**(1), 89–100 (2017). https://doi.org/10.1177/1081286514555451
18. Nobili, A., Prikazchikov, D.A.: Explicit formulation for the Rayleigh wave field induced by surface stresses in an orthorhombic half-plane. Europ. J. Mech. A/Solids **70**, 86–94 (2018). https://doi.org/10.1016/j.euromechsol.2018.01.012
19. Argatov, I., Iantchenko, A.: Rayleigh surface waves in functionally graded materials—long-wave limit. Quart. J. Mech. Appl. Math. **72**(2) 197–211 (2019). https://doi.org/10.1093/qjmam/hbz002

Bending Vibration Systems which are Complementary with Respect to Eigenvalues

Carsten Behn, Christoph Will, Lukas Merker, and Joachim Steigenberger

Abstract In developing prototypes, one fundamental activity is to model appropriate systems which mimic fundamental features of (biological) paradigms. In this way, we set up different models for the investigation of natural frequencies. The aim is to detect object contacts of technical sensors in observing their vibration behavior. For this, we compare the range and the shift of natural frequencies determined from the analysis of the arising two-point boundary-value problems. In particular, we found two systems with complementary spectra of eigenvalues. Considering boundary damping we analyzed these eigenvalues in the first octant of the complex plane. The fundamental result is that these two systems offer no common eigenvalue, they are alternative. This is an interesting and unique observation.

Keywords Bending beam vibrations · Natural frequency · Tactile sensor · Complementary eigenvalues · Animal vibrissa

1 Introduction

In recent years, there is a great interest in tactile sensors, since they can complement other senses and have several advantages: they are superior to optical sensors as in noisy environments (e.g., in the dark), and may also be cheaper in manufacture and

C. Behn (✉)
Schmalkalden University of Applied Sciences, Department of Mechanical Engineering, Schmalkalden, Germany
e-mail: c.behn@hs-sm.de
https://www.hs-schmalkalden.de/en/university/faculties

C. Will · L. Merker
Technische Universität Ilmenau, Department of Mechanical Engineering, Ilmenau, Germany

J. Steigenberger
Technische Universität Ilmenau, Insitute of Mathematics, Ilmenau, Germany

J. Awrejcewicz (ed.), *Perspectives in Dynamical Systems I: Mechatronics and Life Sciences*, Springer Proceedings in Mathematics & Statistics 362,
https://doi.org/10.1007/978-3-030-77306-9_24

use. A tactile sensor from biology is the so-called animal vibrissa found on, e.g., rats and mice. This paragon gets attracted attention in the last decades. They serve for the exploration of the environment, the animals use them, e.g., to detect outer objects, to distinguish between different surfaces, or to recognize surface textures.

The paragon offers an interesting arrangement:

- An animal vibrissa serves for the force transmission/mechanical stimuli in principal. Because it is just a transmitter, a vibrissa itself is made of dead material and, in contrast to ordinary hairs, they are stiffer, have a inherent curvature and a conical shape. Moreover, they are assumed to be hollow due to a multi-layer structure, [1].
- Further on, each vibrissa is embedded in its own follicle-sinus complex (FSC). This FSC serves as a viscoelastic support and exhibits an exceptional arrangement of blood vessels, receptors (e.g., mechanoreceptors) and neural connections [2].
- The FSCs are embedded in the surrounding tissue: intrinsic and extrinsic musculature and the so-called fibrous band [3].
- The hair shaft/vibrissa has also an additional support by the skin.

For the functional understanding and analytical investigations, there are already various mechanical models under investigations in literature for several targeting objectives. With respect to, e.g., object distance determination in measuring and observing the shift of the natural frequencies due to object contacts (changing boundary conditions of the system), we focus on bending beam vibrations in the following. For this, we set up two models of vibrissa-like sensors in the following section.

2 Modeling

The first model serves as a first approach: the support is simply modeled as a clamping. But in contrast to other models from literature, we incorporate the elasticity of the touched skin: a discrete spring-damper combination, see Fig. 1. The second model is completely different from the first one. We neglect the foundation of the FSC and the elasticity of the skin, because the length of the hair under the

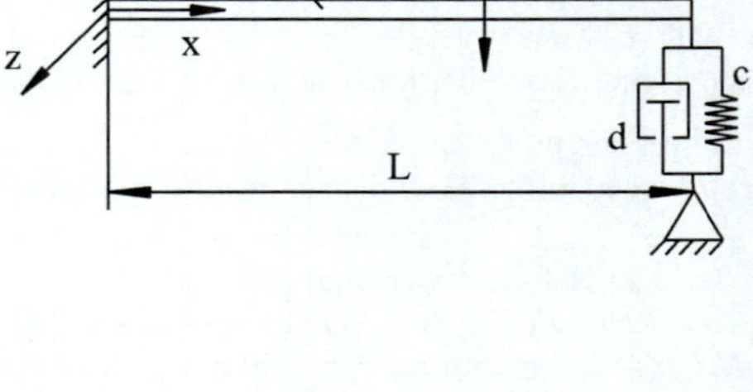

Fig. 1 Model I: one-sided clamped beam with discrete visco-elastic end support

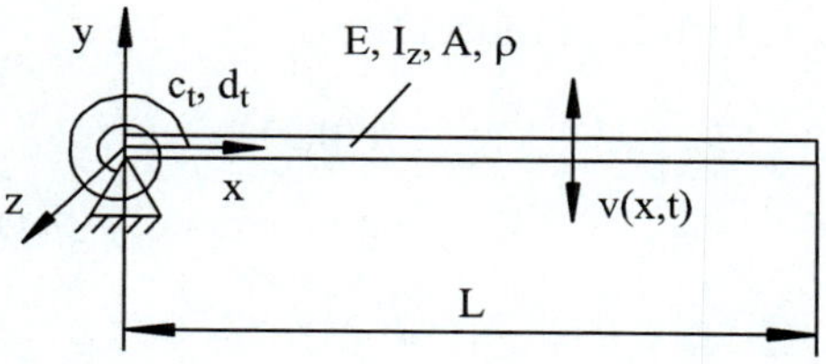

Fig. 2 Model II: beam with a bearing and discrete torsional spring-damper element

skin is much shorter than the free end. Hence, we shrink the support to a rotational viscoelasticity: a bearing and discrete torsional spring-damper element, see Fig. 2.

3 Analyses of the Models

Bending beam vibrations of small amplitude are described by the partial differential equation (PDE), [4],

$$\frac{\partial^4}{\partial x^4} v(x,t) + \frac{\partial^2}{\partial t^2} v(x,t) = 0\,, \quad \forall\ (x,t) \in (0,1) \times \mathbb{R}_+ \tag{1}$$

under particular boundary conditions (BCs). Solutions of this boundary-value problem (BVP) are investigated using separation of variables, see [5].

Remark 1 Referring to [6], we prefer a dimensionless notation throughout by using the following units of measurement matching data of the real background system (L, ρ, A, E, I_z beam parameters; dimensionless variables only here with a tilde which is dropped afterwards):

$$x := \tilde{x} \cdot L\,, \quad t := \tilde{t} \cdot L^2 \sqrt{\frac{\rho\, A}{E\, I_z}}\,, \quad \omega := \tilde{\omega} \cdot \frac{1}{L^2} \sqrt{\frac{E\, I_z}{\rho\, A}}\,,$$
$$c := \tilde{c} \cdot \frac{E\, I_z}{L^3}\,, \quad d := \tilde{d} \cdot \frac{1}{L} \sqrt{\rho\, A\, E\, I_z}\,,$$
$$c_t := \tilde{c}_t \cdot \frac{E\, I_z}{L}\,, \quad d_t := \tilde{d}_t \cdot L \cdot \sqrt{\rho\, A\, E\, I_z}\,.$$

The boundary conditions (BCs) of Model I are:

$$\begin{aligned} v(0,t) &= 0\,,\\ \tfrac{\partial}{\partial x} v(0,t) &= 0\,,\\ \tfrac{\partial^2}{\partial x^2} v(1,t) &= 0\,,\\ \tfrac{\partial^3}{\partial x^3} v(1,t) &= c \cdot v(1,t) + d \cdot \tfrac{\partial}{\partial t} v(1,t)\,. \end{aligned} \tag{2}$$

The BCs of Model II are:

$$\begin{aligned} v(0,t) &= 0\,, \\ \frac{\partial^2}{\partial x^2} v(0,t) &= c_t \cdot v(0,t) + d_t \cdot \frac{\partial^2}{\partial x\, \partial t} v(0,t)\,, \\ \frac{\partial^2}{\partial x^2} v(1,t) &= 0\,, \\ \frac{\partial^3}{\partial x^3} v(1,t) &= 0\,. \end{aligned} \tag{3}$$

Separation of variables,

$$v(x,t) = X(x) \cdot T(t) \;\Rightarrow\; \frac{X^{(4)}}{X}(x) = -\frac{\ddot{T}}{T}(t) =: \lambda^4\,, \quad \lambda \in \mathbb{C}\,, \tag{4}$$

yields the following characteristic equations for λ of both models:

$$\text{Model I}: \lambda^3 \cdot [1+\cosh(\lambda)\cdot\cos(\lambda)] + (c \pm i\cdot d\cdot\lambda^2)\cdot[\cosh(\lambda)\cdot\sin(\lambda) - \sinh(\lambda)\cdot\cos(\lambda)] = 0\,, \tag{5}$$

and

$$\text{Model II}: (c_t \pm i\cdot d_t\cdot\lambda^2)\cdot[1+\cosh(\lambda)\cdot\cos(\lambda)] + \lambda\cdot[\sinh(\lambda)\cdot\cos(\lambda) - \cosh(\lambda)\cdot\sin(\lambda)] = 0\,. \tag{6}$$

Both equations are, because of the $\pm$-sign, in fact *two* equations, i.e., four. Let their left-hand sides (lhs, and rhs, respectively) be symbolized as $eq(\lambda; c, d; +)$ and $eq(\lambda; c, d; -)$, respectively. It is easy matter to verify the following Proposition 1 in observing Eqs. (5) and (6) and using relations from [7].

Proposition 1 *If $eq(\lambda; c, d; +) = 0$ then $eq(\overline{\lambda}; c, d; -) = 0$, where $\overline{\lambda}$ is the complex conjugate of λ. This means that the eigenvalues of a BVP appear as a sequence of conjugate pairs. Moreover, $eq(i\lambda; c, d; +) = -i \cdot eq(\lambda; c, d; -)$ and $eq(-\lambda; c, d; \pm) = -eq(\lambda; c, d; \pm)$ implies that each eigenvalue λ entails the $8-tuple$ $(\pm\lambda, \pm\overline{\lambda}, \pm i\lambda, \pm i\overline{\lambda})$ of eigenvalues.*

In the following section, we present numerical investigations of the solutions of the corresponding characteristic equations, using them to determine the natural frequencies $\omega(c,d) = Re(\lambda(c,d)^2)$, [8].

4 Simulations

At first, we focus on the natural frequencies of both systems, we present the behavior in Figs. 3, 4, and 5.

Both diagrams of Fig. 3 show an interesting feature: if c, c_t are below a certain value c^*, c_t^*, then the natural frequency tends to zero for $d, d_t \to +\infty$, else it

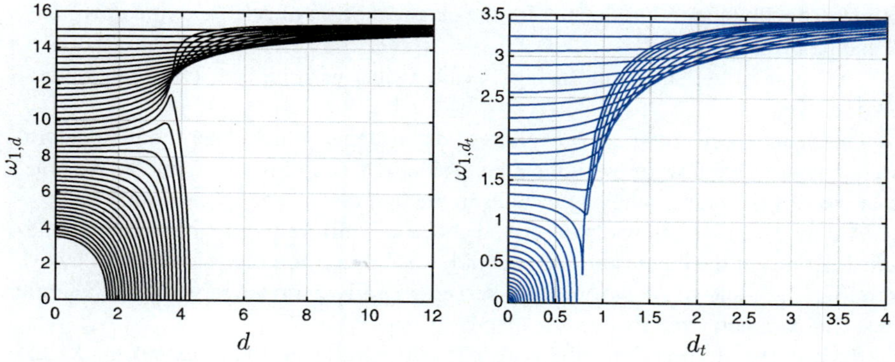

Fig. 3 First natural frequency vs. damping parameter—family parameter c—for Model I (left) and Model II (right)

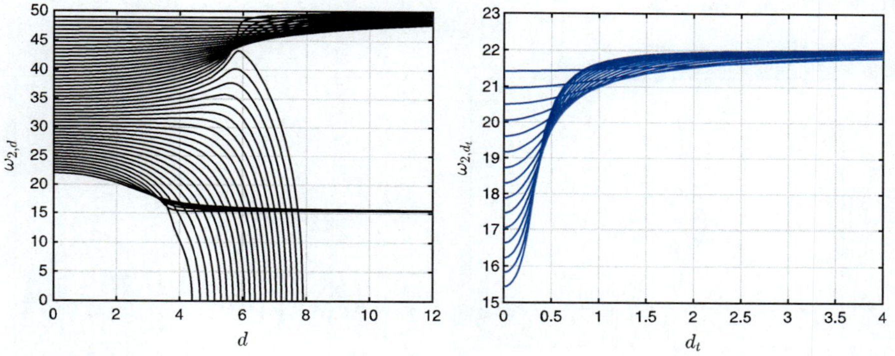

Fig. 4 Second natural frequency vs. damping parameter—family parameter c—for Model I (left) and Model II (right)

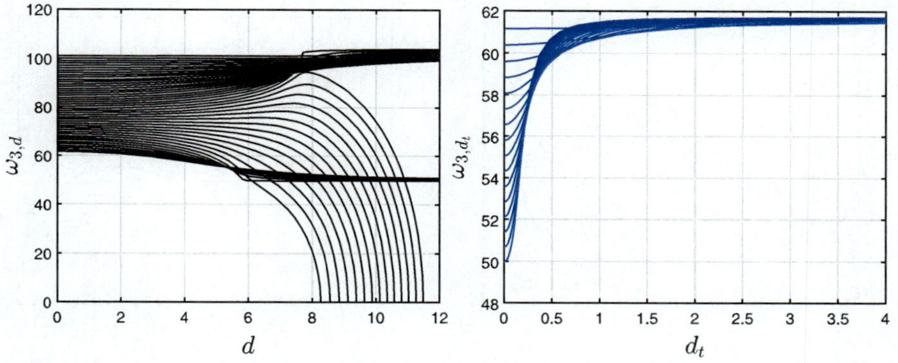

Fig. 5 Third natural frequency vs. damping parameter—family parameter c—for Model I (left) and Model II (right)

tends to the first natural frequency for the classical beam of type "clamped/pivoted", "clamped/free".

The same effect shows up in Figs. 4 and 5, just with higher "classical" natural frequencies.

Let us now focus on the eigenvalues of the systems, which determine the natural frequencies. Again, we present the results side by side in Figs. 6, 7, and 8. We note, that we, at first, present the eigenvalues in the first octant, see Proposition 1.

All these various behaviors described above are mirrored through the corresponding diagrams of the eigenvalues $\lambda(c, d)$, $\lambda(c_t, d_t)$ in the complex plane (first octant), see Figs. 6, 7, and 8. As before, c, c_t serve as family parameters while each single curve in parameterized by d, d_t, respectively.

Each curve starting from the real axis and tending to the diagonal marks the natural frequency as tending to zero. Each curve ending at the real axis marks the natural frequency as tending to a natural frequency of a classical undamped vibrating beam type "clamped/pivoted", "clamped/free". Compare in particular the right parts of Figs. 3 and 6.

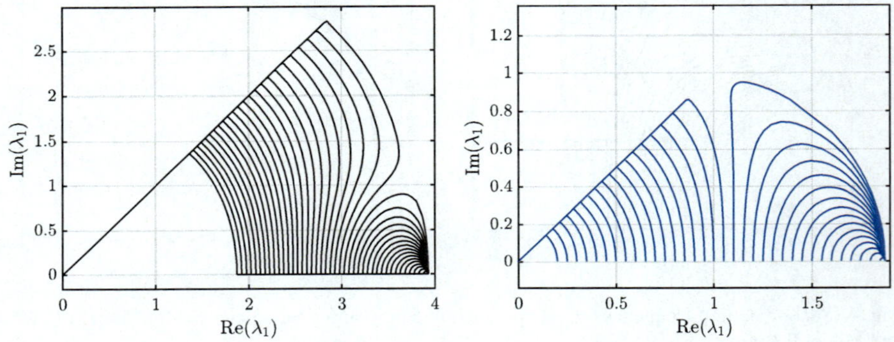

Fig. 6 First eigenvalue for Model I (left) and Model II (right)

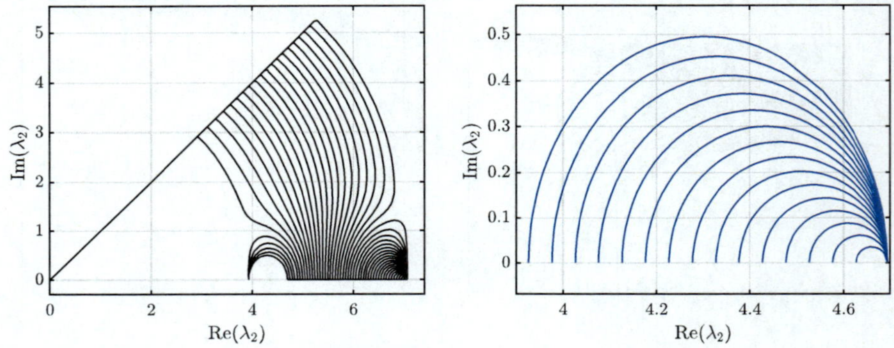

Fig. 7 Second eigenvalue for Model I (left) and Model II (right)

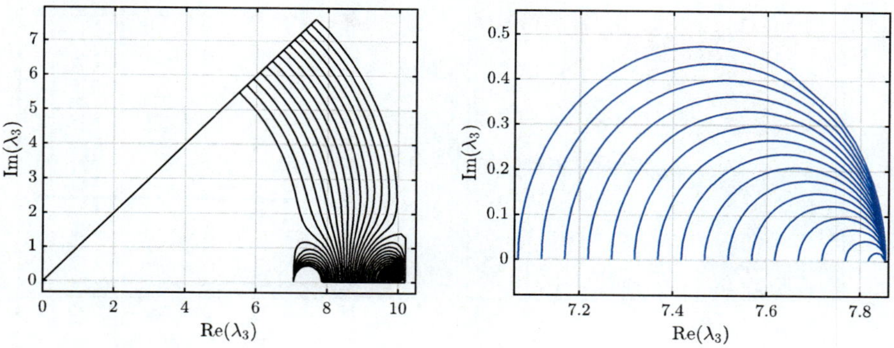

Fig. 8 Third eigenvalue for Model I (left) and Model II (right)

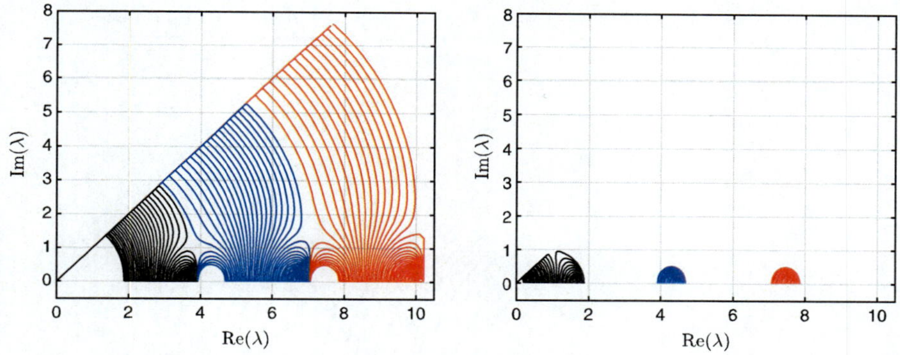

Fig. 9 First three eigenvalues of Model I (left) and Model II (right)

The patterns in complex plane continue unboundedly to the right, see Figs. 9 and 11. Finally, Fig. 12 shows the events of Fig. 11 without confinement to the first octant.

Tacitly, we passed the main point of the headline: "complementarity" becomes obvious in Figs. 9 and 11. Every white place in the left-hand sides of the diagrams fills up with one curve family from the right-hand side diagrams and vice versa—there are no common eigenvalues (besides maybe those on the common boundaries).

Equally, this fact is transferable to the behavior of the eigenfrequencies in inspecting Fig. 10 and, more precisely, in Fig. 13, where the complete vertical axis (ordinate ω) is filled.

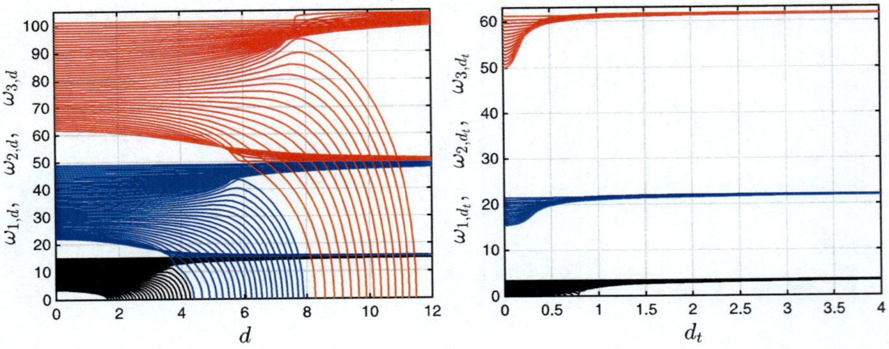

Fig. 10 First three natural frequencies vs. damping parameter—family parameter c—for Model I (left) and Model II (right)

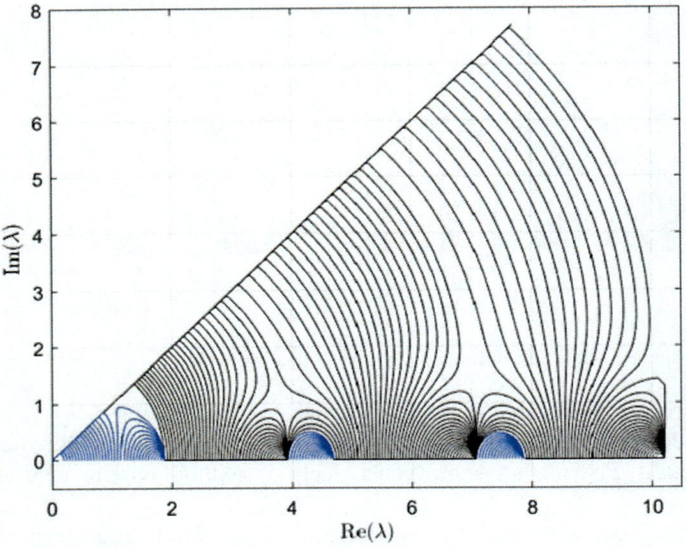

Fig. 11 Superposition of the first three eigenvalues of both systems—Model I (black) and Model II (blue)—in the first octant

5 Conclusions

Until now, the observed complementary is still a remarkable feature. Next work should be concentrate on experimental proof. Possibly, it could be utilized in control problems if both viscoelastic supports are in action.

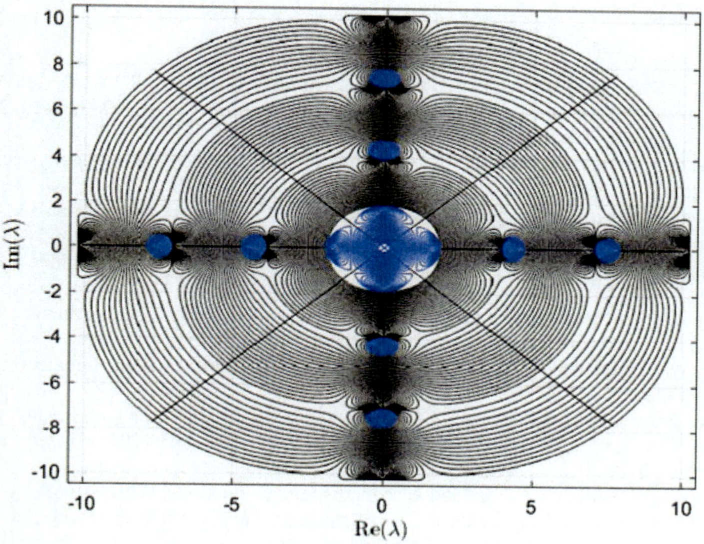

Fig. 12 Superposition of the first three eigenvalues of both systems—Model I (black) and Model II (blue)—in the complex plane

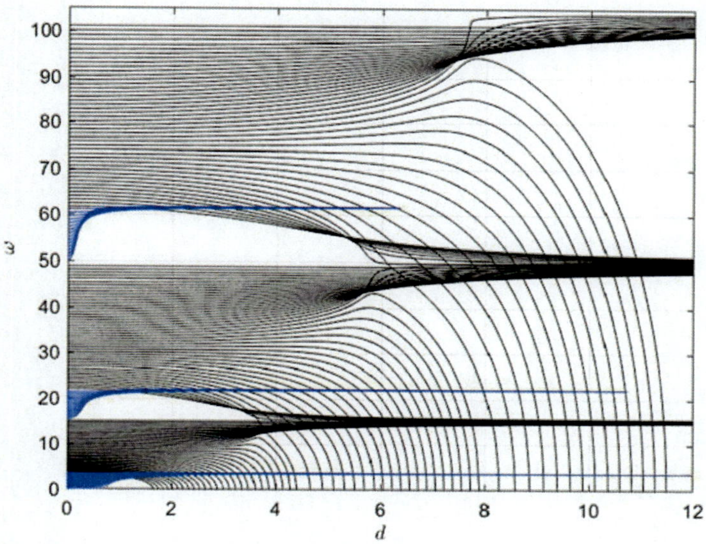

Fig. 13 Superposition of the first three natural frequencies vs. damping parameter—family parameter c—for Model I (black) and Model II (blue)

References

1. Voges, D., Carl, K., Klauer, G.J., Uhlig, R., Schilling, C., Behn, C., Witte, H.: Structural characterization of the whisker system of the rat. IEEE Sensors **12**(2), 332–339 (2012). https://ieeexplore.ieee.org/document/5948319
2. Haidarliu, S., Simony, E., Golomb, D., Ahissar, E.: Muscle architecture in the mystacial pad of the rat. Anat. Rec. **293**(7), 1192–1206 (2010). https://doi.org/10.1002/ar.21156
3. Dörfl, J.: The musculature of the mystacial vibrissae of the white mouse. J. Anat. **135**, 147–154 (1982). https://www.ncbi.nlm.nih.gov/pubmed/7130049
4. Gross, D., Hauger, W., Schnell, W., Wriggers, P.: Technische Mechanik: Band 4 Hydromechanik, Elemente der Höheren Mechanik, Numerische Methoden. Springer, Berlin (2002). (Engl.: Technical Mechanics: Volume 4 Hydromechanics, higher mechanics, numerical methods)
5. Weaver, W., Timoshenko, S.P., Young, D.H.: Vibration Problems in Engineering. Wiley, Chichester (1990)
6. Will, C., Behn, C., Steigenberger, J.: Object contour scanning using elastically supported technical vibrissae. ZAMM J. Appl. Math. Mech. **98**(2), 289–305 (2018). https://doi.org/10.1002/zamm.201600161
7. Abramowitz, M., Stegun I.A.: Handbook of mathematical functions: With formulas, graphs, and mathematical tables. In: National Bureau of Standards Applied Mathematics Series, vol. 55, , corr. edn. United States Department of Commerce, Washington (1972).
8. Baldeweg, D., Will, C., Behn, C.: Tranversal Vibrations of Beams in Context of Vibrissae with Foundations, Discrete Supports and Various Sections. In: Proceedings 58th International Scientific Colloquium,Ilmenau (Germany, September 2014). https://www.db-thueringen.de/receive/dbt_mods_00025169